T0181408

Advances in Intelligent Systems and Computing

Volume 355

Series editor

Janusz Kacprzyk, Polish Academy of Sciences, Warsaw, Poland
e-mail: kacprzyk@ibspan.waw.pl

About this Series

The series "Advances in Intelligent Systems and Computing" contains publications on theory, applications, and design methods of Intelligent Systems and Intelligent Computing. Virtually all disciplines such as engineering, natural sciences, computer and information science, ICT, economics, business, e-commerce, environment, healthcare, life science are covered. The list of topics spans all the areas of modern intelligent systems and computing.

The publications within "Advances in Intelligent Systems and Computing" are primarily textbooks and proceedings of important conferences, symposia and congresses. They cover significant recent developments in the field, both of a foundational and applicable character. An important characteristic feature of the series is the short publication time and world-wide distribution. This permits a rapid and broad dissemination of research results.

Advisory Board

Chairman

Nikhil R. Pal, Indian Statistical Institute, Kolkata, India
e-mail: nikhil@isical.ac.in

Members

Rafael Bello, Universidad Central "Marta Abreu" de Las Villas, Santa Clara, Cuba
e-mail: rbellop@uclv.edu.cu

Emilio S. Corchado, University of Salamanca, Salamanca, Spain
e-mail: escorchado@usal.es

Hani Hagras, University of Essex, Colchester, UK
e-mail: hani@essex.ac.uk

László T. Kóczy, Széchenyi István University, Győr, Hungary
e-mail: koczy@sze.hu

Vladik Kreinovich, University of Texas at El Paso, El Paso, USA
e-mail: vladik@utep.edu

Chin-Teng Lin, National Chiao Tung University, Hsinchu, Taiwan
e-mail: ctlin@mail.nctu.edu.tw

Jie Lu, University of Technology, Sydney, Australia
e-mail: Jie.Lu@uts.edu.au

Patricia Melin, Tijuana Institute of Technology, Tijuana, Mexico
e-mail: epmelin@hafsamx.org

Nadia Nedjah, State University of Rio de Janeiro, Rio de Janeiro, Brazil
e-mail: nadia@eng.uerj.br

Ngoc Thanh Nguyen, Wroclaw University of Technology, Wroclaw, Poland
e-mail: Ngoc-Thanh.Nguyen@pwr.edu.pl

Jun Wang, The Chinese University of Hong Kong, Shatin, Hong Kong
e-mail: jwang@mae.cuhk.edu.hk

More information about this series at http://www.springer.com/series/11156

Ajith Abraham · Azah Kamilah Muda
Yun-Huoy Choo
Editors

Pattern Analysis, Intelligent Security and the Internet of Things

 Springer

Editors

Ajith Abraham
Machine Intelligence Research Labs
  (MIR Labs)
Scientific Network for Innovation
  and Research Excellence
Auburn, WA
USA

Azah Kamilah Muda
Department of Software Engineering
Universiti Teknikal Malaysia Melaka
  (UTeM)
Durian Tunggal
Malaysia

Yun-Huoy Choo
Faculty of Information and Communication
  Technology, Department of Software
  Engineering
Universiti Teknikal Malaysia Melaka
  (UTeM)
Durian Tunggal
Malaysia

ISSN 2194-5357   ISSN 2194-5365   (electronic)
Advances in Intelligent Systems and Computing
ISBN 978-3-319-17397-9   ISBN 978-3-319-17398-6   (eBook)
DOI 10.1007/978-3-319-17398-6

Library of Congress Control Number: 2015936677

Springer Cham Heidelberg New York Dordrecht London

Springer International Publishing AG Switzerland is part of Springer Science+Business Media
(www.springer.com)

Preface

Welcome to Melaka, Malaysia, and to the Parallel Symposiums of the 2014 Fourth World Congress on Information and Communication Technologies (WICT 2014) during December 8–11, 2014. In the past century, our society has been through several periods of dramatic changes, driven by innovations such as transportation systems, telephone. Last few decades have experienced technologies that are evolving so rapidly, altering the constraints of space and time and reshaping the way we communicate, learn, and think. Rapid advances in information technologies and other digital systems are reshaping our ecosystem. Innovations in ICT allow us to transmit information quickly and widely, propelling the growth of new urban communities, linking distant places and diverse areas of endeavor in productive new ways, which a decade ago was unimaginable. Thus, the theme of this World Congress is 'Innovating ICT for Social Revolutions'.

The four day World Congress is expected to provide an opportunity for the researchers from academia and industry to meet and discuss the latest solutions, scientific results and methods in the usage and applications of ICT in the real world. WICT 2014 is Co-Organized by Machine Intelligence Research Labs (MIR Labs), USA, and Universiti Teknikal Malaysia Melaka, Malaysia. WICT 2014 is technically co-sponsored by IEEE Systems, Man and Cybernetics Society Malaysia and Spain Chapters and Technically Supported by IEEE Systems Man and Cybernetics Society, Technical Committee on Soft Computing.

This year, we introduce additional academic activities, Editor-in-Chief's Panel Discussion, Student Symposium, Parallel Symposiums, and Research Product Exhibition. Editor-in-chiefs from world-renowned journals in ICT will be gathered in a special forum to facilitate sharing session for tips and advice on journal publication. The following five Parallel Symposiums were organized:

- Intelligent System and Pattern Analysis
- Emerging Computer Security Issues and Solutions
- Data Quality and Big Data Management
- Innovation in Teaching and Learning
- Requirements Engineering

Many people have collaborated and worked hard to produce a successful WICT-2014 conference. First and foremost, we would like to thank all the authors for submitting their papers to the conference, for their presentations and discussions during the conference. Our thanks to Program Committee members and reviewers, who carried out the most difficult work by carefully evaluating the submitted papers. The themes of the contributions and scientific sessions range from theories to applications, reflecting a wide spectrum of coverage of various data analysis topics covering big data, data quality, pattern recognition, computer security, etc. Each paper was reviewed by at least five reviewers in a standard peer-review process. Based on the recommendation by five independent referees, finally 31 papers were accepted for publication (43 % acceptance rate) in the proceedings published by Springer.

The General Chairs and the Program Chairs along with the entire team cordially invite you to attend the Parallel Symposiums of the 2014 Fourth World Congress on Information and Communication Technologies (WICT 2014).

General Chairs

Ajith Abraham, Machine Intelligence Research Labs (MIR Labs), USA
Ahmad Zaki A. Bakar, Universiti Teknikal Malaysia Melaka, Malaysia

Program Chairs

Azah Kamilah Muda, University Teknikal Malaysia Melaka, Malaysia
Emilio Corchado, Universidad de Salamanca, Spain
Marina Gavrilova, University of Calgary, Calgary

Contents

Bridging Creativity and Group by Elements of Problem-Based Learning (PBL)

Chunfang Zhou

Abstract As recent studies have discussed problem-based learning (PBL) as a popular model of fostering creativity, this paper aims to explore a research question: How can we bridge creativity and group work using elements of PBL to deepen understanding of PBL as a tool for creative learning? A theoretical framework aiming to respond to this question will be provided from a literature review. Accordingly, five elements include (1) group learning, (2) problem solving, (3) interdisciplinary learning, (4) project management, and (5) facilitation. These main elements show PBL is a suitable learning environment to develop individual creativity and to stimulate interplay of individual and group creativity. This study builds a theoretical model, urging a systematic view of PBL in creative facilitation and also indicates its practical significance and potential questions for future investigation.

Keywords Creativity · Group creativity · Problem-based learning (PBL)

1 Introduction

Creativity plays a crucial role in culture; creative activities provide personal, social, and educational benefit, and creative inventions are increasingly recognized as key drivers of economic development [1]. Throughout the world, national governments are adapting their education systems to meet the challenges of the twenty-first century. One priority is to promote creativity and innovation. In the new global economy, the capacity to generate and implement new ideas is vital to economic competitiveness. But education has a greater economic purpose: It must enable people to adapt positively to rapid social change and to live with meaning and purpose at a time when established cultural values are being challenged on many fronts [2].

C. Zhou (✉)
Department of Learning and Philosophy, Aalborg University, Aalborg, Denmark
e-mail: Chunfang@learning.aau.dk

© Springer International Publishing Switzerland 2015
A. Abraham et al. (eds.), *Pattern Analysis, Intelligent Security and the Internet of Things*, Advances in Intelligent Systems and Computing 355,
DOI 10.1007/978-3-319-17398-6_1

1

But what is creativity? In general, the term creativity means to generate new and useful ideas [3]. Early studies of creativity that focused on psychological determinants of the individual, such as genius and giftedness, were followed by later studies that explored such ideas as follows: (a) educating for creativity is a rigorous process based on knowledge and skill, (b) creativity is not confined to particular activities or people, and (c) creativity flourishes under certain conditions, thus creativity can be taught [2]. Accordingly, the topic has exploded with interest. During the past years, creativity has been greatly discussed in the fields of psychology [4, 5], social psychology [6], cultural psychology [7], social culture [8], and even philosophy [9]. The social approaches to creativity, however, have put particular emphasis on group creativity. Following the concept of creativity, group creativity means the creation, development, evaluation, and promotion of novel ideas in groups. This can occur informally during interactions between friends or colleagues, or in more structured groups such as laboratory research scientists and research and development teams [10]. Presently, there is growing attention toward shaping the development of creativity through group work in the learning environment [11].

Problem-based learning (PBL) has been regarded as one of the educational strategies for creativity that provides a group context as the basic condition to develop individual creativity. The term PBL was originally coined by Don Woods [12], based on his work with chemistry students at McMaster University in Ontario, Canada. However, the popularity and subsequent worldwide spread of PBL is mostly linked to the introduction of this educational method at the medical school of McMaster University. Recently, PBL has been introduced into many professional fields of education and appears to have growing interest in higher education [13]. In the PBL context, student learning centers on real-life projects–solving complex problems that do not have a single answer. Students work in collaborative groups to identify what they need to learn to solve the problems. The teacher facilitates the learning process, rather than providing knowledge [14]. The central philosophy of PBL is "student-centered learning," although there are diverse models of application of PBL around the world [15].

A literature review provides a response to the following research question: How can we bridge creativity and group work using elements of PBL to deepen understanding of PBL as a tool for creative learning? Throughout the subsequent sections, this paper explores the link between creativity and learning in the group context and lays the basis for a later discussion of the influences of PBL elements on creativity.

2 Linking PBL Elements and Creativity

2.1 Creativity and Learning: Going Hand-in-Hand in Groups

We can understand group creativity from two aspects: first, to regard it as the "creativity of a group," meaning the outcome of the synergy of individual creativity in a group; second, to view the group as a context for influencing individual creativity. In this sense, it means, the "creativity in a group." However, group creativity happens through the interplay of two such aspects, as illustrated in the following generic model of group creativity [16]:

Figure 1 pictures a dynamic process of individuals engaged into group activities, how they engage in reflective learning through the group process, and how the group performance is improved through individual contributions [17]. Group members bring resources that determine the group's creative potential or what the group is able to accomplish. The contributions of group members need to be combined to yield a group response. The ways in which individual members' contributions are combined constitute the relevant group processes. Finally, the context largely determines which group processes will occur and how individual contributions are combined. Eventually, this determines the quality and creativity of the group response. Thus, according to this framework, the resources of individual group members determine the potential creativity of the group. However, group processes, or the way in which individual contributions are combined, determine whether the group actually achieves its potential. In turn, the social climate and the environment influence group processes [16]. The dynamic process does not always flow in the sequence shown in Fig. 1; however, there is a back and forth exchange between individual-level input and group-level output, and the exchange continues until all individuals yield to the group response due to a satisfying group outcome [17].

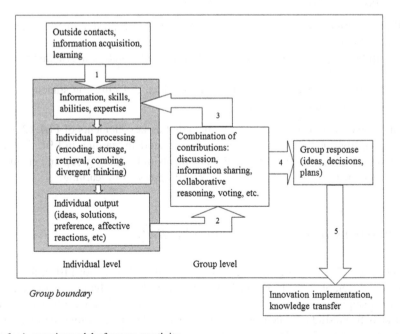

Fig. 1 A generic model of group creativity

A social perspective that links this model to the learning context considers the learner is both "transforming" and "being transformed" when participating in communities of practice. Creativity is regarded as shaping new knowledge [8]. Moreover, researchers have argued that in successful group learning settings learning and creativity go hand-in-hand [18]. Participants build on each other's ideas to reach an understanding that was not available to anyone initially, and group members must also enter into critical and constructive negotiations of each other's suggestions. Meanwhile, well-grounded arguments and counter-arguments need to be shared and critically evaluated through collective talk. These conditions are similar to those needed for collaboration in creative endeavors. Both learning and creativity will take place in groups as the group members participate in shared endeavors, all playing active but often asymmetrical roles in sociocultural activity.

2.2 PBL as a Creative Learning Environment

PBL is well-supported by theories in the learning sciences, ranging from constructivism and cognition to problem solving. Those theories have also been involved in discussions of creativity development in PBL [19]. For example, Tan [20] provided a comprehensive understanding on why and how creativity can be fostered by PBL from different perspectives such as cognitive, sociocultural, psychological, and sociopsychological. According to Zhou et al. [21], there are at least three aspects of PBL that satisfy conditions of creativity:

- Problem orientation and project work—the point of departure in open and real-life problems,
- Group learning context—the process of group collaboration in searching for the solutions, and
- The shifts from teaching to facilitation—the idea of facilitating students' direct learning rather than teaching.

The above three aspects also can be regarded as key elements of PBL [21]. As the project group is the basic form of organizing student learning in PBL, it draws on research focusing on the roles of PBL elements in bridging group and creativity from a systematic view indicated by the social approach to learning.

2.3 Elements of PBL Influencing Creativity

The literature [21] has indicated there are at least five elements of PBL influencing creativity development: (1) group learning, (2) problem solving, (3) facilitation, (4) interdisciplinary learning, and (5) project management. Some other elements have also been studied, such as experience-based learning, active learning, and contextual learning [15, 25]. However, they could be involved in or related to the main pillars mentioned above since they are emphasized by different focuses, but with a common principle advocating for "student-centered learning" [21].

Since group interaction can provide a basis for the exchange of information among group members, effective groups should have individuals with a diversity of knowledge, skills, or new perspectives and should be motivated for a full exchange of ideas. Individuals with different backgrounds may have different value systems, ways of thinking, and attitudes toward collaboration, etc., which may be shaped by different social or cultural environments and influence group process on ideas generation [21]. Group learning lays the basis for the other elements of PBL in relation to creativity development. There are also discussions focusing on relationships between creativity and knowledge, which aim to search further for an appropriate pedagogy to foster creative thinking or greater deep learning [22]. According to Craft [8], whichever approach to learning is dominant in the foundations of one's practice, creativity effectively offers students opportunities to shape new knowledge; for when we learn something new, we are making new connections between ideas and making sense of them for ourselves and we are constructing knowledge; in this sense, we could perhaps describe what we are doing as being creative. However, shaping new knowledge cannot occur without some understanding of what already exists and without opportunities to engage with this and take it to a new place [23]. In other words, creativity cannot appear without knowledge context provided by certain domains and social practice. Because "social practice," to use Wenger's [23] notion, includes both the explicit and the tacit, our communities of practice are places where we develop, negotiate, and share them. To analyze creativity development in settings of group learning, therefore, one should consider the learning stimulus of social practice, from the tacit to the explicit level.

Problem analysis is viewed as a component of PBL, and some universities put more emphasis on it than others. For example, the seven steps to problem solving developed by Maastricht University in the Netherlands are often used in cognitive science [12]. Creativity is included in a number of interdependent and interactive capacities when we solve and analyze problems [20]. There are even researchers who define creativity as the ability to solve problems [24]. The model of creative problem solving (CPS) has been employed broadly in the business context [25] to improve creative products and in education [22] to train students' creative thinking skills. In a real-life problem-solving context, the learning process is creative, dynamic, and iterative. This process involves (a) the identification of problems and problem constraints, (b) identification and clarification of multiple (and possibly conflicting) perspectives of the problem, (c) generation of possible solutions, (d) assessment of the viability of alternative solutions through argument construction and articulation of personal beliefs and assumptions, (e) monitoring of the metacognitive processes involved with the problem-solving activity, (f) testing and recommendation of a solution, and (g) adaptation of a solution [26]. However, a critical aspect of problem solving is that people hold multiple, and sometimes conflicting, perspectives of the nature of the problem, the procedures for solving it, and the appropriate solutions. For this reason, problem-solving methods of instruction typically use learning groups [27]. However, task-related diversity is needed in groups to generate different ideas [28]. In the recent work related to PBL, the discussion of this issue mainly focuses on the role of open, ill-defined, or real-life problems or projects [20] and on the collaborative problem-solving context [15, 17] of creativity.

Interdisciplinary learning is one of the models of PBL, according to Savin-Baden [29] and de Graaff and Kolmos [12]. Interdisciplinary learning may bridge the gap between know-how and know-that and between different forms of disciplinary knowledge. According to studies on communities of practice [23], interdisciplinary projects require the contribution of multiple disciplines. Participating in these kinds of projects exposes practitioners to others in the context of specific tasks that go beyond the purview of any practice. People confront problems that are outside the realm of their competence but that force them to negotiate their own competency with the competencies of others. Moreover, competence and experience are in different relationships at the core and at the boundaries of practices, at the encounters between generations, and in power relationships among participants. The innovation potential of a system lies in its combination of strong practices and active boundary processes—people who can engage across boundaries, but have enough depth in their own practice to recognize when something is really signifi-cantly new. Accordingly, simultaneous participation in communities of practice and project teams creates learning loops that combine application with capability development [23]. In PBL, researchers discuss how to apply this element when designing complex projects for students [15]. Furthermore, researchers have sug-gested this element should be in a curriculum framework of creativity development [8], for much new thinking at the level of "high creativity" does involve the merging of ideas from two or more disciplines [8, 15, 17].

Project management is essential to support learning activities in PBL. Amabile [6] suggested good project management is one of the qualities in work environ-ments that serve to promote creativity from a social psychology approach. Meanwhile, other related qualities have been proposed, such as freedom in deciding what to do or how to accomplish the task, the sense of control over one's own work and ideas, management enthusiasm for new ideas and having the ability to create an atmosphere free of threatening evaluation, sufficient resources of time, pressure and so on. The ability of a manager will be tested to the utmost when complex technical changes demand a high level of corporate activity. A premium is placed upon fixing clear objectives, setting up high-response decision making, communication, and control systems to enable a wide range of resources and disparate talents to be fully harnessed [21]. The social theory of learning indicates the project group works as a community and needs multiple forms of leadership: thought leaders, networks, people who document the practice, pioneers, etc. These forms of leadership may be represented by one or two members of the group or may be widely distributed and will potentially change over time [23]. Although most studies on project manage-ment are concerned with business contexts, especially when creativity is discussed [10, 12], de Graaff and Kolmos [12], Zhou et al. [17], and Smith [29] think it is a way for students to have a social approach to PBL, for students are encouraged to involve themselves in it. Related to the practice of PBL, researchers have observed project management from perspectives of group management, group building and development, and knowledge management to study how students share, collect and exchange knowledge [17].

Facilitation of supervisors is critical to making PBL function well. Many studies emphasized teachers should hold on to the philosophy of student-directed education when solving problems with group work in PBL [18]. According to Hmelo-Silver [14], the PBL facilitator plays an important role in modeling problem solving and self-directed learning skills needed for assessing one's reasoning and understanding, while supporting the learning and collaboration processes, which make students better able to construct flexible knowledge. Facilitation is a subtle skill. It involves knowing when an appropriate question must be asked, sensing when the students are going off-track, and noting when the PBL process is stalled. For example, Dolmans [30] suggested that if group work had a negative outcome because some group members contributed less than the necessary amount of work to the group's activities, supervisors would help the group to perform better as a whole and could develop group spirit by regularly conducting evaluations, making the tutorial group meetings clear, and reiterating his or her expectations about attendance, active participation, etc. Thus, group learning is no guarantee of successful learning, but the stimulation of interactions between students is a prerequisite, which is the same with group creativity development. Some challenges accordingly have been discussed in the shift from teacher-led to student-centered education [14, 30]. PBL places high demands on the problems used and on the skills of the supervisor, to ensure that cooperative learning positively influences students or leads to better learning than individual learning [21].

The above review of the links between PBL and elements of PBL, and creativity is helpful to develop a theoretical model for deeper understanding of these links, a discussion that follows in the Conclusion.

3 Conclusions

The five elements reviewed above are the main pillars for constructing a creative community practice to develop creativity in PBL. As indicated below (see Fig. 2), student groups are the basic way to organize the learning activities in PBL; we therefore define group learning as a main element to develop creativity and the other four as subelements of it. Meanwhile, these elements influence each other within PBL systems, construct this system as a stimulus of social practice, and appreciate conditions of group interactions to generate both individual creativity and group creativity. In other words, creativity development happens in the interplay between these elements and such process is embedded in a situated-learning community built by PBL. In this sense, we view PBL as a community practice due to its theoretical roots of social approaches to learning and its focus on implementing solutions to real-world projects.

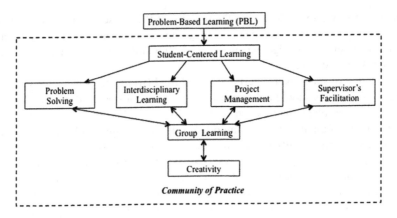

Fig. 2 PBL elements bridging group and creativity

The model developed in Fig. 2, based on the literature review in this paper, has both theoretical and practical significance. Theoretically, it calls for a systematic view when studying the relationship between creativity and learning environment, and advises using diverse channels when linking creativity and PBL. Practically, it provides a guideline for teaching creativity in PBL environments and pays key attention to creativity development by PBL strategy to realize both effective learning and teaching. In the future, some interesting research questions may concern the challenges of teaching staff in building a creative learning environment through roles of PBL elements, examine how students deal with the uncertainty and complexity of learning process in the community of PBL practice, and investigate how supervisors and students construct their social identities of "teacher" and "learner" in creativity development in a PBL environment. Such questions indicate the need to thoroughly unpack the black box of PBL and explore its influences on creativity.

References

1. Mitchell, W.J., Inouye, A.S., Blumenthal, M.S.: Beyond Productivity: Information Technology, Innovation, and Creativity. The National Academies Press, Washington, DC (2003)
2. Craft, A., Jeffrey, B., Leibling, M.: Creativity in Education. Countinuum, London (2001)
3. Sternberg, J.P.: Handbook of Creativity. Cambridge University Press, New York (1999)
4. Guilford, J.P.: Creativity. Am. Psychol. **5**, 444–454 (1950)
5. Sternberg, R.J.: Creativity or creativities? Int. J. Hum. Comput. Stud. **63**, 370–382 (2005)
6. Amabile, T.M.: Creativity in Context: Update to the Social Psychology of Creativity. Crown, New York (1996)
7. Glaveanu, V.P.: Paradigms in the study of creativity: introducing the perspective of cultural psychology. New Ideas Psychol. **28**, 79–93 (2010)
8. Craft, A.: Creativity in Schools, Tensions and Dilemmas. Routledge, New York (1995)

9. Singer, I.: Modes of Creativity, Philosophical Perspective. MIT Press, London (2011)
10. Paulus, P.B.: Group creativity. In: Runco, M.A., Pritzker, S.R. (eds.) Encyclopedia of Creativity, vol. 1, pp. 779–784. Academic Press, New York (1999)
11. Miell, D., Littleton, K.: Collaborative Creativity: Contemporary Perspectives. Free Association, London (2004)
12. De Graaff, E., Kolmos, A.: History of problem-based learning and project-based learning. In: De Graaff, E., Kolmos, A. (eds.) Management of Change, Implementation of Problem-Based and Project-Based Learning in Engineering, pp. 1–8. Sense Publishers, Rotterdam (2007)
13. Newman, M.J.: Problem based learning: an introduction and overview of the key features of the approach. JVME **32**, 12–20 (2005)
14. Hmelo-Silver, C.E.: Problem-based learning: what and how do student learn? Educ. Psychol. Rev. **16**, 235–266 (2004)
15. Zhou, C., Kolmos, A., Nielsen, D.: A problem and project-based learning (PBL) approach to motivate group creativity in engineering education. Int. J. Eng. Educ. **28**(1), 3–16 (2012)
16. Nijstad, B.A., Paulus, P.B.: Group creativity: common themes and future directions. In: Paulus, P.B. (ed.) Group Creativity: Innovation Through Collaboration, pp. 326–346. Oxford University Press, New York (2003)
17. Zhou, C., Kolmos, A.: Interplay between individual and group creativity in problem and project-based learning (PBL) environment. Int. J. Eng. Educ. **29**, 866–878 (2013)
18. Eteläpelto, A., Lahti, J.: The resources and obstacles of creative collaboration in a long-term learning community. Thinking Skills Creativity **3**, 226–240 (2008)
19. Zhou, C.: Group creativity development in engineering education in problem and project-based learning (PBL) environment. Ph.D. Thesis, Akprint, Aalborg, Denmark (2012)
20. Tan, O.S.: Problem-Based Learning and Creativity. Cengage Learning Asia Pte Ltd., Singapore (2009)
21. Zhou, C., Kolmos, A., Du, X., Nielsen, J.F.D.: Group creativity development by solving real-life project in engineering education. In: Excellence in Education 2009–2010: Leading Minds Creating the Future: Proceedings of the Annual Conference of the International Centre for Innovation in Education (ICIE), held in Ulm-Germany (August 24–27, 2009); and Athens-Greece (June 8–12, 2010). ICIE (International Centre for Innovation in Education), Ulm, Germany (2011)
22. Jackson, N., Sinclair, C.: Developing students' creativity, searching for an appropriate pedagogy. In: Jackson, N., Oliver, M., Shaw, M., Wisdom, J. (eds.) Developing Creativity in Higher Education: An Imaginative Curriculum, pp. 118–141. Routledge, London (2006)
23. Wenger, E.: Communities of Practice, Learning, Meaning and Identity. Cambridge University Press, New York (1998)
24. Runco, M.A.: Creativity, Theories and Themes: Research, Development, and Practice. Elsvier Academic Press, London (2007)
25. Isaksen, S.G., Treffinger, D.J.: Celebrating 50 years of reflective practice: versions of creative problem solving. J. Creative Behav. **38**, 75–101 (2004)
26. Lohman, M.C., Finkelstein, M.: Designing groups in problem-based learning to promote problem-solving skill and self-directedness. Instr. Sci. **28**, 291–307 (2000)
27. Nijstad, B.A., Stroebe, W.: Four principles of group creativity. In: Thompson, L.L., Choi, H.S. (eds.) Creativity and Innovation in Organizational Team, pp. 161–177. Lawrence Erlbaum Associates, London (2006)
28. Rogoff, B.: Developing understanding of the idea of community of learners. Mind Cult. Act. **1**, 209–229 (1994)
29. Smith, G.F.: Problem-based learning: can it improve managerial thinking? J. Manage. Educ. **29**, 357–376 (2005)
30. Dolmans, D.: Solving problems with group work in problem-based learning: hold on to the philosophy. Med. Educ. **35**, 884–889 (2001)

Password Recovery Using Graphical Method

**Wafa' Mohd Kharudin, Nur Fatehah Md Din
and Mohd Zalisham Jali**

Abstract Authentication with images or better known as graphical password is gaining its recognition as an alternative method to authenticate users, for it is claimed that images or pictures are easier to use and remember. The same method can be applied to password recovery, with the purpose to ease the process of users in regaining their account in case of forgotten passwords. A total of 30 participants were asked to use a prototype implementation of graphical password recovery and provide feedbacks. The data gained were analyzed in terms of attempts, timing, pattern, and user feedback. Overall, it was found that participants had no problem in using graphical password recovery despite they were new to it. Most of them preferred the choice-based method, even though they agreed that it provided less security. Graphical recovery has potential to be used more widely in current technology, although more works need to be done to balance the issues of usability and security.

Keywords Graphical password · Password recovery

1 Introduction

Normally upon signing up on systems which require usernames and passwords, for example, social media sites (i.e., Facebook and Twitter) and e-mail accounts (Gmail and Yahoo), users are also required to fill up the recovery options for their passwords. The purpose of these recovery options is to make sure users can still login or regain their account in case of passwords forgotten. There are a few options of recovery methods that are being used by almost all systems and sites such as recovery using challenge questions, recovery by e-mail, and recovery by text message.

Recovery by using challenge questions works by asking users to select desired questions from a set of questions and then gives answers to those questions. This

W.M. Kharudin (✉) · N.F.M. Din · M.Z. Jali
Faculty of Science and Technology, Universiti Sains Islam Malaysia, Bandar Baru Nilai,
71800, Nilai, Negeri Sembilan, Malaysia
e-mail: wafamohdkharudin@gmail.com

© Springer International Publishing Switzerland 2015   11
A. Abraham et al. (eds.), *Pattern Analysis, Intelligent Security
and the Internet of Things*, Advances in Intelligent Systems and Computing 355,
DOI 10.1007/978-3-319-17398-6_2

kind of method will help the system to recognize legitimate users later in case if they forget their passwords. Only users who can answer all challenge questions correctly will be granted access to recover or reset their old password. Recovery by e-mail or text message works by the system sending a reset link (or a new password) to their preregistered e-mails or mobile phone numbers.

However, how secure it actually is to be using these methods in case of forgotten passwords? It is no doubt that all the methods provide some sort of protection to users' accounts and data, but then again, is it perfectly secure? Imagine a situation where a hijacker tries to gain access on someone's Facebook account and the hijacker has somehow managed to gain control of the particular user's e-mail account or mobile phone. Then, it will be so easy for the hijacker to log into the user's Facebook account or any other accounts registered under the same e-mail address or phone number. This situation is perfectly possible which could result in breaches of data and even more severe consequences, so it is very crucial to address this issue.

Having said that, we are proposing graphical recovery as an alternative for password recovery. The idea of graphical recovery is to apply graphical methods in graphical recovery technique. The aim of this study was to investigate the usability of graphical recovery. A trial was conducted where 30 participants were asked to use a prototype and later provide feedback. The data collected were then analyzed in terms of number of attempts, timing, pattern, and user feedback.

This paper is arranged as follow. Section 2 discusses the state of the art of graphical authentication, which also highlights the advantages of graphical authentication, especially from psychological aspect. Section 3 describes the methodologies used in this study, while Sect. 4 provides the results gained from the trial. Lastly, conclusion and future works are discussed in Sect. 5.

2 Graphical Recovery

Password recovery process is just as essential as login process, where both can be compromised by malicious attempts. A non-legitimate user might gain access to an account by using some techniques to recover someone's password. One of the most compelling reasons for exploring the use of a graphical method comes from the fact that humans seem to possess a remarkable ability for recalling pictures, whether they are line drawings or real objects [1].

Graphical authentication is not a new thing. Beginning around 1999, a multitude of graphical password schemes have been proposed, motivated by the promise of improved password memorability and thus usability, while at the same time improving strength against guessing attacks [2]. It is reported that there is a growing interest in using pictures as an authentication method, but not much research has been done so far. But the research on this area has now started gaining more attention from researchers; from their classifications up until to their specific applications, both positive and negative findings were reported [3].

From the psychological view, the usability of graphical password schemes is promising as many studies have explained about the 'picture superiority effects' toward verbal and words. The idea behind graphical method is to leverage human memory for visual information, with the shared secret being related to or composed of images or sketches. A number or psychological studies explaining the 'picture superiority effects' toward verbal and words explain the usability of graphical passwords. A study claimed that humans have exceptional ability to recognize images previously seen, even those viewed very briefly [4].

In a study by [5], four experiments were conducted to examine the relationship between perception and memory. The first two experiments were about memory recognition for pictures. Experiment 1 used 1100 pictures taken from the magazines, with Experiment 2 used 2560 pictures obtained from the photographers. Overall, they found that participants scored up to 95 % success for Experiment 1 and for Experiment 2, participants still scored 85 % recognition success even after 4 days time. The last two experiments were about the effect of duration and the effect of reversing and orienting the pictures during viewing. From the results, it was summarized that participants still managed to score above 90 % success rate even the images were reversed. On the whole, they concluded that participants managed to obtain higher success rate for picture recognition. These psychological studies have given an insight to the claim that using images or pictures was superior to using words, with regard to recognizing and memorizing.

In this paper, graphical techniques were grouped into three categories, namely 'choice-based,' 'draw-based,' and 'click-based.' These categories were solely based on the users' actions while carrying out authentication tasks. Briefly, choice-based refers to the action of selecting a series of images from among a larger set of images, draw-based refers to the action of drawing a pattern on an image, whereas click-based refers to the users' action clicking on areas within a given image [6].

The idea of graphical recovery is to allow users to log into their accounts using the conventional usernames and text passwords method as they are used to. But in case of forgotten passwords, instead of recovering their accounts through challenge questions which they might forget the answers of, or by e-mail which may exposed to the risk of being controlled by a hijacker, they can recover their passwords by using graphical method. Graphical recovery is advantageous in the sense that only the right user would know the secrets, it is more memorable as images can trigger user's memory, and it is also more secure as compared to other types of recoveries such as by e-mails or phone.

3 Methodology

In order to test the feasibility of graphical password recovery, a prototype on graphical recovery was developed and a survey was conducted. A number of 30 participants with different backgrounds were asked to use the prototype and then answer a related questionnaire. This activity took approximately 10–20 min to complete depending on the participants' familiarity of using computers. The same participants were tested a week later using the same prototype to reproduce the same secrets, without any prior information.

There were two main modules in the prototype—registration and recovery. In the registration module, participants needed to register their recovery secrets by entering a desired username and password, and then, they needed to choose a picture out of a selected theme, with each theme consisted of a set of images (choice-based). Participants then needed to draw a line on the image they have chosen (draw-based), and finally, they needed to click three times on the same image (click-based). In the recovery module, participants can recover their passwords by using their secrets, with no limitations were made toward the number of trials they were allowed to do.

The development of the choice-based method was made with reference to the scheme [7–9]. The photographs were taken from previous research by [6]. In choice-based, participants were needed to choose an image from three themes which were flowers, places, and animals. These themes were chosen because they were common images which were easy to recognize and remember. Click-based method was made with reference to the scheme by [6, 10], where the tolerance scale of the image was 18×18 pixels. The display size of the images was 320×320 pixels.

All three methods of graphical authentication were combined, making it a hybrid instead of just taking one method with the purpose to make it less guessable. The combination of these three methods was hopefully would result in a hard-to-guess secret, which was not only advantageous in the aspect of security, but also aided memorability and usability.

The setting for this trial was set for the participants to do as following:

1. Register username and password.
2. Register the secrets for graphical recovery.

   (a) Select an image from given theme.
   (b) Draw a line on the image.
   (c) Click 3 points on the image.

3. Test the secrets.
4. A week later—retest the secrets (Figs. 1, 2, and 3).

Fig. 1 The screenshot of the prototype's main menu

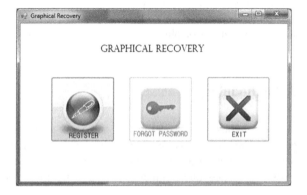

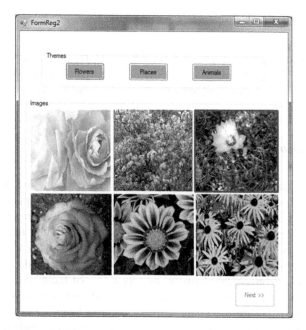

Fig. 2 Screenshot from choice-based secret

Fig. 3 Screenshots of examples from draw-based secret (*left*) and click-based secret (*right*)

4 Results and Findings

A total of 30 participants took part in this study (14 males and 18 females), with an average age of 21 (sample range from 12 to 30). The participants were of various backgrounds, and all of them had more than 3 years of experience using computers.

4.1 Observation

All participants started their registration after being briefed about how the prototype worked. Overall, only 13 out of 30 participants managed to complete both registration and recovery tasks without any failed attempts. The remaining participants needed 2–5 attempts before completing their tasks. These suggest that appropriate training should be provided beforehand for the graphical recovery to be effective.

Based on the observation, it was apparent that the participants were initially quite confused with graphical recovery, as only a few of them were aware about graphical authentication. Only 6 participants claimed that they were aware of it through personal exposure, while the others have never knew about it. Majority of them had problems to understand how it worked, and they drew and clicked on the image during registration without taking into account the memorability of their secrets. As a result, during confirmation process, they needed a couple of (two to four) attempts until they got their secrets right.

4.2 Attempts and Memorability

All participants managed to successfully completed all the tasks required (registration, confirmation, and recovery). For the first step which was choosing an image (choice-based), all participants were able to complete the choosing task with only one attempt. The second step was to draw a single line on the chosen image (draw-based). The number of attempts was significantly higher for this step as they had to precisely draw the exact line as they did for the first time.

The third step required the participants to click 3 points on the same, chosen image (click-based). The number of attempts for this step was also as high as the draw-based. These results were predicted as participants had to carefully click on their secret areas in sequence, which sometimes they did not manage to do.

A week later, the participants were tested again if they remembered their secrets. Most of the participants were able to do choice-based step with only one attempt. However, majority of them needed several (two to five) attempts until they got their secrets for draw-based and click-based right.

4.3 Timing

Each participant's registration and recovery duration were recorded to calculate their average time. The time was measured from they first chosen their secret image until they finished with their drawing and clicking secrets. Table 1 gives the mean and standard deviation (SD) for each registration and recovery process. Note that each process consisted of the three methods—choice-based, draw-based, and click-based.

It was noticeable that participants took longer during registration compared to the confirmation and recovery tasks. It is probably because during registration, it was the first time for most of the participants to be using any sort of graphical authentication or recovery, so it took them time to familiarize themselves with the state of the art of it. As they became clear with the process, it can be seen from the table that the participants took much lesser time for confirmation and immediate recovery.

All participants were tested again a week later to see if they can reproduce their secrets—and from the results in Table 1, they took almost double the time they needed for immediate recovery. All of them did not expect to be tested again; therefore, they did not try to remember their secrets. Majority of them needed two to four attempts until they got their secrets right.

4.4 Pattern

For the choice-based method where participants needed to choose an image, it can be seen that the image chosen was mostly influenced by participants' personal preferences. For example, female participants were most likely to choose image from flowers or animals theme, while male participants tend to choose image from places theme. They would choose the image that they found the most interesting or beautiful. Table 2 shows image popular from each theme.

For the draw-based where they had to draw a line on the chosen image, most of the participants chose to draw from and stop at a point that was sharp (i.e., the tip of a finger). There is admittedly a constraint at this part of study where participants were only allowed to draw a single line. If they were allowed to draw more complex pattern on the image, presumably more interesting patterns can be found.

Table 1 Mean and standard deviations for time taken

N = 30		Time (s)
Registration	Mean	228
Confirmation	Mean	78
Recovery—immediate	Mean	48
Recovery—one week later	Mean	96

Table 2 Popular image from each theme

Theme	Image description	No. of participants
Flowers	Sunflower	7
	Single pink rose	5
Places	Sea port	4
	Flight runway	2
Animals	Lions	4
	Cats	4

For the click-based where participants needed to choose 3 click-points, it was found that most of them liked to click on something sharp or edgy (i.e., the edgy petal of a flower). This would do favor in accuracy aspect, as participants were more likely to accurately clicked on their secret points. Figure 4 shows examples of secrets made by a few participants.

4.5 Participants' Feedback

In general, participants had different perceptions for each method of the whole process. For the choice-based method, 28 out of 30 participants agreed that they could remember their images well and had no problems to perform the task of choosing an image. However, 21 participants thought that it was too vulnerable for security attacks and thus did not offer good security.

For the draw-based, 22 participants agreed that they could remember their secrets well, despite some of them suggested that they should be allowed to draw

Fig. 4 Samples of click secrets. *Different shapes* represent different users

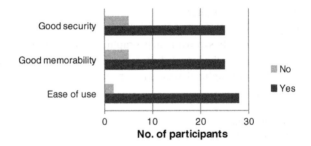

Fig. 5 General feedback from participants based on questionnaire

more complex patterns on the image. However, complex patterns may cause disadvantages in terms of memorability.

For the click-based, 25 participants felt that it was easy to use and agreed that they could remember their click-points well. Majority of them also agreed that click-based provided better security as compared to choice-based. Around 10 participants suggested that they should be allowed to click as many click-points as they like.

Overall, it can be seen that this graphical recovery prototype was well received by the participants, despite some suggestions were made based on their preferences and ease of use (Fig. 5).

5 Conclusion

From the results of survey made to 30 participants, it can be said that many are still not familiar with graphical authentication or recovery. They found it interesting, albeit a little confused with the state of the art of graphical recovery in the beginning. After a brief explanation, all of them were able to carry out all required tasks.

The prototype consisted of a hybrid or 3 combined methods which were choice-based, draw-based, and click-based. This has made participants to develop preferences over which method they liked best. Majority of them preferred choice-based, for its memorability and ease of use. However, they also agreed that choice-based might be prone to attacks such as guessing and shoulder surfing. In regard to security, many of them preferred draw-based and click-based. Nonetheless, participants agreed that this hybrid method which combined all these three graphical methods provided adequate security for their secrets.

Participants were tested again a week after the first survey was conducted to test their memorability on their secrets. Despite taking longer time than the previous week, majority of the participants were able to recall their secrets.

One of the lessons learned from this evaluation came from a number of feedbacks that suggested participants wanted to be allowed to draw any pattern as they liked and clicked as many click-points as they wanted. They believed that this would provide tighter security. However, from the researcher's point of view, this may also lead to less memorability.

This study has proven that graphical recovery has potential to be implemented more widely in the future. Future work will focus on the mechanism of control to be provided in graphical recovery, in order to balance the usability and security of graphical method.

Acknowledgments The authors wish to thank USIM for funding this research. This research is funded under the USIM grant scheme with reference number of PPP/FST/SKTS/30/13612.

References

1. Monrose, F., Reiter, M.K.: Graphical Passwords. Human Centered Systems Group. Department of Computer Science, University College London, London (2005)
2. Biddle R., Chiasson, S., Oorschot, P.C.: Graphical Passwords: Learning from the First Twelve Years. (2010)
3. Ray, P.P.: Ray's scheme: graphical password based hybrid authentication system for smart hand held devices. J. Inf. Eng. Appl. **2**(2) (2012)
4. Standing, L., Conezio, J., Haber, R.: Perception and memory for pictures: single-trial learning of 2500 visual stimuli. Psychon. Sci. **19**(2), 7374 (1970)
5. Standing, L.: Learning 10,000 pictures. Q. J. Exp. Psychol. **25**, 207–222 (1973)
6. Jali, M.Z.: A Study of Graphical Alternatives for User Authentication. School of Computing and Mathematics, Faculty of Science and Technology, University of Plymouth (2011)
7. Passfaces. http://www.realuser.com/personal/index.htm
8. Dhamija, R., Perrig, A.: Déjà Vu: a user study using images for authentication. In: 9th USENIX Security Symposium (2000)
9. De Angeli, A., Coventry, L., Johnson, G., Renaud, K.: Is a picture really worth a thousand words? Exploring the feasibility of graphical authentication systems. Int. J. Hum. Comput. Stud. **63**(1–2), 128–152 (2005)
10. Wiedenbeck, S., Waters, J., Birget, J., Brodskiy, A., Memon, N.: PassPoints: design and longitudinal evaluation of a graphical password system. Int. J. Hum. Comput. Stud. **63**(1–2), 102–127

A Classification on Brain Wave Patterns for Parkinson's Patients Using WEKA

Nurshuhada Mahfuz, Waidah Ismail, Nor Azila Noh,
Mohd Zalisham Jali, Dalilah Abdullah and Md. Jan bin Nordin

Abstract In this paper, classification of brain wave using real-world data from Parkinson's patients in producing an emotional model is presented. Electroencephalograph (EEG) signal is recorded on eleven Parkinson's patients. This paper aims to find the "best" classification for brain wave patterns in patients with Parkinson's disease. This work performed is based on the four phases, which are first phase is raw data and after data processing using statistical features such as mean and standard deviation. The second phase is the sum of hertz, the third is the sum of hertz divided by the number of hertz, and last is the sum of hertz divided by total hertz. We are using five attributes that are patients, class, domain, location, and hertz. The data were classified using WEKA. The results showed that BayesNet gave a consistent result for all the phases from multilayer perceptron and K-Means. However, K-Mean gave the highest result in the first phase. Our results are based on a real-world data from Parkinson's patients.

N. Mahfuz · W. Ismail (✉) · M.Z. Jali
Faculty Science and Technology, Universiti Sains Islam Malaysia,
Bandar Baru Nilai, Negeri Sembilan, Malaysia
e-mail: waidah@usim.edu.my

N. Mahfuz
e-mail: shuhadamahfuz@yahoo.com

M.Z. Jali
e-mail: zalisham@usim.edu.my

N.A. Noh
Faculty of Medicine and Health Sciences, Universiti Sains Islam Malaysia,
Pandan Indah, Kuala Lumpur, Malaysia
e-mail: azila@usim.edu.my

D. Abdullah
Computer Science Department, Universiti Kuala Lumpur, Kuala Lumpur, Malaysia
e-mail: dalilah@unikl.edu.my

Md.J.b. Nordin
Faculty of Information Science and Technology, Universiti Kebangsaan Malaysia,
Bangi, Malaysia
e-mail: jan@ftsm.ukm.my

© Springer International Publishing Switzerland 2015
A. Abraham et al. (eds.), *Pattern Analysis, Intelligent Security
and the Internet of Things*, Advances in Intelligent Systems and Computing 355,
DOI 10.1007/978-3-319-17398-6_3

Keywords Brain wave · Classification · Parkinson's patient

1 Introduction

Electroencephalography (EEG) is a frequently applied medical measurement technique. By measuring the electrical potentials caused by the activation of the neurons in the brain, through electrodes allocated on the scalp, information about the neurological processes in the brain can be extracted. The measured EEG is basically a very restricted indicator of such processes, being influenced by a high level of noise, as well as being a summation of numerous processes in the brain. EEG is used in recording the changes in brain waves when the feeling or emotion changes [1, 2]. EEG has high speed, is non-invasive and causes no pain to the human subjects [3], with minimum expense, and analysis can be performed compared to other medical imaging technique [4].

The data were provided by medical specialists from Universiti Sains Islam Malaysia (USIM) based on the study of Parkinson's patients. This study is to find the "best" classification in the WEKA. WEKA is, in this paper, the preprocessing performed using statistical method. In this preprocessing were required four phases which are first phase between raw data and after statistical method. Second phase is the sum of hertz for raw data and after data processing using statistical method. Third phase is the sum of hertz divided by number of hertz for both raw data and after data processing using statistical method. Last is the sum of hertz divided by total of hertz. The conclusion of this paper shows that BayesNet in the WEKA gave a better result and consistency for all the phases although K-Means gave the highest result in the first phase.

This paper is organized as follows: In the rest of this section, we detail the motivation behind our paper; in Sect. 2, we describe previous work in the area, Sect. 3 details our proposed methodology which explains on the four phases, and Sect. 4 explains in detail our methods.followed by Sect. 5 in which we discuss the results. Lastly, in Sect. 6, we draw conclusions and discuss future research.

1.1 Electroencephalography (EEG)

The study of human brain is not a new thing. In fact, people have been trying to comprehend the science that lies behind the very complex structure of the brain since the ancient times of the Roman Empire, as there was a record about the Greek anatomist Galen who dissected the brains of sheep, monkeys, dogs, swine, among other non-human mammals. Galen concluded that, as the cerebellum was denser than the brain, it must control the muscles, while as the cerebrum was soft, it must be where the senses were processed. Galen further theorized that the brain functioned by the movement of animal spirits through the ventricles [5]. Since then, there have been extensive studies being carried in order to try to understand the brain, which certain neuroscientists' believes might lead to a lot more discoveries about the mind, memory, language, and more importantly, the health status of a person. Today, it is likely that the study on the wave emitted by the brain, known as brain wave, might be the key to diagnose the health status of a person and the kind of disease that a person might be suffering from, which includes Parkinson's disease [6], attention-deficit/hyperactivity disorder (ADHD) [7] and International Affective Picture System (IAPS) [8]. In this literature review, we will be discussing in depth on the classification of the pattern of brain wave that leads to Parkinson's disease.

EEG is an electrophysiology technique that helps to record electrical activity along the scalp. It represents complex irregular signals that may provide information about underlying neural activities in the brain [9]. In neuroscience, EEG is used as a direct medical measurement of the electrical signals in the brain. By measuring the electrical potentials caused by the activation of the neurons in the brain, through electrodes located in the scalp, information about the neurological processes in the brain can be extracted. The measured EEG is basically a very restricted indicator of such processes, being influenced by a high level of noise, as well as being a summation of numerous processes in the brain [10]. Figure 1 shows an EEG cap that is used to obtain raw data of brain wave from a patient.

1.2 Brain Wave

Brain wave, by definition, is the rapid fluctuations of voltage between parts of the cerebral cortex that are detectable with an EEG device. In the brain, there are four basic types of waves that can be distinguished. Each of these waves can be a dominant wave in a period of time.

Fig. 1 EEG cap (*Source* https://cogneuro.byu.edu)

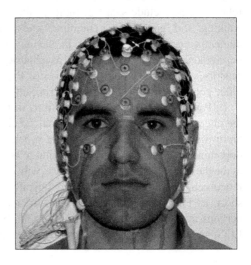

Each wave is identified by amplitude and an interval of frequencies [11]:

(a) Beta waves are in the frequency range of 12–30 Hz. The waves are small and fast, associated with focused concentration and best defined in central and frontal areas. When resisting or suppressing movement, or solving a math task, there is an increase in beta activity.

(b) Alpha waves, ranging from 7.5 to 12 Hz, are slower and associated with relaxation and disengagement. Thinking of something peaceful with eyes closed should give an increase in alpha activity.

(c) Theta waves, ranging from 3.5 to 7.5 Hz, are linked to inefficiency and daydreaming, and the very lowest waves of theta represent the fine line between being awake and in a sleep state. Theta arises from emotional stress, especially frustration or disappointment.

(d) Delta waves, ranging from 0.5 to 3.5, are the slowest waves and occurs when sleeping.

1.3 Parkinson's Disease

Parkinson's disease (PD) is also known as primary parkinsonism or hypokinetic rigid syndrome is a degenerative disorder of the central nervous system. The motor symptoms of Parkinson's disease result from the death of dopamine-generating cells in the substantia nigra, a region of the midbrain; the cause of this cell death is unknown [12]. Basically, what causes Parkinson's disease to develop in the first place is still a mystery, but scientists know that the disease process begins when the brain becomes deficient of a neurotransmitter called dopamine. With diminishing amounts of dopamine, a person with Parkinson's disease will develop several motor symptoms such as movement disorders, tremors, and rigidity [13].

Early in the course of the disease, the most obvious symptoms are movement related; these include shaking, rigidity, slowness of movement, and difficulty with walking and gait. Later, cognitive and behavioral problems may arise, with dementia commonly occurring in the advanced stages of the disease, whereas depression is the most common psychiatric symptom. Other symptoms include sensory, sleep, and emotional problems. Parkinson's disease is more common in older people, with most cases occurring after the age of 50 [13].

In Parkinson's disease, the alterations in basal ganglia physiology may involve the alteration in the pattern of neuronal synchronization particularly involving beta brain rhythms [14]. The level of beta synchronization is in turn modulated by net dopamine levels at sites of cortical input to basal ganglia [15]. Dopamine deficiency as in the case of Parkinson's disease will disrupt the cortico-basal ganglia-thalamocortical circuits, leading to pathologically exaggerated beta oscillations [16]. In short, as Parkinson's disease is a disorder that closely related to the neural and nervous system, therefore by observing the brain wave activity of a person using EEG, we can diagnose if that person has possible motor symptoms of Parkinson's disease. For instance, as explained above, a person with an exaggerated beta wave emitted by the brain most probably is suffering from Parkinson's disease.

2 Previous Work

This is only a preliminary study to find the "best" classification and preprocessing data. Nowadays, classification is widely used in pattern recognition, which includes a number of information processing problems from speech recognition and the classification of handwritten characters to fault detection in technology and medical diagnosis. WEKA is a data analysis software tool which implements a set of machine learning algorithms for data mining tasks [17]. The WEKA workbench is an organized collection of state-of-the-art machine learning algorithms and data preprocessing tools [18]. The choice of the "best" classifier required the performance of a number of experiments from the artificial intelligence networks (ANN) methods.

In the effort to develop classification of emotions, numerous signal processing and artificial intelligence method were utilized to analyze brain waves [19]. In [2], researcher had used the combination of Fast Fourier Transform (FFT), wavelet transform, principal component analysis and, mean and variance to extract features from the EEG data. Neural network is applied to classify four types of emotions (joy, sorrow, relax, and anger) and achieve success rate at 67.7 %. Another possible approach of brain wave classification was presented by [20], and the combination between optimization of neural network and weights of backpropagation as special decoder [3] had computed six statistical features (mean, standard deviation, the means of the absolute values of the first differences of the raw signals, the means of the absolute values of the first differences of the normalized signals, the means of the absolute values of the second differences of the raw signals, the means of the absolute values of the second differences of the normalized signals) from EEG data, and then, backpropagation neural networks are applied to classify human emotions. The results give highest classification rate at 95 %. Heraz et al. [8] investigated on the use of machine learning techniques to predict the three major dimensions of learner's emotions (pleasure, arousal, and dominance). The participants of their study were exposed to a set of pictures from the IAPS using nearest neighbor algorithm. Murugappan et al. [21] had also studied on classifying emotions from EEG signals of brain wave using two simple pattern classification methods K-nearest neighbor (KNN) and Linear discriminant analysis (LDA) for classifying emotions [22] used window or time frame to find the emotional of the patients. WEKA's implementation of linear regression and C4.5 was used to build the emotional models for relaxing and stressful music. Various classification algorithms used in brain–computer interface (BCI) system based on EEG were reviewed in [23]. Common classifications used in BCI systems are linear classifier, neural networks, nonlinear Bayesian classifier, nearest neighbor classifiers, and combinations of classifiers. However, Lip et al. [23] also stressed that some famous kinds of classifier have not be tried in BCI research. Decision trees and whole category of fuzzy classifiers are the most significant ones.

3 Work Process

In this paper, we present four phases. The first phase is the raw data and after data processing using statistical method which are calculations of baseline, standard deviation, and mean as preprocessing. The second phase is the sum of hertz based on delta, theta, alpha, and beta as shown in Table 1 between raw data and pre-processing process. The third phase is the sum of the hertz divided by total of hertz and last is the sum of hertz divided by total of hertz. Then, we use WEKA to find the "best" classification. This is only a preliminary study to find the "best" classification for emotional model for Parkinson's disease (Figs. 2 and 3).

x raw data
T new data after converted into baseline.

4 Method

In the following sections, we present the methods that we have used in this paper.

Step 1: Data gathering
The data from Parkinson were collected from Faculty of Medicine, USIM. The brain wave was collected from the frontier of the head.
Step 2: Preprocessing
Then, collection of data will be removed with all the unclean data which we performed manually as shown in Fig. 1.
Step 3: Design model
The emotional model will be tested for finding the relevant intelligent classifier such as Bayes, multilayer perceptron, and K-Mean in WEKA.

Table 1 Brain wave classification

Type	Frequency (Hz), amplitude (mV)	Normally
Delta	0.5–3 Hz, 20–200 mV	Adults slow wave sleep in babies
Theta	3–7 Hz, 5–100 mV	Young children drowsiness or arousal in older children and adults
Alpha	8–12 Hz	Closing the eyes and by relaxation
Beta	14–30 Hz, 1–20 mV	Active, busy, or anxious thinking, active concentration
Gamma	30–100 Hz, 1–20 mV	Certain cognitive or motor functions

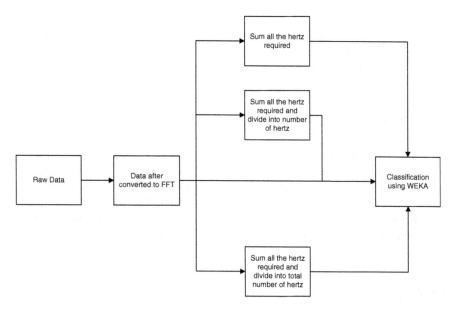

Fig. 2 The data classification with raw data (*x*). *x*-raw data

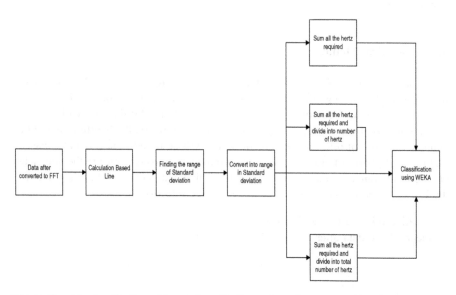

Fig. 3 The data classification with new data (*T*). *T*-new data after converted into baseline

Step 4: Classification
The emotional model will be classified based on Table 1. We did not use
Gamma because the data only on cognitive and motor function are not
reliable for Parkinson's patients.

4.1 Data Gathering

EEG data were recorded from eleven Parkinson's patients after undergoing non-
invasive magnetic stimulation, a form of therapy to alleviate the motor symptoms of
Parkinson's disease. In order to look at the acute effects of the stimulation, we
analyzed the EEG data recorded 20 s post-stimulation. The data sets were taken
from the medical faculty of USIM, Pandan Indah, Kuala Lumpur. The raw data
have been converted into FFT using brain wave software.

4.2 Preprocessing

In the preprocessing, we performed two methods that are statistical method and
classification.

4.2.1 Statistical Method

In the statistical method, we used three steps, which are finding baseline, standard
deviation and ensuring the range in the standard deviation. As for the finding
baseline, the medical doctor will determine the baseline. All the data will be converted into baseline as in Eq. (1). The lacking of the process, the baseline cannot be
counted and lose of information. The mean (ρ) will be found in new data (T) based
on the second part as in Eq. (2). The standard deviation (σ) is to ensure the new data
(T) are in the range based on the Eq. (4). If the new data (T) are not in the range of
the Eq. (4), then the new data (T) will be change into (σ).

j baseline (choose by medical doctor)
ρ mean of the data after converted into baseline.

$$T = \left(\frac{x - j}{j}\right) \times 100 \tag{1}$$

$$\rho = \frac{1}{N}\sum_{i=1}^{N} T_i \tag{2}$$

$$\sigma = \sqrt{\frac{1}{N}\sum_{i=1}^{N} T_i - \overline{T}} \tag{3}$$

$$p - (\sigma * 1.5) \leq T \geq \rho + (\sigma * 1.5) \tag{4}$$

$$T = \begin{cases} 1, & \text{if } T = \sigma \\ 0, & \text{Otherwise} \end{cases} \tag{5}$$

4.2.2 Classification

In the classification, we performed eight preprocessing that consists of four phases in finding the "best" classification in WEKA. In the first part, we took raw data (x) and new data (T). In the second part, we perform the sum of the hertz, in the third part, the sum is divided by the number of hertz, and the last part by sum of all hertz divided by the total hertz.

4.2.3 Attribute

In the attribute, we used as show in Table 2 that used in WEKA. The brain attribute is changes for of sum of the Hertz, sum and divided by the number of hertz, and sum all the hertz and divided by the total hertz.

5 Results

This section shows all the results for classification with 11-fold cross–validation, and choosing domain is hertz in WEKA.

Table 2 Attribute classification

Attributes	Description
Patient	Information about patient
Class	Based on the second capture which are 5 s = 1, 10 s = 2, 15 s = 3, and 20 s = 4
Domain	The domain based on the delta, theta, alpha, and beta
Location	Location of the brain wave cap which are F3, Fz, F4, C3, Cz, C4, P3, Pz, and P4
Brain	Result capture from the brain wave

5.1 Comparison Between Raw Data and After Data Processing Using Statistical Methods

In this experiment, we performed the raw data and new data after performing statistical method. The baseline of new data will be lost. We performed a few methods in the WEKA. Every test in the classification involved 11-fold to get "best" result among classification in WEKA. Table 3 only shows the eight results in the WEKA for the classification Bayes. Table 4 shows the eight results for the

Table 3 Result using WEKA for classification Bayes for raw data and after data processing using statistical method for four phases

Process data	Method	Percentage correct (%)	Percentage incorrect (%)
Classification Bayes			
1st phase experiment			
Raw data (x)	BayesNet	71.68	28.32
New data (T)	BayesNet	62.49	37.51
2nd phase experiment—sum of hertz			
Raw data (x)	BayesNet	48.11	51.89
New data (T)	BayesNet	29.96	70.03
3rd phase experiment—sum of hertz divided by total of hertz			
Raw data (x)	BayesNet	58.09	41.60
New data (T)	BayesNet	35.61	64.39
4th phase experiment—sum of hertz divided by sum of hertz			
Raw data (x)	BayesNet	61.93	38.06
New data (T)	BayesNet	56.23	43.77

Table 4 Result using WEKA for classification function for raw data and after data processing using statistical for four phases

Process data	Method	Percentage correct (%)	Percentage incorrect (%)
Classification function			
1st phase experiment			
Raw data (x)	Multilayer perceptron	62.30	37.69
New data (T)	Multilayer perceptron	62.75	37.25
2nd phase experiment—sum of hertz			
Raw data (x)	Multilayer perceptron	46.40	53.59
New data (T)	Multilayer perceptron	37.37	62.62
3rd phase experiment—sum of hertz divided by total of hertz			
Raw data (x)	Multilayer perceptron	58.33	41.66
New data (T)	Multilayer perceptron	34.68	65.31
4th phase experiment—sum of hertz divided by sum of hertz			
Raw data (x)	Multilayer perceptron	59.91	40.08
New data (T)	Multilayer perceptron	54.46	45.53

classification function, and Table 5 shows results for classification lazy. From the data analysis, it shows that Bayes' network gave a consistent result in the classification, and using raw data can produce better classification. This result can lead to find the "best" classification model in brain wave. The result is shown in Table 6.

Table 5 Result using WEKA for classification lazy for raw data and after data processing using statistical method for four phases

Process data	Method	Percentage correct (%)	Percentage incorrect (%)
Classification lazy			
1st phase experiment			
Raw data (x)	K-Means	75.90	24.09
New data (T)	K-Means	47.50	52.49
2nd phase experiment—sum of hertz			
Raw data (x)	K-Means	10.92	89.07
New data (T)	K-Means	13.72	86.27
3rd phase experiment—sum of hertz divided by total of hertz			
Raw data (x)	K-Means	17.23	92.11
New data (T)	K-Means	13.21	86.78
4th phase experiment—sum of hertz divided by sum of hertz			
Raw data (x)	K-Means	23.73	76.26
New data (T)	K-Means	22.72	77.27

Table 6 Result using WEKA for overall raw data and after data processing using statistical method for four phases

Process data	Method	Percentage correct (%)	Percentage incorrect (%)
Classification lazy			
1st phase experiment			
Raw data (x)	K-Means	75.90	24.09
New data (T)	K-Means	47.50	52.49
2nd phase experiment—sum of hertz			
Raw data (x)	BayesNet	48.11	51.89
New data (T)	BayesNet	29.96	70.03
3rd phase experiment—sum of hertz divided by total of hertz			
Raw data (x)	Multilayer perceptron	58.33	41.66
New data (T)	Multilayer perceptron	34.68	65.31
4th phase experiment—sum of hertz divided by sum of hertz			
Raw data (x)	BayesNet	61.93	38.06
New data (T)	BayesNet	56.23	43.77

6 Conclusion and Future Work

In this paper, we have presented a comparison between four phases by using raw data and preprocessing data. We used classification in WEKA for classifying Bayes, multilayer perceptron, and K-Mean. From the result, it shows that Bayes gave a consistent result in all the phases although K-Mean only gave the highest result in phase 1. This is only a preliminary study to find "best" of classification for Parkinson's diseases. In this paper, we are focusing on raw data because raw data give higher result than processed data. Future work can be focused on processing toward classification on brain wave pattern to gain "best" result for emotional model.

Acknowledgments We would like to thank the Islamic Science University of Malaysia and the Ministry of Higher Education, Malaysia, for supporting this research and providing the grant USIM/RACE/FST/35/50213.

References

1. Teplan, M.: Fundamentals of EEG measurement. Meas. Sci. Rev. **2**(2), 1–11 (2002)
2. Ishino, K., Hagiwara, M.: A feeling estimation system using a simple electroencephalograph. In: IEEE International Conference on Systems, Man and Cybernetics, 2003, Vol. 5. IEEE (2003)
3. Yuen, C.T., et al.: Classification of human emotions from EEG signals using statistical features and neural network. Int. J. Integr. Eng. **1**(3), 71–72 (2011)
4. Hede, S.C.: Signal Detection in EEG Brainwaves-a classification based approach. Dissertation, Technical University of Denmark, Lyngby (2010)
5. Berger, H.: On the electroencephalogram of man. Electroencephalogr. Clin. Neurophysiol. **168** (3931), 562–563 (1969)
6. Noh, N.A., Fuggetta, G.: Direct Electrophysiological Evidence Of Human Cortical Oscillations After Continuous Theta-Burst Stimulation. Magstim TMS summer school, University of Oxford, UK (2010)
7. Hogg, J., Cavet, J., Lambe, L., Smeddle, M.: The use of 'Snoezelen' as multisensory stimulation with people with intellectual disabilities: a review of the research. Res. Dev. Disabil. **22**, 353–372 (2001)
8. Heraz, A., Frasson, C.: Predicting the three major dimensions of the learner's emotions from brainwaves. Int. J. Comput. Sci. **2**(3), 1953 (2007)
9. Rogasch, N.C., Fitzgerald, P.B.: Assessing cortical network properties using TMS-EEG. Hum. Brain. Mapp. **34**(7), 1652–1669 (2012)
10. Freeman, W.J.: Origin, structure, and role of background EEG activity. Part 1. Analytic amplitude. Clin. Neurophysiol. **115**(9), 2077–2088 (2004)
11. Zhang, Y., Llinas, R.R., Lisman, J.E.: Inhibition of NMDARs in the nucleus reticularis of the thalamus produces delta frequency bursting. Front Neural Circ. **3**, 20 (2009)
12. Feinsod, M., Kreinin, B., Chistyakov, A., Klein, E.: Preliminary evidence for a beneficial effect of low-frequency, repetitive transcranial magnetic stimulation in patients with major depression and schizophrenia. Depress. Anxiety **7**(2), 65–68 (1998)
13. Levy, R., Hazrati, L.N., Herrero, M.T., Vila, M., Hassani, O.K., Mouroux, M., et al.: Re-evaluation of the functional anatomy of the basal ganglia in normal and Parkinsonian states. Neuroscience **76**(2), 335–343 (1997)

14. Brown, P.: Oscillatory nature of human basal ganglia activity: relationship to the pathophysiology of Parkinson's disease. Mov. Disord. **18**(4), 357–363 (2003)
15. Jenkinsen, N., Brown, P.: New insights into the relationship between dopamine, beta oscillations and motor function. Trends Neurosci. **34**(12), 611–618 (2011)
16. Moran, R.J., Mallet, N., Litvak, V., Dolan, R.J., Magill, P.J., Friston, K.J., Brown, P.: PLoS Comput. Biol. **7**(8), e1002124 (2011)
17. Hall, M., Frank, E., Holmes, G., Pfahringer, B., Reutemann, P., Witten, I.H.: The WEKA data mining software: an update; SIGKDD Explorations **11**(1), 10 (2009)
18. Frank, E., Hall, M., Holmes, G., Kirkby, R., Pfahringer, B., Witter, I.H., Trigg, L.: Data Mining and Knowledge Discovery Handbook. Springer, Berlin (2005)
19. Wagner, J., Kim, J., André, E.: From physiological signals to emotions: implementing and comparing selected methods for feature extraction and classification. In: IEEE International Conference on Multimedia and Expo, 2005, ICME 2005. IEEE (2005)
20. Azzini, A., Tettamanzi, A.G.B.: A neural evolutionary classification method for brain-wave analysis. In: Applications of Evolutionary Computing, pp. 500–504. Springer, Berlin (2006)
21. Murugappan, M., Ramachandran, N., Sazali, Y.: Classification of human emotion from EEG using discrete wavelet transform. Engineering **2**(4), 390 (2010)
22. Cabredo, R., et al.: An Emotion Model for Music Using Brain Waves. ISMIR (2012)
23. Lips, D., Salden, J., Koper, Y., Abrahams, L., Stupkova, T., Campbell, G.: The influence of binaural beats on brain wave activity. Online Technical Paper. ICC Group Project (2011)

An Ontological Approach for Knowledge Modeling and Reasoning Over Heterogeneous Crop Data Sources

Abdur Rakib, Abba Lawan and Sue Walker

Abstract The past two decades have seen a remarkable shift in the knowledge- and information-sharing paradigm. In the crops domain, for example, the amount of information currently known about underutilized crops, for example, Bambara groundnut their genetics and agronomy are much richer than years before. That paradigm shift offers enormous potential for advancing knowledge representation systems to facilitate access to such data. However, inconsistencies in terminology, improper syntax, and semantics are main obstacles to sharing data and knowledge among disparate researchers. We present a formal framework for representing knowledge using OWL 2 RL ontologies and SWRL rules and to integrate and reason over data from multiple, heterogeneous underutilized crops data sources.

Keywords Ontology · Heterogeneity · Knowledge representation · Reasoning · Semantic Web rule language

1 Introduction

Knowledge representation involves the use of principles and structures to preserve information in a way that facilitates inference. However, due to diverse nature of application domains, inconsistencies in nomenclature, and the variety of knowledge structures involved, similar information can appear to be completely different—the

A. Rakib (✉) · A. Lawan
School of Computer Science, The University of Nottingham Malaysia Campus,
Semenyih, Malaysia
e-mail: Abdur.Rakib@nottingham.edu.my

A. Lawan
e-mail: khyx3alw@nottingham.edu.my

S. Walker
Crops for the Future Research Centre (CFFRC), Semenyih, Malaysia
e-mail: sue.walker@cffresearch.org

© Springer International Publishing Switzerland 2015   35
A. Abraham et al. (eds.), *Pattern Analysis, Intelligent Security
and the Internet of Things*, Advances in Intelligent Systems and Computing 355,
DOI 10.1007/978-3-319-17398-6_4

latter often resulting in greater ambiguity. Such problems are common in various fields of research, such as research activities at the Crops for the Future Research Center (CFFRC), where researchers from various backgrounds must work together to produce reliable knowledge systems that can aid their users in decision making on underutilized crops and related products. To this end, a standard vocabulary needs to be shared and adhered to by researchers in the domain of discourse. Therefore, we should capture and represent knowledge that human experts can use in their decision-making process and that knowledge must be human understandable and machine executable.

Ontologies can be used to formally describe the knowledge of a domain, which gives a clear and coherent view of that domain [3]. Descriptions of the domain concepts and their associations are formalized using logical axioms. This axiom-based formalization of terms and concepts helps to make them less ambiguous and supports the sharing and reuse of formally represented knowledge of a domain. Collaborative ontology development methodology is necessary to enable the knowledge of engineers work closely with domain experts (crop researchers in our context) in the ontology development process. The proposed framework involves the process of gathering knowledge on underutilized crops from heterogeneous sources, such as texts, XML, and relational tables, and then expressing this knowledge using Web ontologies and rule language to develop the domain ontology toward expressive and reasoning-enabled building information model for an intelligent crop base management system.

To standardize the desired ontology into a consistent knowledge base, alignment and merging of several ontologies may be required. Moreover, as not all domain concepts can be expressed using our selected ontology language OWL 2 RL [8], we use user-defined SWRL rules [4] to augment the OWL 2 RL ontologies with additional concepts in the form of horn clauses. We choose OWL 2 RL for its expressive power over RDFS and suitability for the design and development of rule-based systems, whereas SWRL allows user to write rules using OWL concepts that cannot be modeled using OWL 2 RL alone. Thus, the combination of OWL 2 RL and SWRL provides more expressive language having greater deductive reasoning capabilities.

The remainder of the paper is structured as follows. In Sect. 2, we present preliminaries on ontology-based knowledge representation. In Sect. 3, we present a knowledge representation and reasoning framework for underutilized crops domain. In Sect. 4, we present and discuss the Bambara groundnut ontology model adopting the proposed methodology. We evaluate the crop ontology and SWRL rules assertions in Sect. 5. We discuss related work in Sect. 6 and conclude in Sect. 7.

2 Preliminaries

Ontologies have been used in artificial intelligence (AI) as a basis for modeling domains of the real world and sharing and reuse of knowledge. Ontology consists of definition of classes or concepts of a domain, their relationships, objects, instances, and their properties [3, 10]. Another definition by Jacob [5] describes ontology as a representation system that allows detailed specification of the semantics of a knowledge domain. It can be seen in both definitions that ontology is viewed as a vocabulary for a knowledge domain. However, in the second definition, it is more than just a vocabulary of a domain but also a powerful tool that can express hidden knowledge in a domain. Thus, ontology of a domain is any formal explicit specification of such concepts that guides knowledge representation of the domain. This not only gives a common syntax to the domain, but it is possible with the use of ontology language to enrich the data with additional semantics for efficient manipulation and representation of the knowledge.

In [5], Jacob argues that any metadata schema on the Web that specifies the set of conceptual or physical characteristics of resources used by a particular group of users is in itself an ontology. This claim helps to point out the examples of simple and small size ontologies that can be found scattered on the common Web. However, what is certain is that the advancement of the Semantic Web brings about the interest in developing large-scale ontologies and making them a connection point for data integration in large information systems to aid decision-making processes. Moreover, due to the heterogeneous nature of data sources on the Web and the multiplicity of authors, ambiguity is bound to exist on the terms used to represent knowledge. As such, the role of ontologies on the Semantic Web will be to aid data integration and where possible enable the specification of a new relationship from existing ones.

As an example, consider the application of ontology in the domain of interest for this research, the underutilized crops platform. Farmers and agronomists can use the ontology to present, for example, knowledge on pest control, soil nutrients requirements, and other environmental factors, and agrochemical companies may use similar ontologies to present information about fertilizers, pesticides, and their mode of application among other things. This knowledge and information can be combined with the existing crop data to support intelligent applications such as decision support systems that can suggest the optimum crop to be grown given say, an environmental or agronomic data, or the effective pest control to be used given pest information. In addition, the decision support system will contribute toward the utilization of knowledge for underutilized crops and can serve as a starting point for new researchers in the field of underutilized crops seeking first-hand knowledge on the crops covered in the knowledge base. The underutilized crops knowledge base will also enable domain knowledge reuse. For example, interested researchers and software agents from other applications can easily reuse the ontology by adapting the concepts into their knowledge bases.

2.1 DL-Based Ontological Knowledge Representation

The logic behind ontological knowledge representation is known as description logic (DL). Being a decidable fragment of first-order predicate logic (FOL), DL is a collection of logic-based knowledge representation formalisms designed for precise description and reasoning about the concepts in an application domain and the relationships between them. The ability to model a domain and the decidable computational characteristics makes DLs the basis for the widely accepted ontology languages such as OWL. As an example, consider a simple DL (1)–(3) ontology that intuitively describes 'leaf spot is a disease of Bambara groundnut, which is a legume crop with features leaves, stem, and root.'

$$
\begin{aligned}
\text{BambaraGroundnut} &\sqsubseteq \text{Crop} \sqcap \exists \text{isPartOf} \cdot (\text{Legumes} \sqcap \\
&(\exists \text{hasFeatures} \cdot \text{Features} \sqcap \forall \text{features} \cdot (\text{Leaf} \sqcup \text{Stem} \sqcup \text{Root})))
\end{aligned} \tag{1}
$$

$$
\text{LeafSpot} \equiv \text{Disease} \sqcap \exists \text{affects} \cdot \text{Leaf} \tag{2}
$$

$$
\text{BambaraGroundnut}(\text{BambaraGroundnutInd}) \tag{3}
$$

DL-based knowledge representation systems involve two important components: the T-Box and A-Box. The T-Box or terminology-box contains the ontology concepts (owl: Classes) and roles (owl: Properties) also called the terminologies, while the A-Box or assertion-box contains assertions of individual instances from the ontology terms. Example axioms in the T-Box could be the DL axioms (1) and (2) defined above in the simple crop ontology, while a member of A-Box could be the third axiom (3) which asserts the individual 'Bambara groundnutInd' into the ontology as a member of the 'Bambara groundnut' class.

3 A Knowledge Representation and Reasoning Framework

We present a methodological framework for our ontology engineering from different data sources, focusing on texts, XML, and relational tables. Though it can be applied to any other domain, in this work, we focus on underutilized crops using Bambara groundnut as an exemplar crop. Modeling underutilized crops domain requires creating a knowledge model both from existing crop ontologies those focus on the popular crops and various other sources. Information on underutilized crops is usually dispersed among different resources: research papers, implicit knowledge, and the information available from our domain experts at the CFFRC. Many standard terms for popular crops domain already exist in the literature, and to model our domain, a set of standard terms were obtained from AGROVOC [6, 13] and the crop ontology [7], among other sources.

3.1 Ontology Development Methodology

For proper documentation of the ontology engineering process and to ensure a comprehensive modeling of the knowledge domain, we employ the general guidelines advised in the work of Noy and Mcguinnes [10], the METHONTOLOGY [2], DILIGENT [11], and the onto-knowledge methodology. These guidelines help to structure the ontology engineering process by identifying important but non-obvious aspects, such as the target users of the ontologies, supporting tools, and specifying what values can be allowed for properties. Other aspects that are apparent and also common to all methodologies—such as defining domain terms and roles, asserting their hierarchy, and filling the concept slots with individual instances—are performed iteratively for each source of data to populate the underutilized crops ontology. Apart from using the OWL 2 RL, the user-defined rules in SWRL are designed and added iteratively but only at the end when all terms and concepts have been added and the consistency of the ontology asserted by the reasoner (Pellet). Thus, our ontology development methodology is a collaborative one, which involves a team of under-utilized crops domain experts, social engineers from CFFRC, and the ontology knowledge engineers. The major steps can be summarized as follows: (i) ontology requirement specification; (ii) domain knowledge gathering and conceptualization; (iii) model implementation; and (iv) the evaluation of the model.

These steps are performed repeatedly for smaller ontologies, leading to the final larger version, thereby adding two important stages called *versioning* and *assembly*. In 'versioning,' we assign a label to represent each ontology fragment, specifying where it fits to the larger ontology. While in the 'assembly' stage, all the smaller ontologies are put together and the reasoner is invoked to assert the overall classification and check for consistency.

4 The Bambara Grondnut Model

In this section, we present and discuss the Bambara groundnut ontology model adopting the methodology discussed in the preceding section. However, our crop ontology contains both general and some other neglected and underutilized crop.

4.1 Conceptualization

The 'Bambara grondnut' class is the center point of our ontology discussed here, which is a subclass of the 'underutilized crops' concept, sharing similar concepts with other sibling crops and having specific concepts as subclasses. To create the taxonomy, we employ the top-down approach when considering such concepts that are specific to the Bambara grondnut class and also employ the bottom-up approach when considering the concepts that are common to all underutilized crops. In essence, we start the ontology hierarchy with the main class at the center and continuously build it upward or downward depending on the concepts we came across in our domain knowledge acquisition. To speed up the initial process of the ontology development, which is to define the terms and roles relevant to the knowledge domain, an XML-to-OWL [1] and 'relational-to-owl' conversion tools were employed to generate domain concepts from the available XML documents and relational tables containing the domain knowledge. This is because most of the information on underutilized crops available from the CFFRC are either scattered on Web pages stored as XML files, in tabular format as results of data collected by researchers from the field, or as texts in their research documentations. As such, to make the collection of these terms easier, the above tools were utilized.

4.2 Domain Knowledge Gathering

In this work, we utilize the simplicity of XML-Tab plug-in available in Protégé for the XML-to-OWL conversion. However, the XML files have to be generated from the vocabulary of the crop domain stored in text files and relational tables, a summarized Bambara groundnut vocabulary is shown in Table 1.

4.3 The Crop Ontology in OWL 2 RL

For brevity, we omit a detailed description of all the concepts and their relationships used to design the ontology. However, some of the classes of the underutilized crops ontology are shown in Fig. 1. We added the *DomainConcepts* Class as a subclass of owl top class: 'Thing' and this class serves as the ancestor class for our crop ontology. This is followed by the *UnderutilizedCrops* class to contain only those concepts that are categorized as underutilized crops. The focus class for our case is the *Bambara groundnut* class added as the subclass of *UnderutilizedCrops* using the 'class hierarchy' tabs, while the related concepts are added as 'siblings' classes as shown in Fig. 1.

4.4 Alignment and Merging of Crop Ontologies

As a means of dealing with heterogeneous ontologies, the PROMPT tab [9] in Protégé allows for merging, mapping, and aligning multiple ontologies. Here, as an example, we merge two ontologies to achieve a consistent one. Merging ontologies in PROMPT involves a series of varied steps, since it requires user intervention when there is a conflict. In this case, PROMPT was provided with two ontologies: BG-XML1.owl and BG-XML2.owl generated from our two XML files of Bambara groundnut vocabulary, and BG-XML2 ontology was selected as preferred. In Fig. 2, the two ontologies are presented as Arg1 and Arg2 and a conflict is reported as 'two frames with identical name' in the 'reason for selected suggestion' field at the bottom of the page. The conflicting frames (classes) 'Pest-Resistance'

Table 1 Summarized Bambara groundnut vocabulary

Bambara groundnut vocabulary	
Class	Bambara groundnut
Super class	Underutilized crops
Ancestor class	Family legumes
Type	Bunch seed crop
Alias	Vigna subterranea
Region	West Africa
Properties	High nutritional value, pest persistent crop, highly tolerant
Features	Leaf, stem, roots, pods, seed
Soil requirements	**PH level**: 5.0–6.5, **soil type**: loamy (heavy loam, light loams), sandy soil
Rainfall	500–1200 mm seasonal rain
Temperature	**Optimum temp.**: 20–28 °C **base temp.**: 10–12.3 °C **germination temp.**: 30–35 °C
Pests	Spidermites
Purpose/uses	For human consumption
Other information	Growth and development: depends on landrace and environmental condition (e.g., drought, cold, heat, soil moisture (same as soil water), evapotranspiration) BG is drought tolerant (not drought escape or avoidant), i.e., maintains positive turgor at positive turgor at low water potential. BG needs moderate soil moisture. Germination and emergence: takes about 7–15 days. Flowering: takes 30–50 days and depends on: day length, temperature and Landrace. BG is a 'short-day' crop (grows at elevation up to 1600 m). Rainfall: BG needs moderate and evenly distributed rain for successful growth and good yield. If rainfall is evenly distributed and moderate and Stage = Sowing or flowering → Successful growth and Good yield. Harvesting: usually between 90 and 170 DAS (days after sowing). Pests: e.g., spidermites (tetranychus cinnabarinus), can be controlled by pesticides, e.g., phytoseiulus persimilis. Nutrient contents—protein: 16–25 %, carbohydrate: 42–65 %, lipid (oil): 6 %. Minerals: dominant minerals—Ca, K, Mg, Na, P, Cu, Fe, Zn. Growth phases: vegetative phase, reproductive phase (phase has stages). Cultivation: BG is traditionally cultivated by small-scale farmers (majority women farmers) mostly in extreme tropical environments without access to irrigation and/or fertilizers. Life span: averagely 4 months after sowing (120 DAS) or when leaves begin to turn yellow in color

Source CFFRC

from BG-XML2 ontology and 'Pest-Persistence' from BG-XML1 are reported, and since BG-XML2 ontology was set as preferred, 'Pest-Resistance' will be the result of the merge (see the 'Result Classes' highlighted URI in Fig. 2), while 'Pest-Persistence' will be copied into the preferred ontology's named class. Similarly, BinomialName class from the BG-XML1 will be merged into the ScientificName class of BG-XML2 (the preferred ontology) and the result is that ScientificName class stays while BinomialName got merged. Note that, the resulting classes from merge process (URIs in black) shows the results of the merge, while those in color

Fig. 1 A fragment of the crop ontology

(such as PH level and CropYields class) are the result of alignment (copy) by deep copying the classes with its ancestors and subclasses, if no conflict exists.

4.5 SWRL Rules for BG Model

Writing user-defined rules in Protégé 4.3 is not exclusive to SWRL rules as it also supports writing rules in other formats such as Rule Interchange Format—RIF and Rule Mark-up Language—RuleML. A SWRL is written as positive conjunctions (separated by a comma sign ',') on both its head and body with no negation or disjunctions, and also, SWRL allows class, property, or data type predicates to appear in the body or head of a rule with variables representing individuals, data values or their variables as argument. As SWRL rules cannot introduce new terms into an ontology, in order to retain the much desired decidability property, the addition of our user-defined rules was delayed until the final version of the ontology was checked using the Pellet reasoner and found to be consistent. This is also important because it allow us to judiciously utilize the OWL 2 RL syntaxes before employing the SWRL rules in the ontology development. However, considering the main reason of using SWRL rules in our ontology, which is to express complex domain concepts and utilize the SWRL syntax to define and assert domain-specific concepts, it still does no harm to our ontology if we express common OWL 2 RL syntaxes using SWRL.

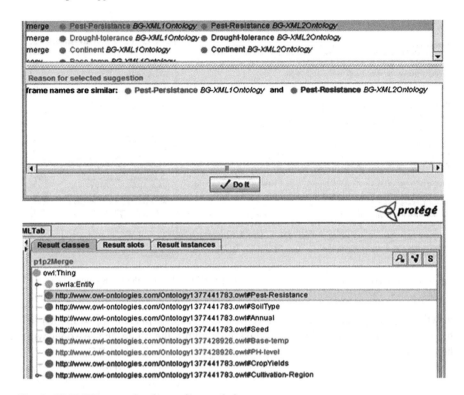

Fig. 2 PROMPT suggestion for conflict resolution

In the rules interface depicted in Fig. 3, we begin with a simple highlighted assertion rule that asserts a relationship between members of Bambara groundnut and those of Bambara groundnutProperties class using *hasProperty* relation. We then continue to assert rules that are not easily expressed in OWL 2 RL—thereby extending the expressive power of the ontology to allow complex domain modeling. For example, the seventh rule—*BambaraGroundnut(?y), Leaf(?z), isFeatureOf (?z, ?y)* → *hasLeafType(?y, "Trifoliate")*—states that if it is true that Bambara groundnut has a feature and the feature is a leaf, then it will assert that the leaf type is 'trifoliate.' Since features such as leaf are not exclusive to Bambara groundnut, then unless the leaf individual is related to Bambara groundnut, the leaf-type 'trifoliate,' cannot be asserted. Rules of these types that are based on certain conditions being true or otherwise are hard to be expressed with OWL 2 RL syntax alone. Details of some of the selected rules are presented as follows:

BambaraGroundnut(?x), DAS(?z), GrowthStage(?y), hasGrowthStage(?x, ?y), hasAverageDaysAfterSowing(?y, ?a), hasCurrentDaysAfterSowing(?z, ?b), great erThanOrEqual(?a, 30), greaterThanOrEqual(?b, 30), lessThanOrEqual(?a, 50), lessThanOrEqual(?b, 50) → *CurrentStage(?y)*—this rule uses the SWRL built-in *greaterThanOrEqual'* and *lessTh anOrEqual'*, to compare the days an individual Bambara groundnut (BG) is planted with the number of days asserted for different

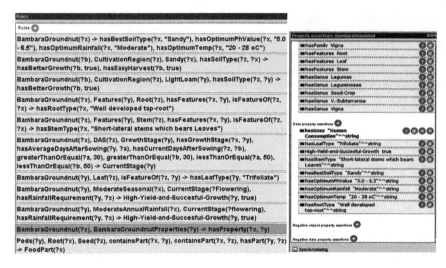

Fig. 3 Rules interface containing some user-defined SWRL rules and their inference

growth stages (e.g., the flowering stage *hasAverageDaysAfterSowing* = *35–50*). If there is a match, i.e., the number of days falls within the range, the reasoner will then assert this growth stage as the current stage of the individual BG.

BambaraGroundnut(*?x*) → *hasBestSoilType*(*?x*, "Sandy"), *hasOptimumPhValue* (*?x*, "5.0–6.5"), *hasOptimumRainfall*(*?x*, "Moderate"), *hasOptimumTemp*(*?x*, "20–28 °C")—this rule is an *optimum condition* test for Bambara groundnut. It can be implemented as a result of a query 'what are the best conditions for BG farming.' The results from the evaluation stage of our ontology are presented in the following section.

5 Reasoning and Query Processing

In this section, we evaluate the ontology and SWRL rules assertions by invoking the Pellet reasoner to classify and check for the consistency of the ontology. Additional knowledge implicit in the crop ontology can then be inferred by this reasoner. Also to verify the conceptual facts and individual assertions, DL queries are used to probe the ontologies. We evaluate 2–3 queries for each SWRL rule, making a total of 56 DL queries. Results of the frequent queries are saved and added as part of the ontology, thereby evaluating automatically once the reasoner is invoked. Using ontologies allows measuring performance at the design as well as run time via a reasoner to compute the ontology classification and ensure consistent knowledge representation. As such a reasoner needs to be active and the ontology classified before writing any DL Queries. Our user-defined SWRL rules are validated by writing DL queries to check their inference or otherwise by the reasoner.

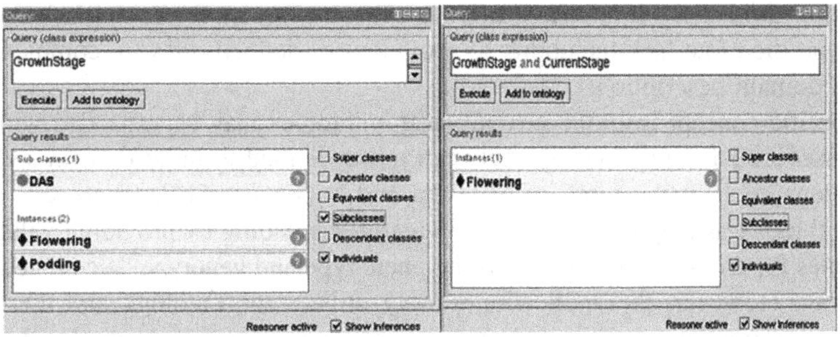

Fig. 4 SWRL rule validation showing query result of the sixth rule

For example, the query result of the sixth rule, which determines the current 'growth stage' of a Bambara groundnut, is depicted in Fig. 4. Results for the rules 2, 3, 8, 9, and 10, which assert data type properties to Bambara groundnut individual, can be seen from the inference provided by the Pellet reasoner in the 'description' panel on the right of Fig. 3. We would like to mention that the Bambara groundnut ontology presented in this paper has 1038 axioms, 122 different classes, and 114 individuals, with 83 object properties and 41 data properties. However, its size will be growing as we receive data from the CFFRC/other sources. In the ontology, all the rules developed are DL safe, thereby decidable. Most of the queries considered in the experiment were originated from the competency questions generated in our ontology engineering stage, due to space constraints we are unable to present those in details.

6 Related Works

In the literature, various crop ontologies have previously been developed to present generalized crops knowledge. Lauser and colleagues in [6, 13] present the Agricultural Ontology model developed using OWL DL profile syntax to represent agricultural and related concepts. The OWL model is open source and is to serve as agricultural-terminology standard for developers of ontologies. A similar approach was observed in the crop ontology tool presented in [7], which is a mainstream crop ontology aimed to provide a standard vocabulary for crops related terms to ontology developers. Provided in RDF format and described using OWL and RDFS, the crop ontology is believed by the authors to be a one-stop platform for creating crop-based ontologies, with an API that can be integrated directly into user applications. In the two approaches, it is observed that general crop domain concepts are represented using ontological knowledge representation to provide among other things; the terminological standards, hierarchical representation of terms, and the relations exist between the terms and/or data type. However, they are not specific to a particular crop domain and therefore may not contain complex descriptions of their application domains. Moreover, the ontologies are not in OWL 2 RL language profile and therefore lack the ability to provide user-defined rules for complex and detailed domain description.

A specific concept from the general crop ontologies may be expanded with in-depth assertions and relationships between terms defined to build domain-specific applications such as the work of Shrestha and colleagues [12], where the 'crop ontology' tool was employed to support integrated breeding by providing validated trait names for crop-breeders to access the phenotype and genotype data related to a given trait. However, the application simply utilizes the ontology and does not extend the crop ontology with additional semantics or syntaxes for complex domain modeling and reasoning. In view of domain-specific crop ontologies, Thunkijjanukij [14] in his doctoral research claims to have developed a pioneer 'Thai Rice Ontology' from scratch, with a specific ontology visualization tool to aid presentation of the ontology. These ontologies, robust as they may be, were developed with older versions of the Web ontology language and, as such, lack the expressive capabilities for domain modeling provided by the latest OWL profiles such as OWL 2 RL. Moreover, none of these works attempt to model an underutilized crop domain that can be used in traditional rule-based application systems. The work presented in this paper is one of the few OWL 2 RL crop ontologies extended with user-defined SWRL rules. Motivated by the need for an efficient knowledge representation of underutilized crops data (from the CFFRC), our work is one of the pioneer ontologies for underutilized crops ontologies.

7 Conclusions and Future Work

In this paper, we proposed a framework for representing knowledge using OWL 2 RL ontologies and SWRL rules and to integrate and reason over data from multiple heterogeneous underutilized crops data sources. Using the Pellet reasoner, we have presented an evaluation that consists of querying the knowledge base to ensure that the query results are consistent with the rules. This includes validation of the ontology and SWRL rules by writing appropriate DL queries. In the future, we plan to develop domain-specific SWRL built-in data types to express our domain knowledge comprehensively and publish the ontology to present the domain knowledge to the general public.

References

1. Bohring, H., Auer, S.: Mapping XML to OWL Ontologies. In: Leipziger Informatik-Tage, vol. 72, pp. 147–156 (2005)
2. Fernández-López, M., Gómez-Pérez, A., Juristo, N.: Methontology: from ontological art towards ontological engineering. In: Proceedings of the AAAI'97 Spring Symposium Series on Ontological Engineering, pp. 33–40 (1997)

3. Gruber, T.: A translation approach to portable ontology specifications. Knowl. Acquisition **5**, 199–220 (1993)
4. Horrocks, I., Patel-Schneider, P.F., Boley, H., Tabet, S., Grosof, B., Dean, M.: SWRL: a semantic web rule language combining OWL and RuleML. Acknowledged W3C Submission, Standards Proposal Research Report: Version 0.6 (April 2004)
5. Jacob, E.K.: Ontologies and the semantic web. Bull. Am. Soc. Inf. Sci. Technol. **29**, 19–22 (2003)
6. Lauser, B., Sini, M.: From AGROVOC to the agricultural ontology service/concept server: an owl model for creating ontologies in the agricultural domain. In: Proceedings of the International Conference on Dublin Core and Metadata Applications: Metadata for Knowledge and Learning, pp. 76–88. Dublin Core Metadata Initiative (2006)
7. Matteis, L., Chibon, P.Y., Espinosa, H., Skofic, M., Finkers, R., Bruskiewich, R., Hyman, J.M., Arnoud, E.: Crop ontology: vocabulary for crop-related concepts. In: Proceedings of the First International Workshop on Semantics for Biodiversity (2013)
8. Motik, B., Grau, B., Horrocks, I., Wu, Z., Fokoue, A., Lutz, C.: OWL 2 Web Ontology Language: Profiles, W3C Recommendation. http://www.w3.org/TR/owl2-profiles/ (December 2012)
9. Noy, N.F.: Ontology management with the PROMPT plugin. In: Proceedings of the 7th International Protégé Conference (July 2004)
10. Noy, N.F., McGuinness, D.L.: Ontology development 101: a guide to creating your first ontology. Technical report, Stanford (2001)
11. Pinto, H.S., Tempich, C., Staab, S.: Diligent: towards a fine-grained methodology for distributed, loosely-controlled and evolving engineering of ontologies. In: Proceedings of the 16th European Conference on Artificial Intelligence, pp. 393–397. IOS Press (2004)
12. Shrestha, R., Matteis, L., Skofic, M., Portugal, A., McLaren, G., Hyman, G., Arnaud, E.: Bridging the phenotypic and genetic data useful for integrated breeding through a data annotation using the crop ontology developed by the crop communities of practice. Front. Physiol. **3**(326), 1–10 (2012)
13. Soergel, D., Lauser, B., Liang, A.C., Fisseha, F., Keizer, J., Katz, S.: Reengineering thesauri for new applications: the AGROVOC example. J. Digital Inf. **4**(4) 1–19 (2004)
14. Thunkijjanukij, A.: Ontology development for agricultural research knowledge management: a case study for Thai rice. Ph.D. thesis, Kasetsart University, Thailand (2009)

A Study on Changes of Supervision Model in Universities and Fostering Creative PhD Students in China

Lingling Luo, Chunfang Zhou and Song Zhang

Abstract This paper aims to explore the changes of supervision model in higher education in relation to fostering creative PhD students in China. The changes are being made from the traditional Apprentice Master Model (AMM) to the modern Collaborative Cohort Model (CCM). According to the results of the empirical work done by questionnaire survey and interviews, this study shows in the background of the Big Science Era and according to theories on systematic view of creativity, the new CCM improves PhD students' creativity to some extent; however, problems exist in the creativity development mechanisms. So this paper also explores the reasons of why the new mechanism of creativity development failed to play fully.

Keywords PhD education · Creativity · Apprentice master model (AMM) · Collaborative cohort model (CCM) · Supervision model

1 Introduction

PhD students are the groups who will gain the highest level of educational degree. To foster creative PhD students is very important for the national strategy of capital development. The supervisors are regarded as the examples of academic research of

L. Luo · S. Zhang
School of Humanities and Law, Northeastern University, 110004 Shenyang, China
e-mail: lll_19500619@126.com

S. Zhang
e-mail: zhangsong767@sohu.com

C. Zhou (✉)
Department of Learning and Philosophy, Aalborg University, 9000 Aalborg, Denmark
e-mail: chunfang@learning.aau.dk

© Springer International Publishing Switzerland 2015
A. Abraham et al. (eds.), *Pattern Analysis, Intelligent Security
and the Internet of Things*, Advances in Intelligent Systems and Computing 355,
DOI 10.1007/978-3-319-17398-6_5

49

PhD students, so they play significant roles in most PhD students' research career and creativity development. Along with the coming of the Big Science Era and the increasing number of PhD students, there are some changes in the traditional ways of supervision that gives the birth of a new supervision model.

From their case studies and based on the theories of systematic view of creativity, Hook and Csikszentmihalyi [1] have generalized a new supervision model that is different from the old model focusing on traditional, close, and one-to-one relationships between supervisors and students. The new model prefers to the following: enough trust and given responsibility; good overlooking and true letting go; and emphasis on group roles, such as influences of peer groups and research communities. Accordingly, the scholars [2] have further generalized such changes of supervision ways are from the Apprentice Maser Model (AMM) to the Collaborative Cohort Model (CCM). This point has been involved into many recent studies [3, 4]. The following four aspects demonstrate the characteristics of CMM.

First, benefits of peer learning on creativity development are emphasized. As the frequent communication can stimulate to learn from each other and to understand process of study, the PhD student can get feedback of self-learning according to the experiences of peers and get peer supports. Supervisors involve their students into learning communities. Second, group supervision is helpful to cooperative innovation that is emphasized. The CCM is gaining increasing popularity internationally and, in some contexts, replacing the conventional model of AMM. Among the motivations advanced for this shift is that the CCM improves completion rates and enhances the quality of research supervision [3, 4]. Third, to make space for creativity of PhD students is emphasized. Making space for creativity is a crucial problem, as we develop higher degree systems. The safest and quickest, but also the most deadening, form of PhD work is where students effectively reproduce the methods of their supervisor. Indeed, students should learn their supervisor's attitudes and methods [5, 6]. Fourth, the influences of cultural capital and social capital of supervisors on students are emphasized. From a sociological perspective, studies on creativity emphasized the importance of relationship between supervisor and PhD student is not on personal relationships, rather on the involvement of PhD students into the supervisors' collaboration network [7].

Recently, China has enlarged the number of PhD students, so how to improve the quality of students; especially, the creativity of students is one of focuses of universities [8–12]. This paper aims to explore the changes of supervision models in PhD study in China. We especially ask two research questions: (1) Are the supervision models of fostering PhD students changing from AMM to CMM in universities in China? And (2) if there are some changes, what are the problems in the practice of change from AMM to CCM? Does the CMM play well with the mechanism of fostering creative students?

2 Methodology

In order to collect data, this study designed 'Investigation Questionnaire of Scientific Research Groups' for supervisors and 'Investigation Questionnaire of Models of PhD Supervision and PhD Study' for students. The latter included 24 questions from the following 5 aspects: (1) ways of supervision, (2) ways of achieving professional knowledge and skills, (3) ways of forming academic norms, (4) ways of fostering creativity among PhD students, and (5) roles of scientific research groups in PhD study. This study also designed interview guidelines for both supervisors and students. In addition, 'Creative Climate Scale for Science and Technology Team (CCSSTT)' was applied for students. The investigation in this scale includes external climate, internal climate, and personal feeling-related climate.

The participants are from 5 scientific research groups from 2 universities in China. From March to June in 2009, researchers of this study interviewed 9 PhD supervisors, 1 co-supervisor, and 14 PhD students. It also included the delivery of questionnaires ($n = 100$, 88 of the 100 are valid) (Table 1).

3 Results

3.1 Results for Research Question 1

The data demonstrated that the supervision model of PhD students is changing from AAM to CMM in China. The relationship between supervisors and students and the form of students' academic norms are at medium level (Tables 2 and 3).

3.2 Results for Research Question 2

As shown in Table 4, the results demonstrate that there are three aspects of significant correlation.

First, the exterior climate (such as information resource of university or faculty, equipment resource, management attitude, innovation policy, and scientific management) has inverse ratio with relationship between supervisors and students ($R = -247$, Sig = 0.020). However, the exterior climate has direct proportion with scientific research group function ($R = 0.293$, Sig = 0.006). This means the higher cores of exterior climate indicate the worse relationships between supervisors and students. In other words, when there are less contact between supervisors and students, the exterior climate plays more roles. The exterior climate has direct influences of scientific research group functions. The better exterior climate indicates the better function of scientific research group.

Table 1 Participants of this study

University	Group name	Specialization	Group level	Group structure		PhD number		Interviewees number	Questionnaire number
Northeastern University	WEN	Mechanical design	National key subject	Group size ($n = 13$)		PhD student ($n = 16$)		Supervisor: 2	21
				Yangtze river scholar ($n = 1$)	Academician ($n = 1$)	Graduated PhD ($n = 70$)		Student: 4	
				Professor ($n = 8$)					
Northeastern University	CUI	Metallurgy and material engineering	National level innovation platform	Group size ($n = 37$)		PhD student ($n = 20$)		Supervisor: 2	13
				Academician ($n = 1$)	Professor ($n = 11$)	Graduated PhD ($n = 25$)		Student: 3	
Northeastern University	HU	Environmental engineering	PhD programme	Group size ($n = 12$)		PhD student ($n = 7$)		Supervisor: 2	11
				Professor ($n = 2$)		Graduated PhD ($n = 7$)		Student: 2	

(continued)

Table 1 (continued)

University	Group name	Specialization	Group level	Group structure	PhD number	Interviewees number	Questionnaire number
Dalian Technology University	LIU	Scientific research and management	International key center for scientific metrology	Group size ($n = 15$)	PhD student ($n = 60$)	Supervisor: 2	30
				Yangtze river scholar ($n = 1$)	Graduated PhD ($n = 100$)	Student: 2	
				Professor ($n = 10$)			
Dalian Technology University	LU	Architecture and design	Programme	Group size ($n = 6$)	PhD student ($n = 6$)	Supervisor: 2	13 (7 master students are included)
				Professor ($n = 1$)	Graduated PhD ($n = 2$)	Student: 2	

Table 2 Results of investigation questionnaire of models of PhD students

$N = 88$ Minimum = 1, Maximum = 6		AMM		CCM
1	Supervising students	Supervisors always directly provide the guidance personally	2.76	The supervision is based on team work
2	Meeting between the supervisors and the students		2.16	
3	Working with the supervisor together	More	3.44 / 2.5	Fewer
4	The relation between the supervisors and the students	Intimate	3.07	Alienating
5	Students' perception for the supervisor	Strict teacher like parents	4.52	Experts
6	Knowledge source	Obtaining knowledge from supervisors	3.3	Mutual inspiration among the team members
7	Ways of knowledge internalization	Obtaining knowledge from supervisors	3.16 / 3.6	Learning from the process of a project
8	Obtaining professional standard	Obtaining knowledge from supervisors	5.06	Learning from the process of a project
9	Professional communication	Communication with teacher	3.17	Increasing the peers' communication
10	Requirements for completing the assignment	Based on supervisors' planning	2.98	The design project by one's own
11	Educational objectives	Get a degree	3.02	The enhancement for research capacities
12	Research methods	Obtaining knowledge from supervisors	3.64 / 4.4	Research design and project management
13	Research data accuracy	The supervisor teaches by personal	3.48 / 4.09 / 4.5 / 4.85	Directly facing social expectations

(continued)

Table 2 (continued)

$N = 88$		AMM		CCM
Minimum = 1, Maximum = 6				
14	Completing academic paper	Face the supervisor's guidance		Directly facing experts' review
15	Submission of manuscript	Deliver to the common journal		Deliver to the anonymous review journal
16	Dissertation Topic	Reading literature		The engagement of supervisors' projects
17	Title choice	The title is assigned by supervisors		Students pick the research topic of interest
18	The fostering of creativity	Finishing the dissertation		Finishing the dissertation by the projects
19	Ways for cultivating creativity	Influenced by the supervisor		Influenced by the research team
20	Supervisor's responsibility	Providing the professional guidance when consulting		Apply for the project and manage the team
21	Career development	Getting the help by the supervisor		Getting lots of help by social network
22	Research teams	No research teams		Stable research team
23	A sense of teamwork	No team spirit		Good team spirit
24	The outcomes of students	Degree certificate		Degree and scientific working experience

Second, the interior climate (including supervisor group academic atmosphere, group culture, interaction of motivation, internal information, and management style of supervisor) has relevance with students' creativity ($R = 0.248$, Sig = 0.020). This means the better interior climate is better helpful to foster students' creativity. The interior climate is also relevant to scientific research group function ($R = 0.33$, Sig = 0.002). The better interior climate is better helpful to play the group roles.

Third, climate related to individual perception (freedom, work pressure, work challenge) has relevance with students' learning specialized knowledge ($R = 0.255$, Sig = 0.017). This means freedom, work pressure, and work challenge can stimulate students to learn specialized knowledge.

Table 3 Five aspects in investigation questionnaire of models of PhD supervision and PhD study

Aspect	N	M	SD	Tendency of AAM or CCM
Relationship between Supervisor and Student	88	2.63	0.98	Medium
Students' specialized knowledge	88	4.10	0.81	Tendency toward learning by collective collaboration
Students' academic norm	88	2.98	1.03	Medium
Students' creativity	88	3.63	1.07	Tendency toward collective collaboration as stimuli
Scientific research group function	88	4.55	1.01	Tendency toward scientific research by collective collaboration

Table 4 Correlation between questionnaire survey and supervision model

N = 88		The method of supervisor's guidance	Students' specialized knowledge	Students' academic norm	Students' creativity	Scientific research group function
Exterior climate A	Pearson correlation	-0.247^a	0.032	-0.136	0.154	0.293^b
	Sig. (2-tailed)	0.020	0.764	0.206	0.153	0.006
Interior climate B	Pearson correlation	-0.158	0.014	-0.043	0.248^a	0.330^b
	Sig. (2-tailed)	0.142	0.896	0.690	0.020	0.002
Climate related to individual perception C	Pearson correlation	0.056	0.255^a	0.009	0.113	0.158
	Sig. (2-tailed)	0.607	0.017	0.932	0.295	0.142

4 Discussion

4.1 Relationship Between Supervisor and PhD Students and Responsibility

According to the data of interviews, the students mentioned now there are less time of face-to-face meetings with supervisors than before. However, they can get more effective communication by information technologies. So it is hard to say the psychological and social distance becomes far. The students also mentioned what they can learn from a good supervisor is much more than knowledge. 'I have been influenced much by tacit knowledge of my supervisor. As a student, I can learn the academic spirits from that: the conscientious and rigorous attitudes, literacy of scholars, and insight of research.' As what Zuckerman [13] found in his early studies, some students of successful scientists reported that scientific knowledge is the least important in what they learned from the Nobel laureates. What are more significant are the relations with their professions and criteria of working as good examples.

Although the PhD supervisors take the responsibilities of academic research and doing contributions to the society, the core task is still to foster and guide undergraduate students. As there are many changes of traditional AMM caused by the new research life, is the 'good overlooking' a lack of responsibility? One point generated from this investigation is that the way of supervision should not be the traditional teaching from hand to hand any longer, rather than the way of 'true letting go.' Roles of behaving as examples have replaced the way of teaching by mouth to a large extent. The supervisors may manage group resource and apply useful resource of Internet. They deliver the specialized knowledge, skills, regulations, and norms of value to the undergraduate students. The estrangement of relationship between supervisor and students is well meaning, which is the trust and the space of self-directed learning and individual growing up given to the students by the supervisors.

Meanwhile, there is another point. Some supervisors overlooked to foster students' ability, as they did not give students the first priority in their work. They only assigned the simple and repeated research activities that led students become labor of scientific research. Some supervisors failed to find the joint points of academic research, student study, and ability fostering. They got half the results with double the efforts. As their attention was distracted, it is very hard to take care of the student well for them.

4.2 Balance Between Recognition of Supervisor and Students' Creativity Constrained by Culture

The results of questionnaire and interviews can confirm that in the research universities, PhD students and master students are the continuous new input of research group. They are successors of knowledge and take on basic research in the groups. The students affirmed the roles of participation of research group on learning knowledge of PhD study and fostering academic norms. Guided by the academic research, students' learning is 'research project-centered.' Besides the compulsory courses, supervisors are not the direct teachers. They introduce students to the professional field and most of them recommend students to practice of research projects. The students can learn and adopt necessary knowledge, academic skills, and norms. They also can master the professional skills and writing norms during participation of project that facilitates them to achieve the criteria of professional work and norms of value.

However, both supervisors and students also mentioned other problem. Firstly, due to the limitations of financial support of research projects, the group depends on much the reputation of academic leader and groups that indicate the strong 'Matthew Effect.' So some supervisors are still using traditional model of AMM to guide PhD students. Secondly, due to the rigorous assessment system of academic research tasks, some supervisors only pay attention to the income of research funds that cause them cannot concentrate on the research directions and achieve the research academic. This further influents the qualification of PhD student.

4.3 Balance Between Students' Acceptance of Culture and Creativity Depending on Supervisors' Attitudes

The responsibility of gatekeeper is to decide what can or what cannot be introduced to the existing knowledge system and then to deliver to next generation. If the experts judge the changes made by the individuals are creative by the criteria of certain field, the person is creative. But if only the research group only emphasizes this, it will decrease students' creativity under the background of introducing students to the formal paradigm of scientific research activities. The good groups generate innovation from students' practice experience that is based on the process from learning knowledge to new knowledge creation. Or the groups should emphasize to develop the individual potential of creativity.

The model of collaborative supervision depends on the group mechanism, which constructs the learning organizations and provides multiple learning objectives to students. Under the leadership of supervisors, the group members own different cognitive styles, knowledge background, and creative motivation. When the traditional AMM change to group-based collaboration supervision model, the supervisors make use of Internet, lead and develop group, and form the group model of fostering PhD student depending on the beneficial environments such as diversity of group members and interaction between members. They stimulate students to better adopt knowledge and creative insights influenced by such model. The PhD students are introduced to study and manage the group and engage themselves into the practical research projects. They can be creative to work and learn confidently and independently.

Among the five groups investigated in this study, four of them are research groups with several PhD supervisors and part-time teachers as the key members. The undergraduate students are floating. Within the groups, the co-supervisors can assist the supervisors to give detailed supervisors and the students also can learn from the teachers and peers besides the supervisor. This avoids causing knowledge gaps when the supervisors are not on the site, which can realize the continuous teaching process. 'Beside the supervisors, the other teachers' ideas and methods can be expressed and shared with students.' However, there are also other points in the investigation. Due to too many layers in the group, sometimes there are weak cooperation and unclear agreements of responsibility between supervisors and co-supervisors that lead to the overlooking of the students. So problems exist, since 'there are so many supervisors, co-supervisors, and assistant professors and unclear responsibility.' This leads to 'situation of passing the buck to each other in supervision group.' 'Free' growing up may delay the students' study and development. In particular to learn the tacit knowledge such as the insight of academic problems and judgment of results, it should be learned from the direct contact with the supervisors. Even students cannot learn in depth from co-supervisors. These barriers to creativity from multiple layers of groups need to be improved by diverse and creative group mechanism.

4.4 Breaking Balance of Relationship Between Society and Culture Depending on Students' Creativity

Specialization exists as the cultural capital. As the social capital, field means a group of people who are the gatekeeper. When the balance between specialization and field is broken, it is usually caused by the appearance of new innovative individuals who own specialized knowledge and skills that engage in professional research and propose creative ideas. Students are usually the innovators guided by the supervisors. In the Big Science Era, the mixed mechanism with educational mechanism and academic research mechanism and problem-guided learning can improve the research ability and creative practice, facilitate students' creativity, break the balance between specialization, and field by either generation of new specialization or new academic leaders. In the universities, students are most creative. The stimulation of individual creativity in the group contexts usually depends on good creative climate. The group climate is crucial influences of creativity development among students. The climate involves process of group creation and different kinds of elements of pressure and environment during the group creation. The group creative climate stimulates members' creative work to reach the goals. This can be another illustration of research group functions.

From the perspective of fostering creativity, the new supervision model indicates the systematic approach to creativity and combines many elements of generating creativity. The adoption of cultural capital becomes the starting points of creative work. Stimuli such as 'letting go' and trust of supervisors, examples of professions, and other resource support make the students to engage into creative learning confidently and independently. The self-influences of achievement of supervisors forms network resource are social capital. It also has accumulation effect that facilitates students to be high level within their field in a short time. As the new supervision model is under construction in mainland China, a lot of problem in relation to creativity research are valuable to be discussed.

The Chinese supervisors usually do some measures in order to stimulate students' creativity. Such measures include encouraging students to apply for research project by themselves, providing opportunities of participation of high-level academic conference, recommending students to go to other countries, and facilitating students to submit articles to good journals. However, the supervisors are still confused that the students are lack of creativity. Now, a serious problem is to improve the theoretical thinking, as some basic phenomenon and basic regulation have been found by the experiments. But how to improve the theories based on what have been found is a puzzle. So the supervisors think to foster the critical thinking skills is very important. However, many Chinese students are lack of the guidance of critical ability. 'Culture of obedience' is integrated into the whole educational process. So the basic conceptualization on changing educational culture is needed.

5 Conclusions

As the literature [1, 7] suggested, we should not ask 'what creativity is,' instead of 'where creativity is.' Creativity is not generalized from personal mind, but from interaction between minds of human being and social–cultural environment. Thus, creativity can be found from a 'system' consist of three elements including 'person,' 'domain,' and 'field' and from the interaction between such three elements. 'Domain' exists as cultural capital. It is a symbol system including specialized knowledge, skills, norms, and value. 'Person' can reach to 'domain' though professional learning and training. 'Field' exists as social capital. It consists of groups of people who have special professional knowledge, skills, and engagement in both research and practice and who may be experts, scholars, and teachers. This study underpins such a systematic view to creativity development among PhD students in Chinese universities. As the shift of supervision model from AMM to CMM is underway, some key issues should be considered including how to build a good relationship between supervisor and students that is helpful to foster creative thinking skills of PhD students, how to stimulate the creative climate of scientific research groups where involving PhD students' creative work, and how to break the cultural barriers to creative teaching and creative learning. Therefore, the future efforts of preparing creative PhD students for the society should be focused on a systematic approach to creativity development calling for closer interaction between 'domain,' 'field,' and 'person' in the changes of supervision model toward CMM.

References

1. Hooker, C., Nakamura, J., Csikszentmihalyi, M.: Supervision model: social capital and the systematic model of creativity. In: N. Paul (eds.) Group Creativity: Innovation through Collaboration (translated by Luo, L. and Li, J.), pp. 305. Liaoning People Publisher, Shenyang (2008)
2. Elke, S.: Undertaking the journey together: peer learning for a successful and enjoyable PhD experience. J. Univ. Teach. Learn. Pract. 7(1), 12 (2010)
3. Denise, W., Dorothy, F., Dorothy, M.: Promoting creativity in Ph.D. supervision: tensions and dilemmas. Think. Skills Creativity 3(2), 143–153 (2008)
4. Krish, G., Rubby, D.: Student experiences of the PhD cohort model: working within or outside communities of practice? Perspect. Edu. 9, 88–99 (2011)
5. Raewyn, C., Catherine, M.: On doctoral education: how to supervise a Ph.D., 1985–2011. Australian Universities'. Review 54(1), 5–9 (2012)

6. Chen, H.: Where is 'Philosophical Ph.D'. in German?—an analysis based on 'Regulation of Cultivating Ph.D. Students' in faculty of philosophy at the beginning of 20th Century. Degree Graduate Student Edu. **3**, 74–77 (2003)
7. Hooker, C., Nakamura, J., Csikszentmihalyi, M.: Supervision model: social capital and the systematic model of creativity. In: Paul, N. (eds.) Group Creativity: Innovation through Collaboration (translated by Luo, L. and Li, J.), pp. 317. Liaoning People Publisher, Shenyang (2008)
8. Zhou, Q., Liu, D.: Roles and responsibility of Ph.D. supervisors—construction of framework of definitions. Degree Undergraduate Student Edu. **9**, 26–29 (2008)
9. Wang, W.: Influencing elements of Ph.D. qualification—an investigation of five research universities. Degree Undergraduate Student Edu. **9**, 16–21 (2008)
10. Gu, J., Wang, Q., Wu, J.: Influences of supervision models on Ph.D. students' creativity—an analysis based on intrinsic and extrinsic motivation theories. Chin. High. Edu. Study. **1**, 35–49 (2013)
11. Fan, K., Shen, W.: What are the good models of Ph.D. supervision?—an analysis of investigation of qualification of Ph.D. students in China. Degree Undergraduate Edu. **3**, 45–51 (2013)
12. Liu, W.: The milestone of creativity assessment research in China: a review of constructing the assessment model for the research scientist's teams. Soc. Sci. Res. J. **5**(1), 133–139 (2005)
13. Zuckerman, H.: Scientific Elite: Nobel Laureates in the United States. Free Press, New York (1977)

Evaluating Different In-Memory Cached Architectures in Regard to Time Efficiency for Big Data Analysis

Richard Millham

Abstract The era of big data has arrived, and a plethora of methods and tools are being used to manage and analyse the emerging huge volume, velocity, variety, veracity and volatility of information system data sources. In this paper, a particular aspect of a business domain is explored where the primary data being stored/accessed are not the data value itself (which is highly volatile), but the frequency of its change. Each data frequency has a chain of related data pertaining to it, whose links must be incorporated into this architecture. The volatility of data necessitates the use of in-memory architectures to reduce access/update times. Given these business requirements, different in-memory architectures are examined, using an experiment with sample data, in order to evaluate their worst case response times for a given test set of data analysis/manipulation operations. The results of this experiment are presented and discussed in terms of the most suitable architecture for this type of data, which is in-memory objects linked via hash table links.

Keywords Big data · In-memory cached architecture

1 Introduction

Big data can often be defined by the nature of its characteristics of the 5Vs of volume, velocity, variety and volatility [1]. Big data are playing an increasingly important role in business. An example, different businesses are analysing big data to provide more informed decision-making and insights into client behaviour [2].

In this paper, we briefly examine the role of big data and the various architectures that have been designed to meet its unique characteristics. Each of these architectural designs has advantages and disadvantages with its design often being best suited to manage a particular facet of big data. As we are focusing on a

R. Millham (✉)
Durban University of Technology, Durban, South Africa
e-mail: richardm1@dut.ac.za

© Springer International Publishing Switzerland 2015   63
A. Abraham et al. (eds.), *Pattern Analysis, Intelligent Security*
and the Internet of Things, Advances in Intelligent Systems and Computing 355,
DOI 10.1007/978-3-319-17398-6_6

particular type of big data that may be particular within a specific business domain, we examine a specific type of data architecture design that is well suited for that type of data, which in our case is in-memory architecture. The nature of the business dictates a particular linking of various related data items. With this type of data and required linkages in mind, we examine different architectures, using a test set of worst-case scenario data analysis/manipulation operations, in order to discover the most time-efficient architecture. This architecture must be generic enough to be implemented for this business aspect with the trade-off being slightly slower speed in return for flexibility rather than being a highly specialised architecture for a very particular instance with optimal speed. The results of our evaluation are presented in order to outline the most suitable architecture for this business aspect with its type of data.

2 Literature Review

Although there are many references to big data and its use within organisations, there is no standard definition of big data or standards governing its use in terms of architecture, analytics, security, etc. Although there are efforts to implement a standardised database to handle big data, the variety, with the requirements of velocity and volume, denotes a customised approach to data storage and collection [3]. Consequently, a wide variety of architectures and security and analysis techniques exist or are emerging. Often, big data is distinguished from other forms of data by its properties of volume, variety, velocity, value, veracity and volatility [1, 4].

Although no standard set of architectures exist to manage big data, one of the focus areas within architecture is on data dynamicity and linkage in order to form new data models that incorporate data linking and referential integrity [3]. This linkage is necessary to ensure data integrity and to reduce data duplication.

In order to manage big data, a number of architectures have been proposed. NoSQL (non-relational databases) rely on "key-value" pairing and may be used to provide distributed, highly scalable data storage for big data [5]. "Key-Value" pairing is used for unrelated data where a key, determined through some function, provides a link to a storage area of its corresponding data [6]. A common architecture for big data is the use of the Hadoop framework. This framework enables large data sets to be distributed across clusters of computers and is highly scalable. Scalability is enabled by having each server in its network have its own local computation and storage and in having this network grow in relation to the growing needs of data—as data grows, more servers are added. A key component of Hadoop is Map-Reduce which distributes data, with its analysis computations, across every server in the Hadoop network. Once each individual server finishes its computations on its local data received by Map-Reduce, these results are sent to Map-Reduce for compilation into final results [7, 8]. Although Hadoop is suitable for large amounts of variable data (data originating from sensors, web clicks, documents, etc.), Hadoop is not suitable for real-time big data use and analysis [5].

Key-value store architectures, which permits clients to insert and access valued per key, are recognised as having the advantages of a simple architecture and high scalability. However, this architecture has the cost of consistency of value storage which prohibits rich ad hoc querying and analytics aspects [9]. Document store architecture allow for storage of more complex data, such as nested documents or list with attribute names to be defined dynamically at run-time. An example of document store architecture is MongoDB. Another architecture, extensible record stores, consists of rows and columns with its rows and columns distributed over multiple nodes. Rows are split across multiple nodes through the use of a shared primary key, and table columns are divided across multiple nodes using "column groups". An example of an extensible record store is Google's Big Table [9]. However, maintaining a record of which rows and columns are split among different data chunks is cumbersome and time-consuming. In addition, unless the partitioning of the data among nodes is done optimally, performance loss may result.

A variable heterogeneous architecture for big data has been proposed where an incoming data stream is first analysed for its format. Depending on the format, the most appropriate architecture is chosen—relational databases for normalised data, key-value storage for simple data with no need of complex querying, etc. In addition, in order to distribute the data amongst servers with minimum communication costs, each data item is calculated in terms of its distance from another. The most centrally located data item, or central centroid, becomes the cloud cluster, and the most closely located data items form this data cluster. This particular analysis has the advantage that outliers, with a great distance to the central centroid are not used in the calculation of average distance, so they do not become part of the cluster, at the expense of a high communication cost [9]. However, the linking and integration of these diverse forms of data amongst these different architectures pose a difficult challenge.

For every origin of big data, there is a target or purpose. An example, for the footing of big data in science, the target was scientific discovery. For the basis of business, the target was personal services [3]. In this study, the foundation of big data in customer relationship, the target is identification of problem areas within complaints.

The data values being stored might range due to a number of considerations. An example, a stock exchange might wish to store the price history of each particular share over a given period time. However, this storage requires considerable data storage and relatively long access times to access a given data item, such as a share price at a particular point in time. Often, a business will store frequently accessed data items, which they commonly used for analysis, in a quickly accessed data structure with links from that item to related items in decreasing order of use. A common example is the current "buy" stock price. Some business analysis models of traders may predict a future rise in a particular stock (based on a detected yet unprocessed "buy" order) and then purchase an amount of stock at the current "buy" price millisecond before the unprocessed "buy" order. The unprocessed buy order is forced to buy that particular stock at a slightly higher buying price.

This scenario is repeated thousands of times per day, which gives this type of trading the moniker of "high-frequency trading" [10].

However, some business models rely on utilising the "herd mentality" to predict stock price. This business focus might be on the most highly traded stocks in order to determine the direction of the market [11]. However, in high-frequency trading, the highly traded stocks, as denoted by frequency figures, might be misleading as the frequency is also dependent on the number of shares traded. These number of shares traded might be stored separately or combined with the stored frequency factor in conjunction with a weighing factor (number of shares traded). However, this frequency factor would consist of two components, the number of times a particular share was traded and the number of shares traded of that share. With an increased number of components comes an increased complexity within its architecture and its algorithms.

Besides data value storage, the manner in which it is stored and accessed must also be considered. As the number of CPUs, CPU speed and memory size increase, many big data workload change from being predominantly I/O bound to CPU bound [12]. Often, there is a trade-off between information representation and data processing efficiency in order to find the optimal cost per bit of information produced. Due to increasing data sizes and energy consumption, cache and memory compression have been areas of focus. Some areas of interest are which modules in the memory hierarchy should contain compressed or non-compressed data and what is the impact of compressed data on performance and energy. Some experiments suggest that, due to the high performance costs of compression and decompression, compression should be reserved for the least-frequently used items [13]. This approach towards compression is adopted by some big data architectures.

In order to manage the volatility and velocity, notably quick access rates, of big data, in-memory architecture is used with periodic backups, during off-peak times, to secondary storage. Although in-memory storage is more susceptible to loss due to disasters such as power outages, the higher access speeds, relative to secondary storage, make it more suitable for volatile data storage [14]. One researcher, to manage the volume and velocity of big data, relied on an architecture of in-memory access to data with customised data structures to hold related data chunks and compression to reduce data size, particularly among less frequently used related items [15]. However, no suitable data structures were specified that would hold these related data chunks yet be flexible enough to manage new varieties of data. In addition, in-memory architecture is suited to data with high velocity and volatility yet poses a danger due to the temporality of its storage nature.

3 Methodology

Although big data have many possible architectures, it was decided to design an architecture that would allow for a variety of different types of data yet allow these data items to be linked, as they often are in a business domain, for related pieces of data. Our data model focuses on developing an architecture, notably for high-frequency data. High-frequency data were chosen in order for businesses that use this data to access need data quickly to make informed decisions within certain domains. Although the stock market business was looked at, it was decided to look at another business using another frequency domain, customer relations, where the frequency factor is the number of complaints per customer. By analysing the frequency factor, businesses can quickly determine their customer relationship direction. The frequency represents an aggregate of individual customer complaints per client. Each individual complaint is categorised into a complaint area, linked with that particular frequency. The complaint area, in turn, is linked to a category of dates. Further analysis of this linked data can be used to yield additional insights into customer relationship management [16].

If further information for analysis is required, each frequency factor has a one-to-many relationship with complaint areas as a customer might have complaints in different areas. If even further analysis is required, each complaint area has a many to many relationship with dates. The area of complaint may have multiple dates upon which the complaint was based. By utilising a unique data per table entry with links, redundant date storage (as in a one-to-many relationship between complaint areas and dates) is eliminated without losing any information [17]. Furthermore, by utilising frequency, complaint area and complaint date as dimensions within a data cube, relationships between these dimensions can be determined through analysis of this data [18]. An example, for a client with high frequency, areas of complaint might be analysed along with the dates of complaint in order to find a correlation between these three dimensions—a problem in a particular area for a particular group of customers during a specific time frame—in order to devise and take remedial action.

In other words, a group of Clients $(CL_1 \ldots CL_n)$ will have their set of complaints: a particular client of this group, CL_i, has a set of complaints $\{CP_1 \ldots CP_n\}$. The frequency of complaints of an individual client CL_iP_i is $(CL_iP_1 + \cdots + CL_iP_n)/T_i$ where T_i is a given time period. Possible complaint areas (CA) are a set $\{CA_1 \ldots CA_n\}$, which are aggregated into a subset $\{CA_{i1} \ldots CA_{in}\}$ that denote grouped complaint areas. Similarly, possible complaint dates are a set $\{CD_1 \ldots CD_n\}$, which are aggregated into a subset $\{CD_{i1} \ldots CD_{in}\}$ that denote grouped complaint dates. Each frequency of an individual client, denoted by F_iCL_i, is linked to a set of grouped complaint areas, denoted by $\{CA_iG_1 \ldots CA_iG_n\}$. Consequently, a client frequency, CL_iF_i, is mapped to a subset of grouped complaint areas, $CL_iF_i\rightarrow(\{F_iCA_iG_1 \ldots F_iCA_iG_n\}$, where $\{F_iCA_iG_1 \ldots F_iCA_iG_n\} \subset \{CA_iG_1 \ldots CA_iG_n\}$. Similarly, a client's present complaint areas are mapped to a particular set of grouped dates. This mapping is denoted as $\{F_iCAG_1 \ldots F_iCAG_n\}\rightarrow\{F_iCD_iG_1 \ldots F_iCD_iG_n\}$ where $\{F_iCD_iG_1 \ldots F_iCD_iG_n\}\} \subset \{CD_iG_1 \ldots CD_iG_n\}$.

Due to the high volatility and high velocity of this type of data, it was decided to adopt an in-cache memory architecture that could store data that could be accessed, inserted and altered in real time. However, as this type of frequency data often had linked data that needed to be accessed for analysis, this architecture also must take into consideration the linking of different types of data. A number of different architectures, based on in-memory access to data, were experimented with and examined for insert/access/alteration speed.

This data architecture provides data linkages of different types of data with the most frequently accessed data being stored and accessed in the first component. Consequently, due to this architecture, there are linkages from the primary component (one to many or many to many) to subsequent components. As a result, if analysis requires more information from an element or set of elements, this information can be derived, in order of use, from following linkages from the element to subsequent components containing this information. In addition, these linkages were designed to be able to form a 3D cube of data, with each component providing a data dimension. Furthermore, these linkages enabled each member of the primary set of data to have multiple elements of linked data in subsequent data sets in order to mirror a business scenario. An example of this type of scenario, given a customer frequency and complaint area, analysis can reveal if these complaints within this area increased, remained stable or decreased over a set period of time and determine what were the areas of complaints and within what time period.

One of the reasons for using objects as data structure entities within our architecture is that objects are easily extendible and can hold a variety of different type of data. As a new data stream becomes available, the object can easily be modified to hold this addition of data. Objects are not restricted to a particular data type—they can simultaneously hold strings, numbers, video and pictures among other data types.

4 Small Example with Results

In order to test the various architectures, a small example data set was used. The data set consisted of originally 20,000 frequencies, 40,000 areas and 80,000 dates. Although this data set is very small (and may not be large enough to be considered to be big data), the experiment was constrained by the equipment that had to be used. This system had a relatively small amount of memory, so it was decided by the researcher to use a smaller data set that would fit in its existing memory rather than utilise a "typical" big data set which would overwhelm the resident memory (and thus defeat the experiment by providing false response rates) and possibly increase CPU overhead unnecessarily (with consequent false response rates possibilities) through paging management of virtual memory.

The system used for this example was an i3 dual-core processor running at 2.53 MHz. It contained 4 GB of RAM and ran using the 64-bit Windows 7 operating System, Service Pack 1. Although this system is not typical of "big data" systems, it was thought that it would provide an accurate relative comparison of different architecture's data manipulation time response rates.

First, a set of test cases needed to be developed in order to test the access/insert/update (data manipulation and business analysis) response times of different architectures. Besides determining the response times for insertion of new data and updating of existing data, response times for data access were also examined. Given the business case scenario, it was decided that typical access queries might involve determining the client with the lowest and highest frequency of complaints, an average frequency of complaints for all clients, and the creation of a data cube for all clients and the creation of a slice of this data cube. The set of test cases are as follows:

1. Determine the maximum frequency and access all linked entities for that frequency
2. Determine the minimum frequency and access all linked entities for that frequency
3. Determine the average frequency and access all linked entities for all frequency used in that calculation
4. Insert 10,000 new frequencies with 10,000 new entities for each linked component (in the relational database case, linked table rows)
5. Update 10,000 new frequencies with 10,000 updated entities for each linked component (in the relational database case, linked table rows)
6. Develop a cube of customer frequency, complaint area and complaint dates
7. Slice this cube for a particular customer with multiple complaint areas but within a certain date range of 3 days

In terms of the update, usually one, or perhaps another linked entity, would be updated with the remaining linked entities remaining the same. However, in our test case, the worst-case scenario is explored where all frequencies, with their linked entities, are updated. An average operation time per architecture is given, from the time to execute for each of these seven test cases, in order to demonstrate which architecture has the fastest overall operations.

4.1 Relational Databases

In order to develop a baseline for access/insert rate, a data model was adopted using a relational database. Though this model is not an in-memory architecture per se, it provides a baseline for comparison with other in-memory architectures. Intermediate tables provided the many-to-many relationship between tables and foreign keys provided a one-to-many relationship between the parent and child tables. Figure 1 denotes the table structure.

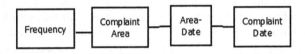

Fig. 1 Relational database tables

Table 1 Query time results for relational database

Operation	Time elapsed (picoseconds)
Maximum	109,000,000,000
Minimum	780,000,000
Average	930,000,000
Insert 10,000 linked entities	2,560,000,000,000
Update 10,000 linked entities	17,261,000,000,000
Create data cube	577,000,000,000
Slice data cube	88,000,000,000
Average operation time	2,942,387,142,857

The relational database used was Microsoft SQL Server, version 10.50.2550. Although different relational databases were available, it was thought that SQL Server would be representative of the relational database architecture.

The above results are given for each test case using the relational database server (Table 1).

The update and insert operations took comparatively longer due to the fact that SQL Server had to insert a row into one entity and then insert another row into another entity with a foreign key, which involved updating its index table for each of the 10,000 entities to be inserted. Similarly, the update operation involved accessing a particular primary entity (frequency), updating different fields of each linked entity and then re-indexing each linked entity for the 10,000 entities to be updated. The creation of the data cube took a comparatively longer time than the statistical operations due to the fact that multiple data items had to be accessed and retrieved. The slice of a data cube consumed relatively less time due to the fact that although all linked data items had to be accessed, relational database built-in optimisation techniques filter out unrequired rows of data items such that only a small portion of these items would have to be amalgamated to form a slice [19].

4.2 Indexed Collections of Objects

In order to speed up all types of operations, different in-cache memory architectures were examined. As frequency changed often, developing an index to access a particular requested frequency, in a quick manner, was impractical. Consequently, each frequency was put into an array element. This frequency element was the "value" with its corresponding array element serving as a pointer (or "key") to a collection of complaint area objects. Each complaint area object had several properties, including a list structure of complaint areas. This list provided a one-to-many relationship between frequency and complaint areas. To access a particular object in the list, the "key" or pointer of the frequency array served as an index to the object in the collection using a pre-defined key-object function. Each list element of the complaint area object had a "key" or pointer that served as an index to

Fig. 2 In-cache memory database using indexed collections

Operation	Time elapsed (picoseconds)
Maximum	604,000,000
Minimum	607,000,000
Average	688,000,000
Insert 10,000 linked entities	0.00,000,003
Update 10,000 linked entities	0.000012850018
Create data cube	3,120,010,000
Slice data cube	3,120,000,000
Average operation time	1,162,715,714

Table 2 Query time results for indexed collection architecture

its corresponding object in the complaint date collection. This list key arrangement provided a many-to-many relationship between complaint area and date. Using a pre-defined key-object function, the list key of the complaint area object is used to access its corresponding complaint date object (Fig. 2).

Using the same set of test cases as were used in the relational database architecture, the above results from this architecture were determined (Table 2).

Access times for the minimum and maximum frequency values have improved significantly. The average frequency is slightly more than the minimum or maximum frequency due to the extra calculation overhead which must be taken into account. The creation and slicing of a data cube consumed approximately the same amount of time. Insertion of new entities in all of the linked entities is much quicker than an update due to the architecture requiring a simple expansion of existing data structures and inserting the new entities into the structure with their new links. An update requires accessing all and each of the frequency elements with their linked data structure entities and updating them and their links. The access and updating of existing elements in linked entities consume more time than simply expanding data structures and adding new elements to these structures (Fig. 3).

4.3 Objects/Lists with Hash Table Links

Using the same set of test cases as were used in the relational database and collection architecture, the following results from this hash table linked architecture were determined (Table 3).

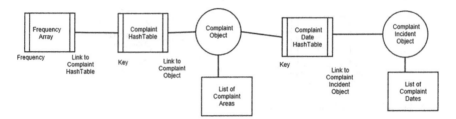

Fig. 3 In-cache memory database using hash table links

Table 3 Query time results for hash table architecture

Operation	Time elapsed (picoseconds)
Maximum	0.000000020012
Minimum	0.000000013008
Average	0.000000013008
Insert 10,000 linked entities	0.000000005
Update 10,000 linked entities	0.0000000400001
Create cube	0.00200943959
Slice data cube	0.00468
Average operation time	0.00095564

Although the hash table architecture does not differ substantially from the in-cache memory collection architecture, the access/update/and insert times are significantly lessened. The common statistical analyses of frequency, along with the creation of a data cube, were relatively quick due to fast access times. The slicing of the data cube took longer due to the need to filter out unwanted data from the cube which was just created. Due to quick access times, the updating of selected rows was very quick. Similarly, due to fast access times, the links to the newly expanded objects (to hold the newly-inserted data) were very quickly updated. The difference in architecture is that rather than relying on indexed object collections, this architecture uses has table data structures to hold "value"-"key" pairs with the "key" being the object index and the "value" being the object itself. Hash tables are data structures that are designed to produce fast associative lookups with average memory access times of $O(n)$ where n represents the number of elements [20].

5 Discussion

This experiment had a number of constraints. Although a comparison of in-memory architecture was performed, the relatively small amount of memory, by big data architectural standards, required the use of a small but representative data set for testing. The set of test cases to be used was limited, but this set was determined to be a common set of operations and business analysis to be performed within this selected particular business domain. The system, although limited in its system resources, would give relative, but not necessarily accurate, response times per given operation for each type of architecture to be tested.

The choice of use of frequency of complaints per client rather than use and storage of individual complaints per customer was selected due to its advantages of quick analysis, management of high volatility of complaints and the reduction of required storage space, with their resultant speed up. By aggregating complaint areas and dates, among other information, to be stored, within appropriate linkages to clients, the amount of storage required is reduced without eliminating any necessary information. This particular business domain could quickly determine the direction of its customer relationship management through accessing the frequency table. If further information regarding a client's customer complaint is required, this information can be retrieved through links from the frequency table.

Because in a large business, customer complaints change so quickly, it was determined that it would be faster to store it, not only as a frequency rather than an individual complaint, but in memory where its high volatility characteristic could be managed in an optimal manner. Consequently, in-memory rather than different big data architectures were considered.

Although a more optimal response time might be achieved using very specialised data structures to hold particular data types within the selected business domain, it was decided to use more generic data structures in order to provide flexibility and to reduce the need to customise the architecture every time that the incoming data type changed.

As this system was not an optimal big data architectural system, it was decided to use response times per operation for different architectures from a relative, rather than absolute, perspective. Although the response rates are given in an absolute manner, these response times are dependent on a number of factors including memory size, CPU size, bus speed, et al. Consequently, when determining the optimal architecture for high-frequency big data, the results must be viewed as in how much they differed from other architectural models performing the same operation rather than in absolute response times.

Viewed in the light of all of these considerations, the experiment revealed that hash tables were time optimal, relative to other architectures, in all of the test case data manipulation/analysis operations. Due to the quick access times of the hash table, items can be quickly retrieved and updated in memory. Hash tables can be quickly expanded in order to accommodate new items. Due to the hash table's quick access times to links to related data, this data can be retrieved quickly.

Acknowledgements I would like to acknowledge the late Mr. Stephen Eta who shared a passion for big data and whose research on the use of the frequency storage dimension within Big Data was tragically cut short by his untimely death.

References

1. Sicular, S.: Gartner's big data definition consists of three parts, not to be confused with three "V"s", Gartner, Inc. **27** (2013). http://www.forbes.com/sites/gartnergroup/2013/03/27/gartners-big-data-definition-consists-of-three-parts-not-to-be-confused-with-three-vs/
2. McGuire, T., Manyika, J., Chui, M.: Why big data is the new competitive advantage. Ivey Bus. J. **76**(4), 1–4 (2012)
3. Demchenko, Y., Ngo, C., Membrey, P.: Architecture framework and components for the big data ecosystem. J. Syst. Network Eng., pp. 1–31, (2013)
4. Millham, R.: Integrating heterogeneous data for big data analysis in cloud infrastructures for big data analytics. IGI Group Hershey, Pennsylvania, USA (2013)
5. Chan, J.O.: An architecture for big data analytics. Comm. IIMA. **13**(2), 1 (2013)
6. Burtica, R., Mocanu, E.M., Andreica, M.I., Tapus, N.: Practical application and evaluation of no-SQL databases in cloud computing. In: Systems conference (SysCon), 2012 IEEE international, pp. 1–6 (2012). IEEE
7. Humbetov, S.: Data-intensive computing with map-reduce and hadoop. The 6th International Conference on Application of Information and Communication Technologies (AICT), IEEE, pp. 1–5 (2012)
8. Yang, H.-C., Dasdan, A., Hsiao, R.-L., Parker, D.S.: Map-reduce-merge: simplified relational data processing on large clusters. In: Proceedings of the 2007 ACM SIGMOD International Conference on Management of Data, ACM, pp. 1029–1040 (2007)
9. Zhang, Q., Chen, Z., Lv, A., Zhao, L., Liu, F., Zou, J.: A universal storage architecture for big data in cloud environment. In: IEEE International Conference on Internet of Things (iThings/CPSCom), and IEEE Cyber, Physical and Social Computing Green Computing and Communications (GreenCom), IEEE., pp. 476–480 (2013)
10. Keim, B.: Nanosecond trading could make markets go haywire. Wired **16** (2012)
11. Raafat, R.M., Chater, N., Frith, C.: Herding in humans. Trends Cognitive Sci. **13**(10), 420–428 (2009)
12. Yang, D., Zhong, X., Yan, D., Dai, F., Yin, X., Lian, C., Wu, G. NativetTask: A hadoop compatible framework for high performance. In: 2013 IEEE International Conference on Big Data, IEEE, pp. 94–101 (2013)
13. Bui, V., Kim, M.A.: The cache and codec model for storing and manipulating data. IEEE Micro **34**(4), 28–35 (2014)
14. Young, M., Tevanian, A., Rashid, R., Golub, D., Eppinger, J.: The duality of memory and communication in the implementation of a multiprocessor operating system. ACM, **21**(5), 63–76 (1987)
15. Matsudaira, K.: Big data without big database—extreme in-memory caching for better performance. ACM Webcast **20** 2013
16. Berry, M.J., Linoff, G.S.: Data Mining Techniques: For Marketing, Sales, And Customer Relationship Management. Wiley, London (2004)
17. Howe, D.R.: Data analysis for database design. Butterworth-Heinemann, London (2001)
18. Gray, J., Chaudhuri, S., Bosworth, A., Layman, A., Reichart, D., Venkatrao, M., Pirahesh, H.: Data cube: a relational aggregation operator generalizing group-by, cross-tab, and sub-totals. Data Min. Knowl. Disc. **1**(1), 29–53 (1997)
19. Sumathi, S., Esakkirajan, S.: Fundamentals of relational database management systems, **47**, Springer, Berlin (2007)
20. Song, H., Dharmapurikar, S., Turner, J., Lockwood, J.: Fast hash table lookup using extended bloom filter: an aid to network processing. ACM SIGCOMM Comput. Commun. Rev. **35**(4), 181–192 (2005)

Engagement in Web-Based Learning System: An Investigation of Linear and Nonlinear Navigation

Norliza Katuk and Nur Haryani Zakaria

Abstract This paper investigates linear and nonlinear navigations in Web-based learning (WBL) systems. The aim of the study was to identify whether the linear and the nonlinear navigations could be the factors that influence students' engagement within WBL environment. An experimental study was conducted on seventy-two students from a university in Malaysia using a Web-based system for learning Basic Computer Networks and a self-report inventory. The results of this study suggested that the types of navigation support affected engagement from certain aspects.

Keywords Navigation · Linear · Nonlinear · Engagement · Web-based learning

1 Introduction

Web-based learning (WBL) emerges as a result of extensive use of information technology (IT) for delivering courses especially in higher learning institutions. One of the WBL important aspects is a proper organisation of learning content and an appropriate navigation support that can facilitate learning. Generally, navigation can be linear and nonlinear and they might not suitable for everyone. The issue is how and what navigation support should be offered to students?

This paper reports on investigation of navigation supports in WBL from students' engagement perspectives. Section 2 explains the basic concepts of

N. Katuk (✉) · N.H. Zakaria
School of Computing, Universiti Utara Malaysia, 06010 Sintok, Kedah, Malaysia
e-mail: k.norliza@uum.edu.my

N.H. Zakaria
e-mail: haryani@uum.edu.my

© Springer International Publishing Switzerland 2015
A. Abraham et al. (eds.), *Pattern Analysis, Intelligent Security
and the Internet of Things*, Advances in Intelligent Systems and Computing 355,
DOI 10.1007/978-3-319-17398-6_7

navigation in WBL and engagement aspects that related to them. Section 3 describes the experimental study and the results. The findings are discussed in Sect. 4.

2 Navigation of Learning Content in WBL

Navigation support is a basic component of any Web-based application that gauges its effectiveness. In simpler words, navigation affects Web usability [1]. In WBL, a good content navigation technique helps students to learn and obtain knowledge effectively [2]. In contrast, a poor navigation technique leads to disorientation [3] and cognitive load [4] that hinder effective learning.

Navigation in WBL comprises two categories: linear and nonlinear [5]. In linear navigation, the process is controlled by the system [6]. Path to access the content has been predetermined either dynamically (following students' needs) or statically (fix path). Students have no control over the sequence or path of learning content. They simply click the provided buttons to move forward or backward. In nonlinear environment, students have greater control over the content compared to the linear. Martin [6] defined nonlinear as a type of navigation that allows students to freely navigate the content and follow their own path. No specific path is determined by the system, which resulted in greater flexibility to browse the content than the linear navigation.

Past studies in linear and nonlinear navigation of WBL have shown mixed results. Connelly et al. [7] found that students make the least mistakes with linear navigation. Further, researchers like Gauss and Urbas [8] suggested that the linear navigation can reduce disorientation among students with low prior knowledge. However, a study by Baylor [9] had a contradict result with Gauss and Urbas. He found that linear navigation is more likely to cause disorientation to students than the nonlinear. His study was about a searching task that required the students to find the location of five sentences within a nine-page Web-based passage.

The other benefits of nonlinear navigation include the following: (i) provides higher interactivity [10], (ii) easy to design and (iii) attractive [11] compared to linear navigation. Al-Hajri et al. emphasised that nonlinear learning may not [12] be suitable to all learners [12]. They also argued that neither linear nor nonlinear will improve students' performance because students are individually different. They suggested that "*only the navigation approaches that accommodate students' differences will result in better performance*". This is true when we refer to Chen et al.'s [13] study. Their study suggested that prior knowledge should be considered when designing a navigation support for WBL. Some studies in the past had also supported that WBL navigation should be designed according to student's individual difference. Table 1 summarises the results of some empirical studies and the factors that influence navigation styles including field dependency, level of knowledge, gender and system experience.

Table 1 Factors influencing the types of navigation in WBL

Components	Linear	Nonlinear
Field dependency (field dependent/field independent) [12]	Field-dependent students prefer a linear content navigation	Field independency relatively enjoys nonlinear content navigation
Level of knowledge [14]	Novice students learn better with linear navigation	Expert students learn better with nonlinear navigation
Gender [15]	Female students prefer linear organisation of learning content	Male students prefer nonlinear learning content organisation
System experience [15]	Students who are not familiar with a particular system prefer linear navigation	Students who familiar with a particular system prefer a nonlinear navigation

2.1 The Effects of Navigation Style on Student's Engagement

Students' engagement (or disengagement) correlates with their motivation to learn. Engagement in the context of this study refers to cognitive engagement; subjective experience persons have when they interact with computer systems [16]. Some studies used the concept of optimal experience or flow [17, 18] to describe the cognitive engagement. In this study, engagement is investigated from four aspects of learning: (i) control, (ii) focus, (iii) curiosity and (iv) intrinsic interests. The following paragraphs describe them generally from the context of learning.

Control refers to the situation in which students feel in control of the learning activities. In the context of WBL, control is a critical component that affects students' motivation, performance and attitudes towards learning [19]. In fact, several studies on learner control in WBL have revealed that giving students control over learning activities leads to an improved academic achievement [20–22].

Besides control, the learning process also requires an optimal level of focus so that a meaningful learning can be obtained. *Attention focus* refers to a situation in which student's mind is absorbed by WBL activities. It actually measures student's level of concentration in the given tasks. Saadé and Bahli [23] defined this condition as cognitive absorption, which plays an important role in generating more positive attitudes towards learning and greater exploratory use of the system.

Webster et al. [24] defined *curiosity* as the situation in which a student is excited and eager to know more about the domain knowledge. It is important to note that the state of curiosity is always inconsistent. Small and Arnone [25] suggested that sufficient and relevant information can increase curiosity. They claimed that motivation could be increased when students are provided with the information that is required for learning, thus encouraging them to explore more about the topic. Consequently, in the context of WBL, insufficient information or knowledge that a student anticipates during a learning process may lead to a significant decrease or even extinction of curiosity.

The last aspect of engagement is *intrinsic interests,* which can be defined as a situation in which a student feels enjoyment with the learning activities. This can be further described by the reasons which can motivate the student to learn. A student with intrinsic interests engages in WBL for the sake of the learning itself without apparent force [26]. Researchers in the area of WBL acknowledge that a proper design of computer systems can help in stimulating intrinsic interests.

Control, focus, curiosity and intrinsic interest are studied with regard to navigation support in this paper. It aims to understand the effects of navigation styles (i.e. linear and nonlinear) on the engagement aspects among adult students in Malaysia. As many Malaysian students use WBL in their courses, the author interested to identify whether specific types of navigation in WBL will provide them with appropriate navigation support. To be specific, the study attempts to answer *"what is the effect of linear and non-linear navigations of learning content on student's control, focus, curiosity and intrinsic interests?"*

3 Experimental Study

An experimental study was conducted to evaluate the effects of linear and nonlinear navigations on the four aspects of engagement (i.e. control, focus, curiosity and intrinsic interests). This section explains the methods for conducting the study and its results.

3.1 Participants

Participants for this study were recruited among students from a higher learning institution in Malaysia between May and December 2010. Emails and advertisement in the university's learning management system (LMS) were used to invite the students to participate in the study. A number of 72 students comprised 33 males and 39 females were recruited. Their mean age was 24.03 ranging from 18 to 45 years with majority of the students were aged between 21 and 25 years (52 students). The students who participated in this study enroled in various programmes.

3.2 Materials

Materials for the study consisted of two Web-based systems and a self-report inventory. The two systems were designed and developed following a course syllabus for the IT fundamental course offered for non-IT students. The original system was known as IT-Tutor with linear navigation. However, for the purpose of this study, the same system has been modified to include the nonlinear navigation. Both systems covered a module named "Basic Computer Networks" that is equivalent to 90 mins of learning. The Web-based systems organised learning content in multimedia formats (i.e. text and images). The systems consist of three main components: quizzes, feedbacks and explanations. The quizzes were used to evaluate students' prior and current knowledge about the domain of learning. The feedbacks notify the students about the answers of the quiz and the associated explanations that they need to know. The explanations are the associated knowledge that gives details of the quizzes. This is the component where linear and nonlinear navigations have been implemented. Figure 1 illustrates the learning process that students need to follow with linear (Fig. 1a) or nonlinear (Fig. 1b) navigation.

Both linear and nonlinear Web-based systems pose a quiz that comprises four questions in every stage of three. When students answer the quiz, both systems analyse the answer and give feedbacks to students. The feedbacks inform the students whether their answers are correct, and if not, they will suggest the content or explanation that the students need to read/learn. After the students access the explanations (either in linear or nonlinear ways), they will proceed to the next stage of the tutorial which is a new set of quiz. This process will be repeated three times in both systems.

Back to the presentation and organisation of the explanations, the linear navigation automatically presents the explanations to the students following the pre-identified contents (they have been generated according to the answers of the quizzes). On the other hand, the nonlinear navigation allows the students to navigate the contents according to their own navigation paths. As shown in Fig. 1a, students will be automatically presented with the explanations after they receive the feedbacks. They can move from one content to another using the given next/back buttons. When they have gone through the explanations, they will move to the next stage of the quiz. Unlike the linear, the nonlinear allows the students either to browse the learning notes independently or to simply move to next stage of the quiz. This is represented by the dotted line in Fig. 1b.

Fig. 1 Linear and nonlinear navigation of the Web-based systems. **a** Linear navigation. **b** Nonlinear navigation

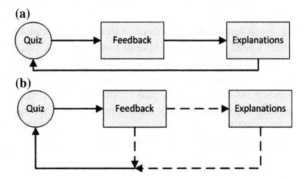

Table 2 The questionnaire for the experimental study [24]

Aspects of engagement	Questions
Control	C1—When using IT-Tutor, I felt in control over everything
	C2—I felt that I had no control over my learning process with IT-Tutor
	C3—IT-Tutor allowed me to control the whole learning process
Focus	F1—When using IT-Tutor, I thought about other things
	F2—When using IT-Tutor, I was aware of distractions
	F3—When using IT-Tutor, I was totally absorbed in what I was doing
Curiosity	CU1—Using IT-Tutor excited my curiosity
	CU2—Interacting with IT-Tutor made me curious
	CU3—Using IT-Tutor aroused my imagination
Intrinsic interests	I1—Using IT-Tutor bored me
	I2—Using IT-Tutor was intrinsically interesting
	I3—IT-Tutor was fun for me to use

The study used a questionnaire as an instrument to measure the effects of linear and nonlinear navigations. This questionnaire was adapted from Webster et al. [24]. The questionnaire comprised twelve items that specifically asked students about their experiences interacting with computer systems as shown in Table 2. A five-point Likert scale (1 is for strongly disagree and 5 is for strongly agree) was used in the questionnaire.

3.3 Procedure

The experimental study was conducted online and in unsupervised mode. The students who agreed to participate were given a URL (a Web link) to access the materials. They were allowed to perform the tasks at their own convenience. Once they accessed the Web link, an information sheet about the study was presented. Then, they were asked about their consent to participate in the study by accepting the given terms and conditions. After that, the students were randomly assigned to either linear or nonlinear system using a binary random number generator. Then, they were required to interact with the systems (answering the quiz and learning the content). When they completed the task, the questionnaire was given to them. Their interactions were logged, and if they had been inactive (no interactions such as moving the mouse and scrolling down the vertical bar) for more than five minutes, they would be logged off from the systems. This was done to protect the reliability of the data in online and unsupervised experimental study.

Table 3 The means (standard deviations) for the linear and nonlinear aspect of engagement

Aspect of engagement	Questions	Linear ($n = 33$)	Nonlinear ($n = 39$)
Control	C1	3.52 (1.093)	3.95 (0.793)
	C2	3.15 (1.176)	2.51 (1.374)
	C3	3.45 (1.034)	3.77 (0.986)
Focus	F1	3.45 (1.227)	3.10 (1.252)
	F2	3.30 (0.984)	3.31 (1.080)
	F3	3.48 (0.939)	3.67 (1.084)
Curiosity	CU1	4.00 (0.707)	4.05 (1.075)
	CU2	3.76 (0.902)	3.97 (1.063)
	CU3	3.73 (0.944)	3.69 (1.004)
Intrinsic interests	I1	2.52 (1.252)	2.54 (1.165)
	I2	3.82 (0.846)	3.62 (1.016)
	I3	3.82 (0.769)	3.79 (1.128)

3.4 Results

The data of this study were analysed using SPSS (version 19). A reliability test was conducted on the twelve items of the questionnaire. The Cronbach's alpha coefficient was 0.774 suggesting that the data had relatively high internal consistency. A normality test following Kolmogorov–Smirnov (K-S) was also conducted on the data, and the result suggests that they were non-normal with $p < 0.002$ for all items. This may due to small sample size of the study. As the data were not-normally distributed, nonparametric tests were applied in the analysis.

Table 3 shows the mean scores for control, focus, curiosity and intrinsic interests for both types of navigation. In terms of the control aspect, the students with the nonlinear agreed that the navigation gave them control over the learning content. Specifically, in question C2 that is "*I felt that I had no control over my learning process with IT-Tutor*", the students with the nonlinear WBL rated significantly lower (2.51) than the linear students (3.15). The difference was statistically proven by the Mann–Whitney U test ($Z = -2.092$, $p = 0.036$, $p < 0.05$). This suggests that the nonlinear navigation gave the students higher level of control over the learning content than the linear in the given WBL. However, the tests suggested that the differences in the other aspects of engagement were not statistically significant.

4 Discussion and Conclusion

There are two important findings of this study. Firstly, it suggests that the linear and nonlinear navigations had different effects on the students from the aspect of control. Specifically, the students with the nonlinear navigation reported higher level of control than the linear students during interaction with the WBL. Secondly, the students' level of focus, curiosity and intrinsic interests was not affected by the navigation types, in which they had similar level of the three aspects, respectively.

The findings of this study should be used with considerable caution due to cultural differences. Further, educational policy and practices in Malaysia are different with other countries which can cause diversity in learning and differentiates them from students of other countries. Lee et al. [27] found the effects of culture and cognitive styles in hypermedia learning among Malaysian and Australian students of higher learning institutions that can be associated with educational practices of both countries. Perhaps understanding the cultural effects of the linear and nonlinear navigations in WBL could be another research opportunity.

References

1. Fang, X., Holsapple, C.W.: An empirical study of web site navigation structures' impacts on web site usability. Decis. Support Syst. **43**, 476–491 (2007)
2. Cress, U., Knabel, O.B.: Previews in hypertexts: effects on navigation and knowledge acquisition. J. Comput. Assist. Learn. **19**, 517–527 (2003)
3. Sutcliffe, A., Gault, B., Maiden, N.: ISRE: immersive scenario-based requirements engineering with virtual prototypes. Requirements Eng. **10**, 95–111 (2005)
4. Tan, G.W., Wei, K.K.: An empirical study of Web browsing behaviour: towards an effective website design. Electron. Commer. Res. Appl. **5**, 261–271 (2006)
5. Hsu, Y.C., Lin, H., Ching, Y.H., Dwyer, F.M.: The effects of web-based instruction navigation modes on undergraduates' learning outcomes. Educ. Technol. Soc. **12**, 271–284 (2009)
6. Martin, F.: Effects of practice in a linear and non-linear web-based learning environment. Educ. Technol. Soc. **11**, 81–93 (2008)
7. Connelly, K., Siek, K.A., Chaudry, B., Jones, J., Astroth, K., Welch, J.L.: An offline mobile nutrition monitoring intervention for varying-literacy patients receiving Hemodialysis: a pilot study examining usage and usability. J. Am. Med. Inform. Assoc. **19**, 705–712 (2012)
8. Gauss, B., Urbas, L.: Individual differences in navigation between sharable content objects—an evaluation study of a learning module prototype. Br. J. Educ. Technol. **34**, 499–509 (2003)
9. Baylor, A.L.: Perceived disorientation and incidental learning in a web-based environment: internal and external factors. J. Educ. Multimedia Hypermedia **10**, 227–251 (2001)
10. Dahlmann, N., Jeschke, S., Seller, R., Vieritz, H.: Accessibility in virtual knowledge spaces for mathematics and natural sciences. In: First International Conference on Automated Production of Cross Media Content for Multi-Channel Distribution, 2005 (AXMEDIS 2005), p. 4 (2005)
11. Bol, L., Garner, J.: Challenges in supporting self-regulation in distance education environments. J. Comput. High. Educ. **23**, 104–123 (2011)
12. Al-Hajri, R., Al-Sharhan, S., Al-Hunaiyyan, A., Alothman, T.: Design of educational multimedia interfaces: individual differences of learners. In: Proceedings of the Second Kuwait Conference on e-Services and e-Systems. ACM, Kuwait City, Kuwait, pp. 1–5 (2011)
13. Chen, S.Y., Fan, J.-P., Macredie, R.D.: Navigation in hypermedia learning systems: experts vs. novices. Comput. Hum. Behav. **22**, 251–266 (2006)
14. Brusilovsky, P.: Adaptive navigation support in educational hypermedia: the role of student knowledge level and the case for meta-adaptation. Br. J. Educ. Technol. **34**, 487–497 (2003)
15. Chen, S.Y., Macredie, R.: Web-based interaction: a review of three important human factors. Int. J. Inf. Manage. **30**, 379–387 (2010)
16. Lin, H.-F.: Examination of cognitive absorption influencing the intention to use a virtual community. Behav. Inf. Technol. **28**, 421–431 (2009)

17. Zhou, T.: The effect of flow experience on user adoption of mobile TV. Behav. Inf. Technol. **32**, 1–10 (2011)

18. Hedman, L., Sharafi, P.: Early use of internet-based educational resources: effects on students' engagement modes and flow experience. Behav. Inf. Technol. **23**, 137–146 (2004)

19. Kopcha, T., Sullivan, H.: Learner preferences and prior knowledge in learner-controlled computer-based instruction. Educ. Tech. Res. Dev. **56**, 265–286 (2008)

20. Corbalan, G., Kester, L., van Merrie¨nboer, J.J.G.: Towards a personalized task selection model with shared instructional control. Instr. Sci. **34**, 399–422 (2006)

21. Shin, E.J., Schallert, D., Savenye, W.C.: Effects of learner control, advisement and prior knowledge on young students' learning in a hypertext environment. Educ. Tech. Res. Dev. **42**, 33–46 (1994)

22. Shyu, H.-Y., Brown, S.W.: Learner control versus program control in interactive videodisc instruction: what are the effects in procedural learning? Int. J. Instr. Media **19**, 85–96 (1992)

23. Saadé, R., Bahli, B.: The impact of cognitive absorption on perceived usefulness and perceived ease of use in on-line learning: an extension of the technology acceptance model. Inf. Manag. **42**, 317–327 (2005)

24. Webster, J., Trevino, L.K., Ryan, L.: The dimensionality and correlates of flow in human-computer interactions. Comput. Hum. Behav. **9**, 411–426 (1993)

25. Small, R.V., Arnone, M.P.: Arousing and sustaining curiosity: lessons from the ARCS model. Training Res. J. **4**, 103–116 (1998)

26. Benabou, R., Jean, T.: Intrinsic and extrinsic motivation. Rev. Econ. Stud. **70**, 489–520 (2003)

27. Lee, C.H.M., Sudweeks, F., Cheng, Y.W., Tang, F.E.: The role of unit evaluation, learning, and cultural dimensions related to student cognitive style in hypermedia learning. In: Sudweeks F., Hrachovec H., Ess C. (eds.) Proceedings of Cultural Attitudes Towards Communication and Technology 2010. Perth, Australia, pp. 400–419 (2010)

Can Single Sign-on Improve Password Management? A Focus Group Study

Norliza Katuk, Hatim Mohamad Tahir, Nur Haryani Zakaria and Mohamad Subri Halim

Abstract This article presents a research concerning password management and single sign-on for accessing Internet applications. Many Internet applications require users to subscribe to their services and authenticate themselves through the use of login credentials. The number of such applications is increasing exponentially, which caused ineffective login credential management among users. This study was conducted with two objectives (i) to identify how users manage their usernames and passwords and (ii) to examine whether users see the benefits of single sign-on. To achieve these objectives, a focus group interview was conducted on students from a local university. The results of the study suggested that the students did not practise proper password management. Further, it suggested that single sign-on may not be the immediate solution to improve the students' password management.

Keywords Single sign-on · Password security · Password management · Security practices · Weak password

N. Katuk (✉) · H.M. Tahir · N.H. Zakaria · M.S. Halim
School of Computing, Universiti Utara Malaysia, 06010 Sintok,
Kedah Darul Aman, Malaysia
e-mail: k.norliza@uum.edu.my

H.M. Tahir
e-mail: hatim@uum.edu.my

N.H. Zakaria
e-mail: haryani@uum.edu.my

M.S. Halim
e-mail: mohamad-subri.halim@hotmail.com

© Springer International Publishing Switzerland 2015
A. Abraham et al. (eds.), *Pattern Analysis, Intelligent Security
and the Internet of Things*, Advances in Intelligent Systems and Computing 355,
DOI 10.1007/978-3-319-17398-6_8

1 Introduction

Password-protected applications are increasing rapidly these days compared to five years ago. Nowadays, a typical user has more than five passwords that are used frequently to access various Internet applications. This causes more challenges to users to manage their passwords effectively, which is not easy due to human memory limitation [1]. Hence, users will likely forget their passwords especially for applications that they use occasionally. In order to cope with this issue, users resort to insecure and poor password management such as writing down the passwords and password reuse. These practices can be a source of attacks, and it is a serious security threat [2] for both users and application providers. Hence, a mechanism that improves this issue is highly needed.

Single sign-on (SSO) is one of them. It is a mechanism that allows users to access a number of applications through the use of a pair of username and password [3]. Authentication is made by a third-party entity that validates users' identity, which benefits both users and Internet application providers. As SSO is a new technology, many users are still not aware of its existence. Hence, this research aimed to identify how users manage their user names and passwords and understand the role of SSO among Internet users.

The remaining of this article describes the study in detail. Section 2 explains literature pertaining password management practices and SSO. In Sect. 3, the methodology for conducting the study is discussed. Then, in Sect. 4, the results are presented. They will be discussed further in Sect. 5. The last section concludes the article.

2 Password Management and Single Sign-on

The number of Internet applications with password protection is increasing. In 2007, it was reported that average users maintained 25 password-protected accounts to access various Internet applications [4]. However, the average has shown a substantial increase recently [5] which indicates the issue of managing passwords is indeed critical.

This section discusses literature on password management practices and SSO as one of the possible solutions to address this issue.

2.1 Password Management Practices

The most common login credential method for Internet applications is a combination of a username and password which is used to authenticate users [5]. Although many researchers suggested that the use of password is no longer a secure authentication method [6], it has been accustomed as part of many existing systems and applications. Hence, the use of login credentials as an authentication method is still feasible. Summers and Bosworth [7] highlighted some of the good practices including the use of alphanumeric words consisting of six to ten characters, and avoiding dictionary words as well as words that are related to ourselves. They also highlighted security policy concerning password management such as frequent changes of user passwords, restrict users from reuse the old passwords, just to name a few.

However, users often practise poor password management. Tam et al. [8] reported in their study that users know what a good password is and the consequences of practising poor password management. However, they tend to practise weak password management because they did not see the immediate negative implications on themselves. Many researchers also agreed that users' behaviours can expose applications in the Internet to various security threats and attacks [8–10]. Therefore, an approach that can help users to improve their password management is highly needed.

One of the possible approaches is SSO which can be used to solve problems related to multiple number of login credentials for accessing different applications [10]. The next subsection explains the SSO concepts.

2.2 Single Sign-on (SSO)

Hardy [11] analogised SSO as a "master key" that allows a user to get access to the systems. Clercq [12] defined SSO as a way to allow users to access multiple secure resources by authenticating themselves to an authentication authority once for all. Authentication authorities are the trusted entities that manage and organise users' credentials as well as access to the resources. Radha and Reddy [13] further classified SSO into three dimensions: (i) the type of networks that the SSO is applied (i.e. intranet, extranet or Internet), (ii) the architecture of SSO implementation (i.e. simple or complex), and (iii) the types of credentials (i.e. single or multiple) and the protocols used for the implementation. Other names similar to SSO are single sign-in (SSI), authentication and authorisation infrastructures (AAIs), privilege management infrastructures (PMIs) and identity management tools [11].

In general, there are three entities involved in SSO: users, application providers and authentication providers. The basic concept of SSO is that users can use a single login credential of their own to access multiple number of Internet applications by supplying their identity to authentication providers. Many leading Internet entities, such as Google, Yahoo, Twitter, PayPal, MySpace and Facebook, provide SSO services to public users to access external applications.

Although SSO has been used since distributed systems were started two decades ago, it was limited to users within an enterprise computing environment. However, it is gradually popular when the Web 2.0 was introduced. Within the social network environment, many leading providers such as Facebook, Yahoo!, LinkedIn,

MySpace and Google provide the authentication services for both users and application providers. Further, they also have higher chances to attract new users of the social networking providers to use their applications or services. In general, the SSO has eased the user identity management for both application providers and users.

Within the Internet environment, SSO standards such as SAML Oasis [14] and OpenID [15] have been proposed and developed by community groups. OpenID is the standard that has been used by major social networking providers. It aimed to provide users and application providers with secure cyber identity management. Any Web entities can be an OpenID provider for free. It guarantees the security and privacy of users' passwords that they will not be revealed to the application providers in any way. Some of the benefits of using OpenID for users are as follows: (i) acceleration of login process of favourite websites, (ii) efficient maintenance of login credential, (iii) gaining greater control over personal cyber identity and (iv) minimisation of security risks [15].

Apart from this, the standards that govern the implementation of SSO guaranteed the security of users' passwords, which makes SSO an efficient approach towards identity and login credential management.

3 Methodology

Data for this research were collected through focus group interviews. It is a suitable way for studying human behaviour especially in the area of human–computer interaction. Data obtained from the study were qualitatively analysed and presented in a descriptive analysis form.

3.1 Respondents

The respondents were recruited among undergraduate students of Universiti Utara Malaysia who took the final-year project paper of information technology (IT) degree programme. Students who enrolled in this course were encouraged to actively participate in the school research projects as a way to expose them to research. Eleven students (10 males and 1 female) participated in the study. The students comprised 5 Malaysians, 5 Yemenis and 1 Russian aged between 20 and 23.

3.2 Materials and Procedure

A set of open-ended questions for the interview was constructed based on our literature analysis of current and past studies concerning password security and SSO.

The questions were divided into three parts as below: (i) demographic information, (ii) password security practices and (iii) knowledge on SSO. The focus group interviews were conducted in 3 sessions. Each session comprised either 3 or 4 students with a same facilitator for all the sessions. The focus group interviews were conducted between forty minutes, in a room where the participants and the facilitator sat in a circle facing each other. The facilitator posed the same questions to each of the participants. The students took their turn to answer the questions, and they were allowed to discuss the topic within the group. Light refreshments were also provided at the end of the session. The focus group interviews were voice-recorded with consent from the participants. The interviews were transcribed, and the analysis of the data is presented in the next section.

4 Results

4.1 Access to Applications with Authentication

We firstly investigated the number of Internet applications with login credentials (i.e. a user name and a password each) for authentication that the students subscribed. In average, the students subscribed to five Internet applications. It ranges between two to ten applications including emails (Gmail, Yahoo, Hotmail, University email, etc.), social network sites (i.e. Facebook and Twitter), video-on-demand, online shopping, file sharing and voice-over-IP applications.

4.2 Password Practices

We were interested to know how the students managed their usernames and passwords. Hence, the facilitator asked the students on their techniques to remember them. As expected, five of them reuse the same user names and passwords to access different applications. A student wrote his usernames and passwords on a piece of paper and saved them in his mobile phone, respectively. In order to remember the password easily, a student used his best friend's phone number, his parents' names, his pet's name, his date of birth and his citizen identification number, respectively. One student used his favourite movie titles for passwords.

The facilitator further asked them whether the approaches they used for remembering their passwords were working well. All of the students reported that they had no problem so far.

4.3 Techniques to Improve Password Management

The students were asked to suggest ways to improve password management particularly for users with many login credentials. A student suggested the use of biometric technology for authentication. Another student suggested the use of a SSO mechanism.

Further, the facilitator asked the students concerning the feasibility of SSO. Ten students thought that SSO is feasible to be implemented. Although SSO is feasible, one student suggested that it is not suitable for banking and financial systems.

In terms of managing the SSO, the students suggested that the government should play a role in initiating an independent and trusted body to manage SSO for the citizen. Such body could be Malaysian Communications and Multimedia Commission (SKMM), for example.

4.4 The Role of SSO and Students' Awareness

In order to understand the students' views regarding the role of SSO, the facilitators first informed them the needs to have good password management. Then, they were asked on the methods or strategies to secure users' login credentials for Internet applications. Majority of them suggested that Internet application providers should strengthen their security protections. Only one student recommended that Internet users should manage their login credentials properly by protecting and securing them from other people. He also indicated that the use of different usernames and passwords for different Internet applications is sufficient.

The facilitator further asked the students whether they know the existence of SSO technologies that manage users' credentials in the Internet. Four of the students knew about it. The students who knew about SSO were further asked to explain the basic concept of it. All of them were able to explain the concept correctly. The students who knew about SSO anticipated that it is secure. However, those who did not know about SSO thought that the technology does not provide users with high level of security and privacy. The students also anticipated some security weaknesses especially cybercrime involving identity theft and breach of privacy.

5 Discussions

The previous section discussed the results of focus group interviews regarding password management and SSO. Generally, we found that the students had used minimum of two and up to ten Internet applications at one particular time. This shows that many applications are now being online and users are required to authenticate themselves to get an access to these applications. Although authentication through login credentials can provide security protections, this has obviously created another issue to users that is the challenge to manage a number of login credentials at one particular time.

Managing a number of passwords is difficult. Users that have security precautions in their mind might create strong alphanumeric passwords. However, if the number of login credentials keeps increasing, users are more likely to face problems to recall their passwords correctly especially for those that are used seasonally. This issue is common due to the reason that human memory is limited in their capacity [9]. Hence, remembering a number of alphanumeric passwords is a great challenge especially when the passwords have no relationships with other things or events in their memory.

A quick and intuitive solution to this problem is users tend to practise techniques that will ease them to recall those login credentials. Among those practices are repeating the same login credentials in many applications, writing them on paper, saving them on mobile phones and using easy-to-remember passwords. These are among the practices that we discovered in this study.

We also found that the students were comfortable with their practices because of two reasons. First, the practices helped them to manage their login credentials easily with low chances of forgetting them. Second, they did not experience any security issues as immediate consequences of their practices. These two reasons have made the students think that their password management practices were adequate.

Further, the students thought that their practices have no security implications. In reality, this could be a source that leads to security attacks to the Internet applications. Many Internet applications are securely protected through the use of encrypted passwords. However, there are also some other Internet application providers that have not implemented such security protections. When users replicated passwords in both secure and unsecure applications, this will open a door to hackers. Hackers can observe and analyse users' passwords, capture the unencrypted passwords and use them to access the secure applications. This kind of security attack does not affect the users directly; however, it does have impact on the Internet application providers and their resources.

The interview also reveals that students have limited knowledge on SSO and its current implementation. Some of them suggested that the government should play a role in initiating centralised digital authentication systems. This could be beneficial for local Internet applications. However, it is important to note that many users use Internet applications hosted in other countries around the world. Hence, the government role is limited in this scenario.

It is interesting that the students have mix views in terms of SSO security. A quarter of the students knew about its existence, and they were certain with SSO security. In contrast, the students with no prior knowledge on SSO indicated that SSO does not provide strong security and privacy protections. They also anticipated that SSO implementation could lead to cybercrimes.

Another interesting finding is on the students' thought that Internet application providers should implement strong security protections for their applications.

Through this thought, we can see that users highly rely on Internet application providers to provide security protections. Users do not realise the importance of their role in ensuring the security of the applications.

The main issue that should be taken into consideration is on users' behaviours and attitudes. Although Internet application providers implement all possible security protections, they are still exposed to security attacks due to their users' behaviours. In this case, the use of SSO is not a direct solution to this problem. This is true when users still replicate their passwords for applications with SSO and without SSO. The fundamental issue remains there and unresolved. In a nutshell, SSO is a good solution to reduce the number of login credentials that users have; however, it does not improve users' password practices especially the behaviours on password replication.

From the discussion in the preceding paragraphs, we highlighted some actions that need to be taken to ensure that computing recourses pertaining Internet applications are properly protected against security threats and attacks. The actions include the following: (i) the need for users' security education especially on proper password management practices, (ii) the need for a security policy that enforces secure password behaviours among users, (iii) the need for SSO implementation and (iv) the implementation of alternative authentication methods other than login credentials. It is suggested that the above strategies are implemented as a group of actions rather than individually.

6 Conclusion

Weak password management among users could be a security threat for Internet applications. Although Internet applications implemented strong security protections, their users' password practices can be a source of security attacks especially users who practise password replications. Passwords that are used for fully secured applications can be easily captured by hackers. They can simply analyse or steal passwords from the same user in unsecured applications.

The SSO implementation is a good solution to reduce the number of login credentials that users have at one time. However, it is unable to improve users' password management practices especially password replication. The only way to this is through education and awareness programmes. This study would like to emphasise that the core issues to this are users' behaviours. Organisations and governments should take part and play their roles to educate users.

The findings of this research also indicate the need to study the alternative authentication methods that are more friendly to users than the existing multiple login credentials. This could be an opportunity for researchers to explore a new method for Internet authentication.

Acknowledgments This work was supported in part by a grant from Universiti Utara Malaysia (LEADS—S/O Code: 12397).

References

1. Zhang, J., Luo, X., Akkaladevi, S., Ziegelmayer, J.: Improving multiple-password recall: an empirical study. Eur. J. Inf. Syst. **18**, 165–176 (2009)
2. Cameron, K., Jones, M.B.: Design rationale behind the identity metasystem architecture. ISSE/SECURE 2007 Securing Electronic Business Processes, pp. 117–129. Springer, Berlin (2007)
3. Riedel, M., Mallmann, D., Streit, A.: Enhancing scientific workflows with secure shell functionality in UNICORE grids. First International Conference on e-Science and Grid Computing, pp. 8–139. IEEE, New Jersey (2005)
4. Florencio, D., Herley, C.: A large-scale study of web password habits. In: Proceedings of the 16th international conference on World Wide Web, pp. 657–666. ACM (2007)
5. Bang, Y., Lee, D.-J., Bae, Y.-S., Ahn, J.-H.: Improving information security management: an analysis of ID–password usage and a new login vulnerability measure. Int. J. Inf. Manage. **32**, 409–418 (2012)
6. Ciampa, M., Revels, M., Enamait, J.: Online versus local password management applications: an analysis of user training and reactions. J. Appl. Secur. Res. **6**, 449–466 (2011)
7. Summers, W.C., Bosworth, E.: Password policy: the good, the bad, and the ugly. In: Proceedings of the winter international symposium on information and communication technologies, pp. 1–6. Trinity College Dublin (2004)
8. Tam, L., Glassman, M., Vandenwauver, M.: The psychology of password management: a tradeoff between security and convenience. Behav. Inf. Technol. **29**, 233–244 (2010)
9. Kumar, N.: Password in practice: an usability survey. J. Glob. Res. Comput. Sci. **2**, 107–112 (2011)
10. Ciampa, M.: Are password management applications viable? An analysis of user training and reactions. Inf. Syst. Educ. J. **9**, 4 (2011)
11. Hardy, G.: The truth behind single sign-on. Inf. Secur. Tech. Rep. **1**, 46–55 (1996)
12. Clercq, J.D.: Single sign-on architectures. Proceedings of the International Conference on Infrastructure Security, pp. 40–58. Springer, Berlin (2002)
13. Radha, V., Reddy, D.H.: A survey on single sign-on techniques. Procedia Technol. **4**, 134–139 (2012)
14. SAML. (2013). Welcome to SAML Oasis.org. Available: http://saml.xml.org/
15. OpenID. (2013). The Benefits of OpenID. Available: http://openid.net/get-an-openid/individuals/

The Relevance of Software Requirement Defect Management to Improve Requirements and Product Quality: A Systematic Literature Review

Nurul Atikah Rosmadi, Sabrina Ahmad and Noraswaliza Abdullah

Abstract Software is an intangible computer's component, and their requirements are the greatest challenge to handle but yet the most important. In order to ensure the requirements are in good quality, defect management is one of the promising efforts to adopt. This paper aims to provide a literature review regarding defect management for software requirements and the relevance of the defect management to improve requirements and product quality. The paper is structured based on a systematic literature review method which is constructed from significant questions. The findings on the literatures show that many efforts have been done to improve defect management effort in several stages of software development life cycle. However, the efforts are scarce in the requirement engineering phase. This paper provides a foundation study to pursue research in improving requirements and eventually product quality through defect management.

Keywords Software requirement engineering · Defect management · Requirement defects · Requirement negotiation · Requirement management and requirement analysis

1 Introduction

Developing a reliable software system which fulfils various stakeholders' needs is never an easy task. It needs specific skills, wise decisions and structured plan. In terms of requirement engineering which involve various stakeholders, the process

N.A. Rosmadi (✉) · S. Ahmad · N. Abdullah
Faculty of Information and Communication Technology, Universiti Teknikal Malaysia
Melaka, Durian Tunggal, Melaka, Malaysia
e-mail: atikahrosmadi@gmail.com

S. Ahmad
e-mail: sabrinaahmad@utem.edu.my

N. Abdullah
e-mail: noraswaliza@utem.edu.my

© Springer International Publishing Switzerland 2015
A. Abraham et al. (eds.), *Pattern Analysis, Intelligent Security
and the Internet of Things*, Advances in Intelligent Systems and Computing 355,
DOI 10.1007/978-3-319-17398-6_9

of eliciting software requirements is crucial to ensure that the software developed is based on the right and quality requirements. The efficiency of the elicitation process and the quality of the requirements obtained depend on the approaches and techniques used to elicit the requirements from various resources. Eliciting and analysing software requirements is a critical stage because defects originated in this stage will propagate to the subsequent stages and therefore effect the entire software development process. According to Boehm [1], the cost of fixing errors increased exponentially, and in the later, the errors were detected in the development life cycle because the artefacts within a serial process were built on each other. Also, the average cost arose exponentially after the defects were detected because the development continues upon an unstable foundation. This meant that there was a possibility that the software being developed had been based on unwanted or wrong requirements.

Besides, requirement engineering process basically mould the shape of the entire software development life cycle as the process is concerned on determining the goals, functions and constraints for software systems [2]. This is also supported by Pandey et al. [3] as he stated in an article that requirement engineering process covers the importance of the entire system and software development life cycle. However, defects are inevitable since the interpretation of requirements from the sources to the documents mostly based on natural language. Requirements are usually communicated in natural language [4] which leads to several problems such as imprecision, ambiguity, incompleteness, conflict and inconsistency [5]. These take time to resolve. In industrial RE, natural language is the most frequently used representation in which to state requirements that are to be met by information technology products or services. The use of natural language to specify requirements has many benefits, although it also creates some problems. It was claimed [6] that a major and well-recognized problem is the inherent ambiguity of natural language. Ambiguity is a phenomenon [7], which straddles all three dimensions: specification, agreement and representation. The level of ambiguity decreases when the project is making progress and requirements become more complete, more accepted and more formal. The problem is that there may be unconscious and unrecognized ambiguity. According to an empirical study [6], ambiguities are usually resolved over time without any effort on the part of the stakeholders. This is a serious problem because the contextual knowledge of customers and software developers usually differs. Thus, implicit assumptions are likely to be wrong when a system is more complex. In addition, ambiguities that were not recognized were likely to be misinterpreted more often [8]. However, the use of special languages and notations may overcome some natural language weaknesses. One of the well-known notations is the unified modelling language (UML) [9]. It uses a combination of graphical diagrams and words to capture software requirements. By reducing the natural language scope, these formal languages reduce the ambiguities that exist in natural language. To further reduce the ambiguities, mathematical notations are used. The disadvantage of special notations or languages is that people have to learn how to use and interpret them. This takes some time and effort.

Even though there has been much research conducted to help practitioners understand and improve the way requirements are specified and formulated, there is still a large gap between the formal methods advocated by researchers and the informality that dominates in industry. There are reasons why requirements are initially specified in natural language [10]; it is the primary communication language which is shared by the stakeholders, and many formal methods do not offer support for the management and analysis of erroneous, incomplete or partially specified requirements. Therefore, the research community should expect that its use cannot be avoided. Hence, a prevention action to avoid, reduce or divert the impact of erroneous in requirement statements would be excellent.

This systematic literature review will analyse and identify the relevance of software requirement defect management to improve requirements and product quality. The result of this systematic literature review exposed the importance of defect management in preventing the defects, thus improving the product quality.

This paper consists of four sections. Following introduction, Sect. 2 describes the review method used to produce the systematic literature view. Next, Sect. 3 describes the results of the review and Sect. 4 concludes the paper.

2 The Review Method

2.1 Introduction to Review Method

N. Salleh stated that a systematic literature review establishes a connection between the existing knowledge and the problem to be solved [11]. This paper adopted a systematic literature review method to systematically accumulate, organize, evaluate and synthesize all existing research evidence related to our research area in order to provide testimony and confidence. It is all based on the previous research/ articles/journals by other researchers on the defect management effort in the area of software requirement engineering. The systematic review has 3 steps: (1) planning, (2) conducting and (3) reporting. Figure 1 provides more information on the activities within the three steps.

In the planning phase, the objectives were identified as to discover defect management for software requirements, to discover the role of negotiation as a prevention action to handle defects in software requirement and to improve the current approaches. In order to achieve the objectives, the review protocol was established. The review protocol specified the questions to be addressed, the database to be searched and the methods to be used to identify, assemble and assess the evidence [12]. This is followed by the conducting phase in which the relevant questions were sort out and prioritized accordingly in order to extract the right data. Then, the studies were assessed and relevant data were synthesized. In the reporting phase, all the relevant studies and significant data were reported accordingly.

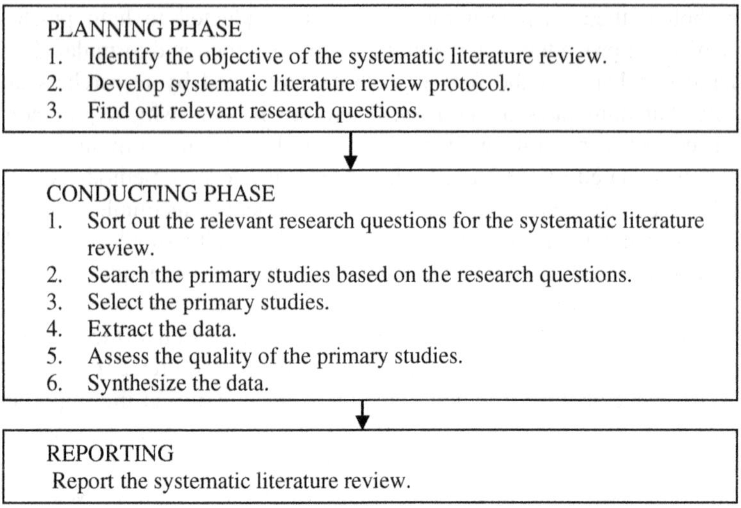

```
┌─────────────────────────────────────────────────────────────────┐
│ PLANNING PHASE                                                  │
│ 1.  Identify the objective of the systematic literature review. │
│ 2.  Develop systematic literature review protocol.             │
│ 3.  Find out relevant research questions.                      │
└─────────────────────────────────────────────────────────────────┘
                                ↓
┌─────────────────────────────────────────────────────────────────┐
│ CONDUCTING PHASE                                               │
│ 1.  Sort out the relevant research questions for the systematic│
│     literature review.                                         │
│ 2.  Search the primary studies based on the research questions.│
│ 3.  Select the primary studies.                                │
│ 4.  Extract the data.                                          │
│ 5.  Assess the quality of the primary studies.                 │
│ 6.  Synthesize the data.                                       │
└─────────────────────────────────────────────────────────────────┘
                                ↓
┌─────────────────────────────────────────────────────────────────┐
│ REPORTING                                                      │
│ Report the systematic literature review.                       │
└─────────────────────────────────────────────────────────────────┘
```

Fig. 1 Steps in systematic literature review

2.2 Research Questions

The questions were focused on to find the evidence on the relevance of defect management in software requirements and their limitations. The research questions are as follows.

1. Research question 1: "What is software requirement defect and does it exist?"
2. Research question 2: "How does defect management help to reduce requirement defects and therefore improve requirements and product quality?"

Table 1 shows the relationship between the research questions and the research motivations.

Table 1 Research questions and motivations

Research questions		Motivations
RQ1	What is software requirement defect and does it exist?	Discover the definition and types of software requirement defects
RQ2	How does defect management help to reduce requirement defects and therefore improve requirements and product quality?	Find out efforts to reduce software requirement defects and improve requirements and product quality through defect management

2.3 Search Strategies

After the research questions being finalized, the search process took place. The sources of the search were digital libraries and databases which were searched through search string and refining search string. The list of the digital databases is the most popular and familiar databases to ease and broad the set of related search papers. Below is the list of the digital databases that being used to search papers in our study:

1. IEEE Xplore (ieeexplore.iee.org)
2. ScienceDirect (sciencedirect.com)
3. Springer (Springerlink.com)
4. Scopus (scopus.com)
5. Google Scholar (scholar.google.com)
6. Elsevier (Elsevier.com)
7. ACM Digital Library (dl.acm.org)
8. Cornell University Library (arXiv.org)

The search strings are based on the research questions and the keywords of the research field such as software requirement engineering and defect management. Search papers were using the title and author name. The search was limited by the year of 1996–2014 including journal papers and conference proceedings. Language for the search was limited to English only.

2.4 Identification of Relevant Literature

The search was based on the guideline to write the systematic literature review constructed upon the research questions [13]. The listed research questions were answered by the related works searched in order to find out the current finding on the research title. The keywords from the primary studies or keywords that we already known were used to find more articles related to the research. The synonym words and alternative words related to the studies also applied in order to optimize the related work search.

The general keywords that we used in searching the related articles or journal for this systematic literature review were "systematic literature review", "software requirement engineering", "requirement engineering" and "defect management". Narrowing the search, we also used the keywords "requirement negotiation", "defect", "error", "software problem", "requirement elicitation phase", "requirement elicitation" and "defect classifications". The search was resulted in the various reliable journals and conference proceedings covering issues in the software requirement engineering defect management. In order to obtain as many citations as possible, we also used library facilities as to access full-text articles to non-subscribed databases such as Springer.

Upon the completion of the searching process, we filter the findings and narrow them to related works only. The next phase of review was based on the references of relevant papers to the research questions, and the filtered papers were added to the final list for further analysis.

2.5 Study Selection

Following the identification of relevant papers, a quality assessment was performed to prevent bias, internal validity and external validity. Mendeley software package for desktop and online (http://mendeley.com) were used to store and manage the references. The inclusion and exclusion criteria are listed as in Table 2.

2.6 Significant Journal Publications in Research

In this literature review, we have included 30 primary studies that related to the defect management in software requirement engineering. The distribution over the years is presented to identify the interest of studies in this field (Fig. 2).

Table 2 The inclusion and exclusion criteria

Inclusion criteria	Exclusion criteria
• Focusing on software requirement defects to improve software quality	• Tutorials
• Defect management	• Studies not related to research questions
• Software requirement quality and software requirement engineering	• Studies that are unclear
• Empirical studies on defect management in software development	
• Systematic literature review	
• Defect/error/faults classifications	

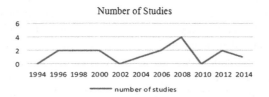

Fig. 2 Graphical representation of number of studies involved

3 The Review Results

In this section, we present the evidence of the related work search. The evidence was based on the previous research articles, journals and books from other researchers. The time limit for the search had been set to any time since we need to discover as many results related to the defect management in requirement engineering as possible. The online search was mostly done through Google Scholar. The papers from IEEE database and Springer database were requested through library service in Universiti Teknikal Malaysia Melaka.

The data from the paper being gathered were tabulated in a table based on the research questions created initially. Almost 80 % of the references were searched through Google Scholar, Science Direct, Springer, Scopus, Elsevier, ACM Digital library, Cornell University Library and IEEE Xplore digital library. The other 20 % are based on books, e-book and doctoral thesis. The primary data searches found were too general and mostly related to defect in coding and other deliverables but very scarce in requirements. Therefore, we narrowed the search to requirements and also looked at negotiation as the mechanism to provide the prevention action.

There are two questions which focused to discover literature review on software defect management in requirement engineering process. The data collected were divided into 2 parts; defects and defects management.

3.1 Review Based on Research Questions

Research question 1: "What is software requirement defect and does it exist?"
Defects usually referred to software defects which appear in the code of a software system. A software defect is an error, failure or fault in software [14] that produces an incorrect or unexpected result or causes it to behave in unintended ways. It is a deficiency in a software product that causes it to perform unexpectedly [15]. Software defects are expensive in quality and cost. Moreover, the cost of capturing and correcting defects is one of the most expensive software development activities [16]. It is even worst when the defects are originated from requirements. Software requirements are critically important because defects originated in requirement engineering stage will propagate to the subsequent stages and therefore affect the entire software development process. Defects in requirements exist in the requirement statements usually presented in a requirement document.

A software requirement document is a formal documentation to record the requirements following the completion of requirement engineering process [3]. It also includes the identification of requirements and software and system requirement specifications. Eliciting requirements is crucial to determine the functionality and constrains of the system to be developed. Dealing with multiple stakeholders and sources, the interpretation of the requirements into a document has high potential to contain defects. It is especially so when the process involves human factors which is reported to contribute 34 of the risks in the research of global software development [17]. IEEE defined defect as fault or an incorrect step, process or data definition in a computer program [18]. A defect is a problem that occurs in an artefact and may lead to a failure [19]. Any blemish, imperfection or undesired features in the product are defined as defect [20].

Definitions of defects and its acronyms are stated below [21], and it is consistent with software engineering textbooks [12] and an IEEE standard [18]:

1. **Error**—defect in the human thought process made while trying to understand given information, solve problems or to use methods and tools. In the context of software requirement specifications, an error is a basic misconception of the actual needs of a user or customer.
2. **Fault**—concrete manifestation of an error within the software. One error may cause several faults, and various errors may cause identical faults.
3. **Failure**—departure of the operational software system behaviour from user-expected requirements. A particular failure may be caused by several faults, and some faults may never cause a failure.

Defects are always being mentioned to be appearing in the line of codes. However, a study had discovered that defects found in software requirements are in higher percentage compared to the defects in the line of codes [22]. In software requirements, the defects that appear will propagate and affect the sequential stages of software development life cycle. Requirement defects may appear in requirement statements while transforming the stakeholder's wish list into requirement specification document. Davis [23] and Yilmatzurk [24] came out with lists of quality attributes. The quality attributes if read in reverse will give a list of requirement defect types as presented in Table 3.

Table 3 List of requirement defect types

Defect types	Quality attributes
Infeasibility	Feasibility
Imprecise	Precise
Incomplete	Complete
Winding	Concise
Incorrect	Correct
Misinterpret	Interpretable
Externally inconsistent	Externally consistent
Internally inconsistent	Internally consistent
Unmodifiable	Modifiable
Redundant	Not redundant
Unorganized	Organized
Unusable	Reusable
Ambiguous	Unambiguous
Incomprehensible	Understandable
Unnecessary	Necessary

The list gives us an indication of qualities to achieve and at the same time the defect types that might exist in the requirement documents. This can be a fundamental guideline to show the difference between requirement defects and software defects. However, the types of software defects are not included in this paper. Software defects are basically active in nature in which they can be detected and removed during software run-time. Unlike requirement defects, the defect types are passive and need to be identified from the requirement documents.

Referring to Margarido et al. [19], stated below is the list of requirement defect classifications taxonomy. Types of defect that being mentioned to appear and common in the requirement statements are as follows (Table 4).

The defects listed are understandably possible to be detected and removed from the requirement specification document. In order to handle all defects listed in Table 3, a complete document must be presented. However, in a notion to introduce prevention action through negotiation, the defects are expected to be avoided during the requirement elicitation process in which a proper requirement specification document is not there yet. Therefore, this research has scope the list of defects to comprehensibility, consistency, completeness, feasibility (resources and dependency) and correctness. The list is already identified [25] to be feasibility exists in the high-level requirement statement during elicitation process.

Research question 2: "How does defect management help to reduce requirement defects and therefore improve requirements and product quality?"
An early-stage defect analysis in software development can reduce the time, cost and resources required for work [20]. The defects basically need to be identified, addressed and later removed from the documents or any other deliverables to improve overall quality. Therefore, defect management effort will be an excellent attempt to perform those tasks. A study has been done [20] and suggested three conventional defect prevention strategies which are product approach of defect

Table 4 Types of defects

Defects	Explanation
Missing or incomplete	The requirement is absent in document
Incorrect information	The information contains the false requirement
Inconsistent	The requirement is inconsistent with the overall document and in conflict with another requirement that is correctly specified
Ambiguous or unclear	Information or vocabulary has more than one interpretation
Misplaced	The requirement information is misplaced either in the section of the requirement specification document or in the functionalities, packages or system
Infeasible	The requirement is not implementable
Redundant or duplicate	Requirement duplicate of another requirement or part of it already present in the document
Typo or formatting	Orthographic, semantic, grammatical error or missing word
Not relevant	Unnecessary information or out of project scope

prevention, process approach of defect prevention and automation of development process [26].

Based on the above-mentioned strategies, this research is looking into the process approach. The prevention will take place in the earliest stage of requirement engineering process which is in requirement elicitation. At this stage, stakeholders are communicating to solicit potential requirements for the system to be developed and high-level requirement statement will be produced. Therefore, defect prevention will commence to ensure defects that are not introduced into the requirement statements. This effort will eventually improve the quality of software requirement document. Requirements are the foundation of the software development process which provides the basis for the costs and schedules estimation as well as developing design and testing specifications [27]. Therefore, requirement defect may cause poor project management and estimation, the development of architecture will be based on the wrong foundation, the design will be inappropriate, and the test cases will be absolutely out of the picture. This is supported by Alves et al. [28] who stated that many important decisions are made during the requirement engineering phase in the software product line. Therefore, a good quality of requirements will eventually improve the product quality.

Even though prevention action is rather new in ensuring software requirement quality, a study has been done and it is estimated that by using certain techniques, 15 % of all the requirement-related defects can be prevented [29].

Therefore, defect management plays an important role to identify, address and remove the defects from the requirement statement. On the other hand, when prevention action is adopted, the defect management effort is responsible to detect the probability of defects occurring and avoid them altogether. Defect prevention in defect management helps in reducing the requirements-related defect by preventing the defects from occurring and it is eventually save the cost of rework to fix the defects later.

4 Conclusion

Defect management is common to handle software defects, but in the area of requirements, research on requirement defects is scarce. This systematic literature review gives an insight into the relevance of defect management for software requirements and the importance of it. The review revealed that defects in requirement statement do exist and the types of the defects are different from defects in the other phases in the software development life cycle. Since requirements are the first thing that a system development needs to deal with, this research is promoting prevention action to the defect management. It helps in reducing the requirements-related defects and saving the cost of rework in fixing the defects later. Based on the evidence gathered from the review, there are many opportunities to improve the approach of defect management in order to produce better software requirements. On top of that, defect classification and taxonomy can be developed to provide traceability mechanism for better defect management.

References

1. Boehm, B.: Software Engineering Economics. Prentice-Hall, New Jersey (1981)
2. Laplante, P.A., Agresti, W.W., Djavanshir, G.R.: Guest editor's introduction, special section on IT quality enhancement and process improvement. IT Prof. 10–11 (2007)
3. Pandey, D., Suman, U., Ramani, A.: An Effective Requirement Engineering Process Model for Software Development and Requirements Management. In: International conference on advances in recent technologies in communication and computing, pp. 287–291, 2010
4. Boehm, B., Egyed, A., Kwan, J., Port, D., Shah, A., Madachy, R.: Using the winwin spiral model: a case study. Computer **31**(7), 33–44 (1998)
5. Kotonya, G., Sommerville, I.: Requirement Engineering: Processes and Techniques. Wiley, New York (1998)
6. Kamsties, E.: Understanding ambiguity in requirements engineering. In: Aurum, A., Wohlin, C. (eds.) Engineering and Managing Software Requirements, pp. 245–266. Springer, Berlin (2005)
7. Galliers, R.D., Swan, R.D.: Against Structured Approaches: Information requirements analysis as a socially mediated process. In: Proceedings of the Thirtieth Hawaii International Conference on System Sciences, Information Systems Track, pp. 179–187, 1997
8. Boehm, B., Grunbacher, P., Briggs, R.O.: Developing groupware for requirements negotiation: lessons learned. J. IEEE Softw. **18**, 46–55 (2001)
9. Hickey, A.M., Davis, A.M.: A unified model of requirements elicitation. J. Manage. Inf. Syst. **20**(4), 65–84 (2004)
10. Dag, N.O., Regnell, J.B., Gervasi, V., Brinkkemper, S.: A linguistic-engineering approach to large-scale requirements management. Softw. IEEE **22**, 32–39 (2005)
11. Salleh, N.: Systematic Literature Review for SE Research. In: Software Engineering Postgraduate Workshop, Nov 2013
12. Walia, G.S., Carver, J.C.: Development of Requirement Error Taxonomy as a Quality Improvement Approach: A Systematic Literature Review, Department of Computer Science and Engineering MSU-070404, 2007
13. Kitchenham, B., Charters, S.: Guidelines for Performing Systematic Literature Reviews in Software Engineering, Keele University and Durham University Joint Report, EBSE 2007-001, 2007
14. Naik, K., Tripathy, P.: Software Testing and Quality Assurance: Theory and Practices. Wiley, New York (2008)
15. Adikari, S., McDonald, C., Lynch, N.: Design Science-Oriented Usability Modelling for Software Requirements. In: 12th International Conference, HCI International 2007, Beijing, China, 22–27 July 2007, pp. 373–382
16. Mantyla, M.V., Itkonen, J.: How are software defect found? The role of implicit defect detection, individual responsibility, documents, and knowledge. Inf. Softw. Technol **56**, 1597–1612 (2014)
17. Verner, J.M., Brereton, O.P., Kitchenham, B.A., Turner, M., Niazi, M.: Risk and risk mitigation in global software development: a tertiary study. Inf. Softw. Technol. **56**, 54–78 (2014)
18. IEEE Standard Classification for Software Anomalies, 1044–2009

19. Margarido, I.L., Faria, J.P., Vidal, R.M., Vieira, M.: Classification of defect types in requirements specifications: literature review, proposal and assessment. In: 2011 6th Iberian conference on information systems and technologies (CISTI), vol. 15–18, pp. 1–6, June 2011
20. Suma, V., Gopalakrishnan Nair, T.R.: Defect Management Strategies in Software Development. Intec Web Publishers, Germany (2012)
21. Lanubile, F., Lonigro, A., Visaggio, G.: Comparing models for identifying fault-prone software components, In: Seventh international conference on software engineering and knowledge engineering, pp. 312–319, 1995
22. Kumaresh, S., Baskaran, R.: Defect analysis and prevention for software process quality improvement, Int. J. Comput. Appl. **8** (2010)
23. Davis, A., Overmyer, S., Jordan, K., Caruso, J., Dandashi, F., Dinh, A., Kincaid, G., Ledeboer, G., Reynolds, P., Sitaram, P., Ta, A., Theofano, M.: Identifying and measuring quality in a software requirements specification. Proceeding of 1st International Software Metrics Symposium, Baltimore, Maryland, United States, IEEE Computer Society, pp. 141–152, 1993
24. Yilmatzurk, N.: Good quality requirements in unified process. In: Aurum, A., Wohlin, C. (eds.) Engineering and Managing Software Requirements, pp. 373–401. Springer, Berlin (2005)
25. Ahmad, S.: Measuring the Effectiveness of Negotiation in Software Requirements Engineering. The University of Western Australia, Australia (2012)
26. Tian, J.: Quality Assurance Alternatives and Techniques: A Defect-Based Survey and Analysis, ASQ by Department of Computer Science and Engineering, Southern Methodist University, SQP vol. 3, No. 3/2001
27. Javed, T., Maqsood, M.E., Durrani, Q.S.: A study to investigate the impact of requirements instability on software defects. ACM Softw. Eng. **29**(4), 1–7 (2004)
28. Alves, V., Niu, N., Alves, C., Valenca, G.: Requirements engineering for software product lines: a systematic literature review. Inf. Softw. Technol. **52**, 806–820 (2010)
29. Lauesen, S., Vinter, O.: Preventing requirement defects: an experiment in process improvement. Requirement Eng. **6**, 37–50 (2001)

Finding the Effectiveness of Software Team Members Using Decision Tree

Mazni Omar and Sharifah-Lailee Syed-Abdullah

Abstract This paper presents steps taken in finding the effectiveness of software team members using decision tree technique. Data sets from software engineering (SE) students were collected to establish pattern relationship among four predictor variables—prior academic achievements, personality types, team personality diversity, and software methodology—as input to determine team effectiveness outcome. There are three main stages involved in this study, which are data collection, data mining using decision tree, and evaluation stage. The results indicate that the decision tree technique is able to predict 69.17 % accuracy. This revealed that the four predictor variables are significant and thus should consider in building a team performance prediction model. Future research will be carried to obtain more data and use a hybrid algorithm to improve the model accuracy. The model could facilitate the educators in developing strategic planning methods in order to improve current curriculum in SE education.

Keywords Software team performance · Decision tree · Software methodology · Personality types · Personality diversity · Academic achievement

1 Introduction

Teamwork is vital to encourage members to become more responsible and stimulates collaboration and creative thinking [1], specifically in educational environment. Furthermore, teamwork activities often enhance learning experiences [2] when

M. Omar (✉)
School of Computing, Universiti Utara Malaysia (UUM), 06010 Sintok, Kedah, Malaysia
e-mail: mazni@uum.edu.my

S.-L. Syed-Abdullah
Department of Computer Sciences, Faculty of Computer and Mathematical Sciences,
Universiti Teknologi MARA, Arau Campus, 02600 Arau, Perlis, Malaysia
e-mail: shlailee@perlis.uitm.edu.my

© Springer International Publishing Switzerland 2015
A. Abraham et al. (eds.), *Pattern Analysis, Intelligent Security*
and the Internet of Things, Advances in Intelligent Systems and Computing 355,
DOI 10.1007/978-3-319-17398-6_10

members actively develop better relationships among them and thus improving team performance. In completing certain tasks collaboratively, members inevitably encounter differences and hence must adjust to realize their commonalities to suit their learning objectives. This situation invites members to build the capacity to be more tolerant in resolving differences and increase team dynamism. In addition, learning to achieve consensus among members is a valuable lesson for them, as this is one of the crucial parts of decision making in human life. However, finding the effectiveness of software team members is a challenging task because it is dealing with complex activity and tasks. Due to this, there is a need to ascertain relationships of effective software team members in order to assist decision maker to form a balance of team. Thus, this paper presents steps taken in finding the effectiveness of software team members using decision tree.

2 Related Works

Among popular classification data mining techniques are logistic regression, decision tree, artificial neural network (ANN), support vector machine (SVM), and rough set. Decision tree is a classification technique used in predictive data mining. There are various decision tree learning algorithms that can be used, such as ID3 [3], C4.5 [4], and CART [5]. Trees are usually represented graphically as hierarchical structures, thus making them easier to visualize and interpret.

Literature review indicated that decision tree was widely used in educational data mining due to their simplicity and ease of visual interpretation, which enable the researcher to identify interesting patterns for future analysis. Decision tree has been widely used in predicting student performance [6–8]. Nevertheless, variables and accuracy obtained by using this technique are varied. Fang and Lu [9], in their study, showed that decision tree technique is able to obtain higher accuracy (83.3–85.9 %) compared to linear regression (66.7–71.9 %) when predicting academic achievement for engineering students. Nghe, Janecek, and Haddawy [10] also conducted comparative studies on technique to predict academic performance, indicating that decision tree was more accurate compared to Bayesian network. Moreover, accuracy of decision tree was increased from 71–73 to 93–94 % when the number of outcome classes was reduced from four to two.

Al-Radaideh et al. [11] also achieved good prediction accuracy with 87.9 % when using decision tree to predict suitable education track for students. However, in respective studies, [12] and [13] received only 47 and 38 % prediction accuracy when using decision tree to predict performance of students in higher education. This shows that prediction accuracy of this technique is highly depended on nature of data used in a particular study. It was noted that most applications of decision tree are based on Waikato Environment for Knowledge Analysis (WEKA) tool to analyze patterns of the available data sets, as this tool is freely available, thus saving more time in analyzing data.

3 Method

Guidelines from knowledge discovery in databases (KDD) [14] were used in this study to uncover data patterns and provide useful knowledge derived from the data collected. The three main stages involved in this study are as follows:

3.1 Data Collection

Empirical data from [15–18] was used in this study. The data set consists of 120 data sets collected during two comparison studies and one case study conducted in an academic setting, where participants were software engineering (SE) students. Four predictor variables were selected; this is based on the experiment results carried out in [18]. The four predictor variables were used—prior academic achievements, personality types, personality diversity, and software methodology, whereas the outcome variable is team effectiveness. The descriptions of variables selected are defined in Table 1.

Table 1 Description of predictor and outcome variables

Description of predictor variables used	Category level
1	**Prior academic achievements** – Refers to advanced programming (TIA1023) course that have taken by the students before enrolling the software engineering course	Grade A = 4 Grade B = 3 Grade C = 2 Grade DF = 1
2	**Personality types** – Refers to Jung Myers–Briggs personality types [19], which are: • Introvert (I) versus extrovert (E) • Sensing (S) versus intuitive (N) • Thinking (T) versus feeling (F) • Judging (J) versus perceiving (P)	Introvert (I) = 1 Extrovert (E) = 2 Sensing (S) = 1 Intuitive (N) = 2 Thinking (T) = 1 Feeling (F) = 2 Judging (J) = 1 Perceiving (P) = 2
3	Team personality diversity – The team personality diversity was determined based on [20]. In this study, diversity from 0 to 3 is considered as homogeneous, while 4–8 as heterogeneous team	Between 0 and 8
4	Type of methodology – Software methodology used by the team during their software project, whether using agile methodology or formal methodology	Agile = 1 Formal = 2
Description of outcome variable	Category
1	Team effectiveness (Q) – Refers to grade achieved by the team assessed by the software project client	Effective = 1 (TRUE) Ineffective = 0 (FALSE)

3.2 Data Mining

Data mining is a process of analyzing and discovering previously unknown and potentially interesting patterns of available data [14]. Data mining plays an essential role in this study to discover and represents patterns to the user in a human-understandable form [14]. In this study, decision tree is selected because it is inductive learning algorithms that include the following:

- Decision tree is able to analyze nominal and ordinal data.
- Decision tree is suitable for small data sets because it is free from data normality assumptions and is suited for discovering knowledge derived from the empirical data.
- Decision tree is faster to construct and easier to visualize.

Constructing the model using decision tree was relatively straightforward when using WEKA tool. WEKA is a machine learning workbench that can be used to apply decision tree algorithms to data sets. This tool was selected because the J48 class in WEKA tool represents function based on C4.5 [4] algorithm. It is a decision tree algorithm commonly used for classification and is well suited for this study. Figure 1 illustrates the steps in generating decision tree using WEKA.

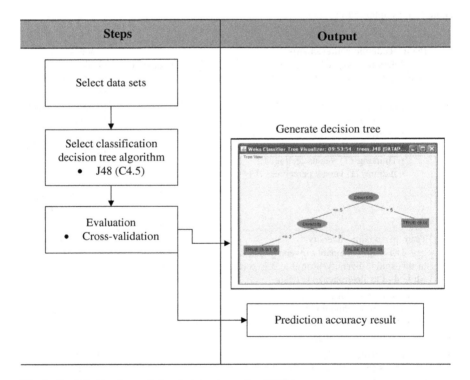

Fig. 1 General steps for applying decision tree using WEKA

```
IF Diversity <=5 AND Diversity >3
   THEN class = FALSE (ineffective)
ELSE
   class = TRUE (effective)
```

Fig. 2 Example of rules generated from decision tree

From the decision tree generated, rules can be constructed as shown in Fig. 2. However, decision trees make rule generation rather complex [21]. In such cases, additional tools, such as Wekatext2Xml, are used for generating the rules. However, the decision tree generated in this study is simple, and the focus is on the model accuracy; hence, no additional tool was required.

3.3 Evaluation

The third stage involved in this study is the evaluation of the patterns in data mining stage. Cross-validation was chosen because the data set was too small to split into training sets and validation data sets as recommended by [22]. In this study, the data size was limited by the data collection time and the difficulty in obtaining accurate data. For the purpose of evaluation the decision tree, cross-validation technique was used to determine the prediction accuracy. The detailed results of decision tree patterns and prediction accuracy are discussed in Sect. 4.

4 Results and Discussion

Decision tree provides a good visual representation that can help discover patterns among variables when determining team effectiveness based on the location of the variables. The pattern of decision tree in this study is illustrated using WEKA tool and is depicted in Fig. 3. The decision tree shows the predictor variables and the outcome variables. The predictor variables are represented their names in node form and their splitting value. The outcome variables are represented by TRUE value (indicating effective team), or FALSE value (indicating ineffective team).

In this study, prior academic achievement is represented by advanced programming (gTIA1023) at the top of the decision tree, thereby indicating that this predictor is the most significant attribute in predicting team performance. Team members who obtained less than or equal to D grade (≤ 1) indicated ineffective team membership. Therefore, all team members must achieve at least C grade (>1) for them to be effective. The next important predictor is the type of methodology (type) used by the team, which is either agile or formal methodology. This is followed by a combination of team personality diversity (diversity) and personality types

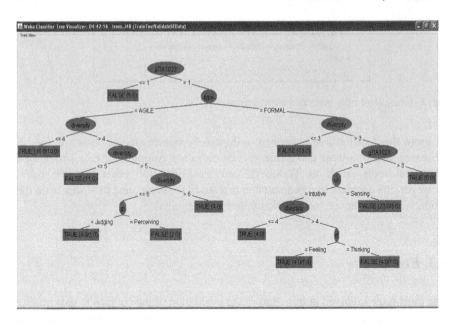

Fig. 3 Patterns of team members effectiveness using decision tree

(jp, sn, tf). The patterns produced by applying the decision tree technique were analyzed to discover the predictor combinations that contribute to effective team.

Overall, effective team must have all members with at least C grade in advanced programming course (gTIA1023). The pattern generated shows that the combination of methodology type, team personality diversity, prior academic achievements, and personality types plays significant role in determining the team effectiveness. Moreover, the analysis indicates that the effectiveness of agile teams depends on team personality diversity, whereby homogeneous team diversity was able to be effective team. As described in Table 1, a homogeneous team refers to team with less than four team personality diversity.

Among the formal teams, grade plays an important role in determining the team effectiveness. For the team to be effective, it was observed that the team must have members who scored at least B grade, indicating that it is difficult for weak students to contribute to an effective team. In this context, a weak student is a student that has not acquired sufficient knowledge and understanding of programming skills. Therefore, teams working according to formal methodology do not encourage weak students to explore or improve their programming skills. However, using agile methodology, it is possible to have a combination of weak and advanced students in a team and still able to achieve the stated goal in the project. This may indicate that suitable training program can induce team members to perform better and thus can work collectively.

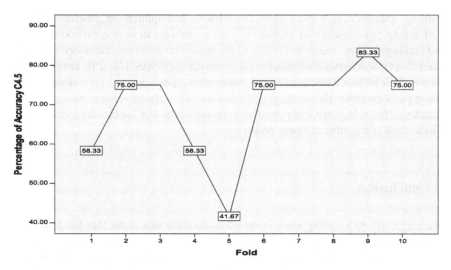

Fig. 4 Decision tree (C4.5) prediction accuracy using 10-fold cross-validation

In order to test the modeling, the k-fold cross-validation method was used to validate the prediction accuracy of the decision tree technique. The result is illustrated in Fig. 4.

It is noted in Fig. 4 that there is evident fluctuation in the prediction accuracy. The highest accuracy for $k = 10$ is 83.33 %, and the lowest is 41.67 % for $k = 10$. The fluctuations of the accuracy values reflect that the training and test data sets were not equivalent. Therefore, there is a need to use the average accuracy for each fold. Accordingly, the average accuracy result by using decision tree is 69.17 %.

The results show that decision tree can only predict 69.17 % accuracy of the team effectiveness. This is lower than prediction accuracy in [18] using rough set technique. However, decision tree provides a good data visualization pattern and shows important of each predictor variables to determine effective team. It is noted that previous programming grade is a key predictor to ensure that team members can interact effectively and thus can produce high-quality software. This result consistent with other studies that demonstrates cognitive ability is significant to determine students' performance [7, 8, 23]. Software team must consist of members that have good programming skills in order to assist other members to learn and share knowledge among them. This study also clearly demonstrated that type of training or intervention program is able to encourage team members to be effective. A good training program can bring out human potentials which ensuring each of the team members has equal chance to learn and perform better. Therefore, more research is needed to be carried out in order to measure the effectiveness and suitability of a certain training program.

This study also revealed that team personality diversity is one of the most important predictor variables to determine team effectiveness. In teamwork environment, diverse personalities of team members are required to promote creative learning. This is because more ideas, suggestions, and opinion are generated during the learning process and thus encouraging team members to have good coordination and communication. Previous studies highlighted that extrovert members are most likely to perform better compared to other personality types [24, 25]. However, in teamwork, a balanced combination of personality types, introvert (I)–extrovert (E), sensing (S)–intuitive (N), feeling (F)–thinking (T), judging (J)–perceiving (P), is essential. This is because team dynamism is one of the key aspects for members to interact and solve programming problems.

5 Conclusion

The results of data mining techniques using decision tree show that the four predictor variables, which are prior academic achievements, type of methodology, team diversity, and personality types, are significant to predict software team performance. Future studies will be carried out to collect more data in order to build a reliable prediction model. The prediction model can be improved by integrating hybrid algorithm techniques, such as fuzzy rough set, which may allow high prediction accuracy to determine team effectiveness. The model could facilitate the educators in developing strategic planning methods in order to improve current curriculum for SE students when developing teams.

Acknowledgments The authors wish to thank the Ministry of Education Malaysia for funding this study under Fundamental Research Grant Scheme (FRGS), S/O project:—12818 and Dana Kecemerlangan UiTM, code project:—600-RMI/ST/DANA 5/3/Dst (102/2009).

References

1. Bushe, G.R., Coetzer, G.H.: Group development and team effectiveness: using cognitive representations to measure group development and predict task performance and group viability. J. Appl. Behav. Sci. **43**(2), 184–212 (2007)
2. Deeter-Schmelz, D.R., Kennedy, K.N., Ramsey, R.P.: Enriching our understanding of student team effectiveness. J. Mark. Educ. **24**(2), 114–124 (2007)
3. Quinland, J.R.: Induction of decision trees. Mach. Learn. **1**, 81–106 (1986)
4. Quinland, J.R.: C4.5: Programs for Machine Learning. Morgan Kaufmann Publishers, USA (1993)
5. Breiman, L., Friedman, J.H., Olshen, R.A., Stone, C.J.: Classification and Regression Trees. Taylor & Francis, USA (1984)
6. Kabakchieva, D.: Predicting student performance by using data mining methods for classification. Cybern. Inf. Technol. **13**(1), 61–72 (2013)
7. Pal, A.K., Pal, S.: Analysis and mining of educational data for predicting the performance of students. Int. J. Electron. Commun. Comput. Eng. **4**(5), 1560–1565 (2013)
8. Mishra, T., Kumar, D., Gupta, S.: Mining students' data for prediction performance. In: Fourth International Conference on Advanced Computing and Communication Technologies (ACCT), pp. 255–262. IEEE, Rohtak (2014)

9. Fang, N., Lu, J.: Work in progress-a decision tree approach to predicting student performance in a high enrollment, high-impact, and core engineering course. In: 39th ASEE/IEEE Frontiers in Education Conference, pp. 1–3. IEEE, San Antonio (2009)
10. Nghe, N.T., Janecek, P., Haddawy, P.: A comparative analysis of techniques for predicting academic performance. In: 37th ASEE/IEEE Frontiers in Education Conference, pp. 7–12. IEEE, Milwaukee, WI (2007)
11. Al-Radaideh, Q.A., Al-Shawakfa, E.M., Al-Najjar, M.I.: Mining student data using decision trees. In: 2006 International Arab Conference on Information Technology (ACIT'2006), pp. 1–5. Jordan (2006)
12. Ramaswami, M., Bhaskaran, R.: A chaid based performance prediction model in educational data mining. Int. J. Comput. Sci. Issues IJCSI 7(1), 10–18 (2010)
13. Al-Radaideh, Q.A., Ananbeh, A.A., Al-Shawakfa, E.M.: A classification model for predicting the suitable study track for school students. Int. J. Res. Rev. Appl. Sci. 8(2), 247–252 (2011)
14. Fayyad, U., Piatetsky-Shapiro, G., Smyth, P.: From data mining to knowledge discovery in databases. In: Fayyad, U., Piatetsky-Shapiro, G., Smyth, P., Uthurusamy, R. (eds.) Advances in Knowledge Discovery and Data Mining, pp. 1–34. AAAI/MIT Press, Cambridge, Mass (1996)
15. Mazni, O., Sharifah-Lailee, S.-A., Naimah, M.H.: Analyzing personality types to predict team performance. In: CSSR 2010, pp. 624–628. IEEE, Kuala Lumpur (2010)
16. Mazni, O., Sharifah-Lailee, S.A.: Identifying effective software engineering (SE) team personality types composition using rough set approach. In: International Conference on Information Technology (ITSIM 2010), pp. 1499–1503. IEEE, Kuala Lumpur (2010)
17. Sharifah-Lailee, S.-A., Mazni, O., Mohd Nasir, A.H., Che Latifah, I., Kamaruzaman, J.: Positive affects inducer on software quality. Comput. Inf. Sci. 2(3), 64–70 (2009)
18. Mazni, O., Sharifah-Lailee, S.-A., Naimah, M.H.: Developing a team performance prediction model: a rough sets approach. In: ICIEIS 2011, Part II, CCIS 252, pp. 691–705. Springer, Berlin, Heidelberg (2011)
19. Myers, I.B., McCaulley, M.H., Quenk, N.L., Hammer, A.L.: MBTI Manual: A Guide to the Development and Use of the Myers Briggs Type Indicator, 3rd edn. Consulting Psychologists Press, USA (1998)
20. Pieterse, V., Kourie, D.G., Sonnekus, I.P.: Software engineering team diversity and performance. In: 2006 Annual Research Conference of the South African Institute of Computer Scientists and Information Technologists on IT Research in Developing Countries, pp. 180–186. ACM, Somerset West (2006)
21. Whitten, I.H., Frank, E.: Data Mining—Practical Machine Learning Tools and Techniques, 2nd edn. Morgan Kaufmann, USA (2005)
22. Hubert, M., Engelen, S.: Fast cross-validation of high-breakdown resampling methods for PCA. Comput. Stat. Data Anal. 51(10), 5013–5024 (2007)
23. Borkar, S., Rajeswari, K.: Attributes selection for predicting students' academic performance using education data mining and artificial neural network. Int. J. Comput. Appl. 86(10), 25–29 (2014)
24. Rhee, J., David, P., Basu, A.: The influence of personality and ability on undergraduate teamwork and team performance. SpringerPlus 2(16), 1–14 (2013)
25. Gilal, A.R., Jaafar, J., Mazni, O., Tunio, M.Z.: Impact of personality and gender diversity on software development teams' performance. In: International Conference on Computer, Communications, and Control Technology (I4CT), pp. 261–265. IEEE, Langkawi (2014)

Data Completeness Measures

Nurul A. Emran

Abstract This paper presents a review of the literature on data completeness with the aim to learn the different forms of completeness measure proposed to date. By learning the features of the completeness measures in the literature, an understanding of the similarities (and differences) of those measures will be provided, and the gaps in the current completeness measure proposals will be identified. Definitions of data completeness and comparison of several types of completeness measures proposed to date will be presented. In particular, for each proposal, the definition of the reference data set which is used in completeness measurement and the method used to measure completeness are examined. This paper concludes by pointing out the gaps in the current literature that will be addressed in the future work.

Keywords Data quality · Completeness · Population-based completeness · Data completeness measures · Measuring completeness

1 Introduction

Data completeness is not a new problem as studies in data completeness can be seen in the literature as early as 1970s. During this period, among scholars in the database community [1–7] as well as among statisticians [8–11], the data completeness problem was well known as the problem of missing information.

The early work on completeness for the database community largely dealt with the problem of representing missing values (as opposed to 'empty' or undefined values) within the relational tables, where *nulls* were usually assigned for the missing values in the tables [7]. Various representations of null have been used, for

N.A. Emran (✉)
Computing Intellingence Technologies (CIT) Lab, Centre of Advanced Computing Technology (C-ACT), Universiti Teknikal Malaysia Melaka (UTeM), 76100 Hang Tuah Jaya, Melaka, Malaysia
e-mail: nurulakmar@utem.edu.my

© Springer International Publishing Switzerland 2015   117
A. Abraham et al. (eds.), *Pattern Analysis, Intelligent Security and the Internet of Things*, Advances in Intelligent Systems and Computing 355, DOI 10.1007/978-3-319-17398-6_11

example, the @ symbol [12], ω [1] and the use of variables such as x, y and z [5]. About 14 manifestations of nulls have been identified in the ANSI/SPARC interim report [7], but the two types of null that can be frequently seen in the literature are (1) the *unknown* nulls, where the values are missing because they are unknown, and (2) the *non-existence* nulls, where the values are missing because the attribute of the relation is inapplicable. For example, the attribute 'property owned' of the citizens in Rome city council records is assigned a null in the case where a citizen does not own any properties.

The studies in completeness that deal with how to distinguish the types of null just mentioned have been driven by the need to determine whether the completeness problem exists or not. The presence of nulls is regarded as legitimate in the 'non-existence' case but not the 'unknown' case. Therefore, if 'non-existence' nulls are present, no completeness problems arise. But, the presence of unknown nulls signifies the existence of a completeness problem.

In order to distinguish the types of null, different representations have been proposed for each. In 1986, Codd proposed different representations of nulls according to their type, where the unknown nulls are represented by A (denoting applicable, but absent), while I (denoting inapplicable) is used to represent the non-existent nulls [2]. The work assumes that database administrators could provide the information on both types of null and perform the updates (i.e. from I to A), which could be a burden in cases where the volume of data is large.

Furthermore, distinguishing the types of null is not straightforward especially for attributes which are not mandatory. For example, as every registered citizen must have 'nationality' information, a null 'citizen nationality' attribute in a citizen record means that a citizen's nationality is 'unknown'. However, it is hard to distinguish whether the nulls that are present in the 'property-owned' attribute of citizen records are 'unknown' or 'non-existence' as having a property is an optional characteristic (rather than mandatory) for the citizens. In this case, additional information is required to distinguish the types of null. For example, the 'property-type' attribute might be referred to where, if the 'property-type' attribute has a value, then a null in the 'property-owned' attribute is an 'unknown' (rather than a 'non-existence' null). On the contrary, if the 'property-type' attribute has a null, then a null in the 'property-owned' attribute might be a 'non-existence'-type null.

The early studies on completeness also dealt with how to evaluate queries where nulls are present. One of the questions that arises in the query evaluation is, should a query expression such as $X > Y$ be evaluated as TRUE or FALSE if either X or Y is null? Codd suggests the three-valued logic in treating a query expression involving nulls where the MAYBE state is assigned for such an expression [12]. However, Grant argued that the three-valued logic is only applicable for the unknown nulls but not for the non-existence nulls, and he suggested that expressions involving the non-existence nulls should be evaluated as either TRUE or FALSE only, as there is no uncertainty (denoted by a MAYBE state) issue that arises for the non-existence nulls [3]. In 1979, Vassiliou extended Codd's work by introducing the four-valued logic to distinguish both types of null using the denotational semantic [6] but, this approach has been criticised by Zaniolo in terms of its computational cost, as for certain queries the cost could be high [7]. Perhaps, because there is no clear consensus on the adoption of either the three-valued logic or the four-valued logic, the two-valued logic remains a common way to treat nulls in query evaluation in most database systems today. In addition, no special symbols are used to distinguish the type of null as the cells containing nulls in relational tables are usually left empty.

For statisticians, missing values have been regarded as a nuisance that could prevent them from making reliable analyses. The focus of the studies in statistics is largely deal with the effort to reduce the impact of missing values that appear in statistical surveys or in questionnaire results. The missing values are usually represented by responses like 'don't know', 'refused', 'unintelligible', and the difficulty to distinguish whether the missing values are legitimate or not has been reported as one of the complexities in handling missing values [13].

Deletion of the entire records containing missing values has been identified as one of the methods that has been applied to handle missing values in statistics. By deletion, the missing values are ignored and excluded from the studies, causing only the complete survey items to be retained in the studies. However, in the case where the proportion of missing values is large relative to the number of items in a survey, deletion of missing values is not always preferable as much valuable information will be lost [11] and selection bias will be introduced. Thus, the imputation method was proposed in which the values that are missing are substituted. Through imputation, no survey items will be discarded but we need to make sure that the imputed values will not cause bias in the study. Statistical procedure has been proposed in order to improve estimation of the missing values, where studies conducted by Dempster et al. [9] have been recognised as the earliest work that formalises an expectation-maximisation (EM) algorithm that is used to estimate the missing values [13]. Imputation techniques which are based on maximum likelihood (ML) and multiple imputations have been used to estimate the missing values [13, 14].

The following are the studies in which completeness is a key factor in several application domains:

- In biology, missing values in microarray data sets can be caused by experimental faults such as when the measurements needed to derive the values cannot be performed due to technical reasons [15] (i.e. the presence of dust on the slides of the specimen or the probe may not be fixed properly [16]). Using complete microarray data sets is crucial for scientists, as missing data points can hinder downstream analyses, such as unsupervised clustering of genes, detection of differentially expressed genes, construction of gene regulatory networks [15, 17]. Imputation methods from statistical methods have been applied to estimate the missing values so that analysis and scientific interpretation that rely on the data sets can be performed. In studies conducted by Jörnsten et al. [18] a new imputation method called *LinCmb* has been proposed to estimate missing values in microarray data sets where the method is adaptable to the frequency of missing values and to the distribution of the missing values in the data set (whether the values are heterogeneous or homogeneous).

- In the government sector, missing values in time series data affect the measurement of the return premiums generated by liquidity differences in bonds for US government notes and bond portfolios [19]. The EM algorithm as introduced by statisticians (refer to [9]) has been used to estimate the missing values.
- In social sciences, missing values have been identified in surveys caused by several factors such as lack of cooperation from the respondents, or the deletion of inappropriate values in the survey [20]. Researchers in this domain study the types of missing value based on the distribution of the missing values in the data set. By identifying the types of missing value (i.e. generic/random, univariate and monotonic), suitable techniques to handle the missing values are identified. Statistical methods have been applied to impute the missing values.

The studies presented above deal with nulls (that are illegitimately present) in order to produce reliable and complete analyses. Completeness has been defined as the absence of illegitimate nulls in the studies presented in this section. To continue exploring the meanings of completeness, in the next section, definitions of completeness as documented in the literature will be presented. Section 3 provides details on the types of completeness measures drawn from the literature. Finally Sect. 4 concludes the paper and highlights the possible future work.

2 Definitions of Completeness

A survey conducted by Wang and Strong showed that completeness is an important data quality criteria for data consumers [21]. However, even though many agree that completeness is important, a standard definition of completeness that can be accepted by the people who are concerned about completeness is not available. We extract a range of completeness definitions from the literature on databases as well as on data quality, proposed from 1985 to 2009 as shown in Table 1, in order to examine the various definitions of completeness.

Based on the definitions, it is found that the common question that arises is the question of 'what is missing?'[1] and that there are several types of missing 'units' proposed. We can see in Table 1 some missing units appear in multiple definitions (such as 'value', 'tuples' and 'information') and some missing units only appear in a single definition of completeness (such as 'elements' and 'contents').

We can also see from the definitions that the 'view(s)' of 'what is missing' was included by some researchers (notably [22–29]). For example, Fox et al. [24] addressed missing values from the entity attributes view, while Wolf et al. [28, 29] addressed missing values from the view of a tuple. As more than one view is available, we know that completeness is a multi-dimensional concept that requires some way to measure the missing units from the data set concerned.

---

[1]'What is missing' in every definition of Table 1 is in italics.

Table 1 Definitions of completeness in the literature

Researcher	Completeness definition
Ballou and Pazer [22]	'All *values* for a certain variable are recorded'
Fox et al. [24]	'The degree to which a data collection has *values* or all attributes of entities that are supposed to have values'
Pipino et al. [27]	Column completeness: 'a function of the missing *values* in a column of a table'
Wolf et al. [28, 29]	'Let $R(A_1, ..., A_n)$ be a database relation. A tuple $t \in R$ is said to be complete if it has *non-null values* for each of the attributes A_i; otherwise, it is considered incomplete. A complete tuple t is considered to belong to the set of completions of an incomplete tuple $\hat{t}$ [denoted $C(\hat{t})$], if t and $\hat{t}$ agree on all the non-null attribute values'
Motro [30]	'Completeness as constraints correspond to predicates whose interpretation must contain all the *tuples* that represent real-world relationships'
Li [25]	'Query answer that could be computed if we could retrieve all the *tuples* from the relations in the query'
Kahn et al. [31]	'The extent to which *data* are not missing and are of sufficient breadth and depth for the task at hand'
Wang and Strong [21]	'Breadth, depth and scope of *information* in the data'
Jarke and Vassiliou [32]	'The percentage of the real-world *information* entered in the sources and/or the warehouse'
Motro and Rakov [33]	'Measures of the true proportion of the *information* that is stored'
Pipino et al. [27]	Schema completeness: 'the degree to which *entities and attributes* are not missing from the schema'
Wand and Wang [34]	'The ability of an information system to represent every *meaningful state* of the represented real-world system'
Bovee et al. [35]	'Having *all required parts* of an entity's information present'
Naumann and Rolker [36]	'Quotient of the number of *response items* and the number of real-world items and also coverage, scope, granularity, comprehensiveness, density, extent'
Pernici and Scannapieco [26]	'The degree to which the *elements* of an aggregated element are present in the aggregated element instance'
Ballou and Pazer [23]	'Presence of all defined *content* at both data element and data set levels'

Unfortunately, how completeness is measured was described by only some of the studies that provide definitions of completeness (notably [23, 24] and [27]). To learn how completeness is measured, we surveyed the literature where completeness measures were proposed.

3 Measuring Completeness

Driven by the 'what is missing?' question and the types of the missing unit, completeness measures are divided into the following types:

- Null-based completeness (NBC): the focus of the measures is on the 'values' that are missing from the data set under measure, where the missing values are represented by nulls.
- Tuple-based completeness (TBC): the focus of the measures is on the 'tuples' that are missing from the data set under measure.
- Schema-based completeness (SBC): the focus of the measures is on the missing 'schema elements' (e.g. attributes and entities) from the schema under measure.
- Population-based completeness (PBC): the focus of the measures is on the missing 'individuals' from the data set under measure relative to a reference population.

A simple ratio method is usually applied to measure completeness [27] where completeness of a data set under measure can be defined as:

$$\text{completeness}(D, R) = \frac{|(D \cap R)|}{|R|} \in [0, 1], \qquad (1)$$

where D is the *data set under measure* and R is the *reference data set*.

Within each proposal of completeness measures, the following characteristics will be observed:

- the view(s) of 'what is missing' from the data set under measure (e.g. missing values from a tuple view, missing tuples from a relation view),
- the reference data set used,
- how the reference data set is defined,
- the method used to measure completeness.

3.1 Null-Based Completeness (NBC)

The first proposal for a measure of NBC was made by Fox et al. [24]. In explaining their completeness measures, they described a datum as a triple $\langle e, a, v \rangle$, where v is the value of the attribute a that belongs to an entity e [24]. Nulls were viewed from two levels of granularity: single datum level and at data collection level. According to the study, different granularity levels require different types of measurement. At the single datum level, a *binary measure* was proposed which checks whether a datum has a value or not. At the collection level, the study described the completeness measure as an *'aggregate' measure* that computes the fraction of the data that are null. The reference data set for this study is defined implicitly, which is a data set that has no nulls (where all attributes are mandatory). In this study, a null is represented by an empty slot for v in each datum and all nulls under observation are regarded as illegitimate.

Another NBC measure, by Pipino et al. [27] viewed nulls from the column level of a relational table. 'Aggregate' measure is used by taking the ratio of the number of nulls in a column of a table to the number of tuples in the column and subtracting it from 1. In this study, the reference data set is defined implicitly from the column under measure (i.e. it consists of a column with the same number of tuples, none of which are nulls). The values from the column were treated as a bag (rather than a set), where all individual null values are counted in the NBC measurement.

Ballou and Pazer [23], however, view NBC from a vector of 'categories of data' (e.g. teaching data, research data). Each data category is assigned with a weight that shows its importance relative to the other categories. Each category consists of one or more data sets, and by using the 'aggregate' measure, the aggregated completeness is computed to determine the overall completeness of the category. The overall NBC has been defined as an aggregate function where the average of the product of the individual category completeness measurements and their weights is computed. An NBC measure called *structural completeness* was defined by the authors to measure completeness of a data set, as shown below.

$$\text{Structural completeness} = \frac{(\text{values that are recorded})}{(\text{values that could have been recorded})},$$

where the 'values that are recorded' are the non-null values in the data set under measure, and the 'values that could have been recorded' are the reference data set that has no nulls (where all attributes are mandatory). As in the measures by Fox et al. (see [24]) and Pipino et al. (see [27]), the reference data set is defined implicitly from the data set under measure. The authors suggest that different granularities of views (such as an entire database, a table or a record) and different formats of data sets (i.e. spreadsheet or file) can be supported within a category. However, no further discussion has been provided on how the different requirements for NBC were treated.

Scannapieco and Batini [37] extended the NBC measures by introducing more granularity levels from which NBC can be viewed (beyond the column level proposed in [27]). Four granularity levels were introduced, namely the value, tuple, attribute and relation levels. NBC is measured at each granularity level by the binary measure (called as Boolean function) or by the 'aggregate' measure that computes the percentage of the non-nulls within the data set under measure. The reference data set was implicitly defined from the data set under measure, with an assumption that the data set under measure should be free from nulls, which is similar to the NBC measures presented earlier.

A study by Sampaio and Sampaio which was conducted in the context of Web query processing also focuses on nulls (called missing instance values), but within XML documents that are queried from a Web source [38]. Similar to the NBC measures presented so far, 'aggregate' measure is used in this study where NBC is determined based on the information on the number of missing instance values within the query results relative to the total number of instance values (which is the reference data set). In this study, nulls were viewed from the query results (which is the data set under measure), and the reference data set is defined implicitly from the query results, with an assumption that the query results should be free from nulls. The study, however, assumes that the Web source could provide information on the missing instance values that may be supported by an established collaboration with the Web sources (e.g. within the context of a cooperative Web information system [39] and a shared workspace system [40]).

Naumann et al. [41] proposed an NBC measure called a *density measure*. Within the context of a virtual data integration of multiple data sources, the notion of a union schema that consists of the schema of relations (of all data sources) was used. Each data source that takes part in the integration contributes one or more relations. A universal relation was introduced, which consists of the union of relations of data sources whose schemas are present in the union schema. The density measure is an 'aggregate' measure which is based on counts of non-nulls in the data set under measure, viewed from attribute view, source view and query view. The density of an attribute was defined as 'the ratio of non-null values in the attribute to all values in the attribute'. The density of a source was defined as 'the average density of all attributes that appear in the global schema', while the density of a query was defined as 'the average density of all attributes that appear in the query'. Similar to the other NBC measures presented in this section, the reference data set (that comes from the universal relation) is implicitly defined as a data set that has no nulls (with attribute(s) that is/are mandatory).

3.2 Tuple-Based Completeness (TBC)

Most studies on TBC have been performed within the context of the relational model.

The TBC measure proposed by Motro and Rakov [30, 33] is not only useful for detecting missing tuples, but it also helps to determine whether the tuples are accurate. TBC, in their proposal, was viewed from a database level and is defined as an 'aggregate' measure as follows:

$$\text{Completeness(of the database relative to the real world)} = \frac{|D \cap W|}{|W|},$$

where D is the actual stored database instance while W is the ideal, real-world database instance. From this definition, an important insight into completeness is gained which is completeness can be affected by the presence of errors in the data set. W in the definition represents not only a reference data set that is complete, but also a reference data set that is accurate. Nevertheless, because W is very unlikely to be acquired, the measure used the sample of W which came from alternative databases or judicious sampling (where the verification of the samples is made by humans) [33].

Fox et al. also proposed a TBC measure (in addition to NBC) where, in their study, a tuple is defined as a collection of triples $\langle e, a, v \rangle$ that belong to the same entity (denoted as e), where v is the value for the entity's attribute a [24]. The authors stated that a tuple is missing if the triplet is missing entirely. An 'aggregate' measure is used to measure TBC, where the authors stated that the fraction of triples that are missing from the 'data collection' is computed against the number of triples in the data collection. The reference data set in this measure is the data collection; however, no further description of how the data collection was obtained was documented.

Naumann et al. proposed a TBC measure called a *coverage measure* (in addition to the density measure as described earlier in Sect. 3.1) [41] which is an 'aggregate' measure. As in the density measure, the reference data set used in the coverage measure is derived from the universal relation. TBC was viewed from the data source level, where the coverage measure was defined as the ratio of the number of records in the data source to the number of records of the universal relation.

Scannapieco and Batini dealt with TBC which was viewed from a relation level [37]. Like other TBC measures presented earlier in this section, an 'aggregate' measure is used in which, the completeness of a data set under measure (a relation) was defined as the ratio of the number of tuples it contains the number of tuples of the reference data set (called a reference relation). By assuming the number of tuples of the reference data set is known, this study extended TBC by proposing ways to determine completeness of a data set under measure based on the knowledge on the completeness of the relations from which the data set under measure is derived (called defining relations by Scannapieco and Batini). This extension of the TBC measure is particularly useful within the context where the reference data set for the data set under measure is unknown or not available (that prevents a TBC measure), and the information about the completeness of its defining relations is available to determine completeness of the data set under measure. Taking the example given by the authors, given that D is the data set under measure, $R1$ and $R2$ are the defining relations of D, where $D = R1 \cup R2$. Consider further that the completeness ratio of $R1$ is 0.5 and the completeness ratio of $R2$ is 0.5. In the case where $R1$ and $R2$ are disjoint, completeness of D is defined as the sum of the completeness ratio of $R1$ and $R2$, which is equal to 1.

However, the extension to TBC proposed by Scannapieco and Batini may introduce TBC measurement complexity as knowledge about how those defining relations are related (i.e. overlap or disjoint) and about how the data set under measure is derived from the defining relations (i.e. through union, intersection and Cartesian product operators) is required. In addition, we also need to know whether the same reference data set is used to measure completeness of the defining relations or not, as how TBC is determined for the data set under measure is not the same for both cases. For example, if different reference data sets are used to measure completeness of $R1$ and $R2$, completeness of D cannot be determined by simply adding the completeness ratio of $R1$ and $R2$, even though $R1$ and $R2$ are disjoint.

In a study of TBC by Fan and Geerts, the notion of *master data* was used as the reference data set in determining TBC from the view of a query level [42]. The authors stated that if a query is submitted against the reference data set, the result of the query (which is a set of tuples) is complete. The study addressed the problem of determining TBC for queries submitted against a database that consists of not only the tuples that fully overlap with the reference data set but also other data sets which may be incomplete. Nevertheless, this study does not attempt to compute how many tuples are missing from the query answer (which is the data set under measure), but rather it defines the characteristics of queries that will yield complete answers based on the information of the queried data sets. TBC is not measured based on a mathematical equation such as the simple ratios, but is determined based on the evaluation of the content of the queried data sets and the queries.

3.3 Schema-Based Completeness (SBC)

SBC is called 'model completeness' by Sampaio and Sampaio who defined it as 'the measure of how appropriate the schema of the database is for a particular application'. From an XML point of view, Sampaio and Sampaio defined SBC as the number of missing attributes relative to the total number of attributes [38]. The definition of SBC given by Pipino et al. states that SBC is 'the degree to which entities and attributes are not missing from the schema' [27].

Both definitions tell us that the SBC measures are the 'aggregate' measures, where attributes and entities are the views from which SBC can be assessed. For SBC proposals, there is a notion of database schemas with complete entities and attributes that are used as reference, but the details of how SBC is actually measured in practice are missing from the literature. Another limitation of the SBC literature is that the explanations of where these reference database schemas come from and how they are defined are missing.

3.4 Towards Population-Based Completeness (PBC)

To the best of our knowledge, the first recorded use of the term 'population' in connection with completeness is in a proposal by Pipino et al. [27]. The authors did not provide a formal definition of the PBC measure, but hinted at the presence of this useful concept through an example. In the example, the authors stated that 'if a column should contain at least one occurrence of all 50 states, but only contains 43 states, then we have population incompleteness' [27]. From the example, we observe that there is a data set under measure (from state column) in which its completeness is determined by the number of missing 'individuals' (the states) from a 'reference population' (a set of 50 states). There is a notion of reference population that is used to represent a population that consists of complete individuals.

Another example, provided by Scannapieco and Batini regarding Rome's citizens [37] might represent a form of PBC. This is because there is a notion of a reference population, which is a citizen population based on the personal registry of Rome's city council, and the data set under measure that consists of a set of Rome citizens derived from a company that stores Rome citizen information for the purpose of its business (in which its completeness is of concern). Based on the literature, we say that two elements must be present for PBC measurement which are the data set under measure and the reference population.

However, details of how PBC measurement is made in practice are missing from both proposals, especially in terms of how the reference populations are acquired and used. The elaboration of the concept of PBC therefore remains an open question for research in terms of the current literature.

4 Conclusions

This paper provides background on studies that deal with completeness (e.g. in the database community and in statistics), with some examples of completeness problems in several application domains. We surveyed definitions of completeness with the aim of learning the common question that arises from the definitions and the views from which completeness has been considered. Based on the 'what is missing?' question (that is found to be the common question asked), several types of missing 'units' (i.e. values, tuples, attributes) are present. We set to observe several characteristics of completeness measures proposed in the literature such as the view(s) of 'what is missing' from the data set under measure, the reference data set used, how the reference data set is defined and the method used to measure completeness. From the survey of completeness measures we conclude that:

- Given the missing 'unit' of concern, a data set under measure can be viewed at different granularity levels (i.e. tuple level, relation level and source level) and an 'aggregate' measure is the most common method used to measure completeness. Most of completeness measures are objective, mathematical measures but only some of the measures provide a formal definition of the completeness measure formula (notably [23, 30, 37, 41]).
- Every measure requires a reference data set, that is, considered to be complete. However, in most studies, the reference data sets have been assumed to be available and we cannot learn much from the literature about how reference data sets are defined or acquired. Motro and Rakov [33] pointed out the fact that a true reference data set is difficult to construct and suggested one method based on judicious sampling of alternative databases. While this could be a sensible way to address the complexity of establishing a reference data set, no details were provided about how the sampling can be done.
- For NBC, a separate reference data set is unnecessary as it can be determined from the data set under measure itself, where the reference data set has been implicitly stated in most completeness measures of NBC. Most other completeness measures, however, require reference data sets that are separate from

the data set under measure. While the need for a reference data set in completeness measures has been clearly stated in the completeness measure proposals, the complexity of acquiring (or providing) the reference data sets needed in completeness measurements has not been addressed by those proposals, either for reference data sets that are separate from the data set under measure or for reference data sets that are retrievable from the data sets under measure.

The gaps in the completeness literature, as stated above, can be seen for completeness measures of all categories. However, we found that the study of PBC is very limited as compared to other categories of completeness measures. Yet, we expect that PBC has much more to offer than its limited coverage suggests. Therefore, defining the meaning of PBC and exploring its contribution in measuring completeness are among the questions that we have explored in our earlier work (see [43–46]). Nevertheless, studying the practical issues (in PBC and other forms of measures) that might arise in acquiring (or providing) reference data sets, and the ways to deal with those issues remain open problems in data completeness studies, which will be addressed in our future work.

Acknowledgments The authors would like to thank the financial assistance provided by the Universiti Teknikal Malaysia, Melaka (UTeM) and The Ministry of Education Malaysia during the course of this research.

References

1. Codd, E.F.: Extending the database relational model to capture more meaning. ACM Trans. Database Syst. (TODS) **4**, 397–434 (1979)
2. Codd, E.: Missing information (applicable and inapplicable) in relational databases. SIGMOD Rec. **15**, 53–53 (1986)
3. Grant, J.: Null values in a relational data base. Inf. Process. Lett. **6**, 156–157 (1977)
4. Lipski, W.: On semantic issues connected with incomplete information databases. ACM Trans. Database Syst. (TODS) **4**, 262–296 (1979)
5. Imielinski, T., Lipski, J.: Incomplete information in relational databases. J. ACM **31**, 761–791 (1984)
6. Vassiliou, Y.: Null values in data base management a denotational semantics approach. In: Proceedings of the 1979 ACM SIGMOD International Conference on Management of Data, pp. 162–169. ACM (1979)
7. Zaniolo, C.: Database relations with null values. J. Comput. Syst. Sci. **28**, 142–166 (1984)
8. Reich, D.E., Gabriel, S., Atshuler, D.: Quality and completeness of SNP databases. Nat. Genet. **33**, 457–458 (2003)
9. Dempster, A.P., Laird, N.M., Rubin, D.B.: Maximum likelihood estimation from incomplete data via the EM algorithm (with discussion). J. Roy. Stat. Soc. **39**, 1–38 (1977)
10. Royston, P.: Multiple imputation of missing values. Stata J. **4**, 227–241 (2004)
11. Schafer, J.L., Olsen, M.K.: Multiple imputation for multivariate missing-data problems: a data analyst's perspective. Multivar. Behav. Res. **33**, 545–571 (1998)
12. Codd, E.F.: Understanding relations (installment #7). Bull. ACM SIGMOD **7**, 23–28 (1975)
13. Schafer, J.L., Graham, J.W.: Missing data: our view of the state of the art. Psychol. Methods **7**, 147–177 (2002)
14. Little, R.: Missing-data adjustments in large surveys. J. Bus. Econ. Stat. **6**, 287–296 (1988)

15. Aittokallio, T.: Dealing with missing values in large-scale studies: microarray data imputation and beyond. Briefings Bioinform. **11**, 253–264 (2010)

16. Kim, H., Golub, G., Park, H.: Missing value estimation for dna microarray gene expression data: local least squares imputation. Bioinformatics **21**, 1410–1411 (2006)

17. Tiffin, N., Andrade-Navarro, M.A., Perez-Iratxeta, C.: Linking genes to diseases: it's all in the data. Genome Med. **1**, 1–7 (2009)

18. Jörnsten, R., Wang, H., Welsh, W., Ouyang, M.: DNA microarray data imputation and significance analysis of differential expression. Bioinformatics **21**, 4155–4161 (2005)

19. Warga, A.: Bond returns, liquidity, and missing data. J. Financ. Quant. Anal. **27**, 605–617 (1992)

20. Little, R., Rubin, D.: Missing data: our view of the state of the art. Sociol. Methods Res. **18**, 147–177 (1989)

21. Wang, R.Y., Strong, D.M.: Beyond accuracy: what data quality means to data consumers. J. Manage. Inf. Syst. **12**, 5–33 (1996)

22. Ballou, D.P., Pazer, H.L.: Modeling data and process quality in multi-input, multi-output information systems. Manage. Sci. **31**, 150–162 (1985)

23. Ballou, D.P., Pazer, H.L.: Modeling completeness versus consistency tradeoffs in information decision contexts. IEEE Trans. Knowl. Data Eng. **15**, 240–243 (2003)

24. Fox, C., Levitin, A., Redman, T.: The notion of data and its quality dimensions. Inf. Process. Manage. **30**, 9–19 (1994)

25. Li, C.: Computing complete answers to queries in the presence of limited access patterns. VLDB J. **12**, 211–277 (2003)

26. Pernici, B., Scannapieco, M.: Data quality in web information systems. J. Data Semant. **1**, 48–68 (2003)

27. Pipino, L.L., Lee, Y.W., Wang, R.Y.: Data quality assessment. Commun. ACM **45**, 211–218 (2002)

28. Wolf, G., Khatri, H., Chokshi, B., Fan, J., Chen, Y., Kambhampati, S.: Query processing over incomplete autonomous databases. In: Proceedings of the 33rd International Conference on Very Large Databases (VLDB), VLDB Endowment, pp. 651–662 (2007)

29. Wolf, G., Kalavagattu, A., Khatri, H., Balakrishnan, R., Chokshi, B., Fan, J., Chen, Y., Kambhampati, S.: Query processing over incomplete autonomous databases: query rewriting using learned data dependencies. VLDB J. **18**, 1167–1190 (2009)

30. Motro, A.: Integrity = validity + completeness. ACM Trans. Database Syst. **14**, 480–502 (1989)

31. Kahn, B.K., Strong, D., Wang, R.: Information quality benchmarks: product and service performance. Commun. ACM **45**, 184–192 (2002)

32. Jarke, M., Vassiliou, Y.: Data warehouse quality design : A review of the DWQ project. In: Proceedings of the International Conference on Information Quality (IQ), MIT, pp. 299–313 (1997)

33. Motro, A., Rakov, I.: Estimating the quality of databases. In: Proceedings of the Third International Conference on Flexible Query Answering Systems (FQAS), pp. 298–307. Springer, Berlin (1998)

34. Wand, Y., Wang, R.: Anchoring data quality dimensions in ontological foundations. Commun. ACM **39**, 86–95 (1996)

35. Bovee, M., Rajendra, P.S., Mak, B.: A conceptual framework and belief-function approach to assessing overall information quality. Int. J. Intell. Syst. **8**, 51–74 (2003)

36. Naumann, F., Rolker, C.: Assessment methods for information quality criteria. In: Proceedings of the International Conference on Information Quality, pp. 148–162. ACM (2000)

37. Scannapieco, M., Batini, C.: Completeness in the relational model: A comprehensive framework. In: Ninth International Conference on Information Quality (IQ), MIT, pp. 333–345 (2004)

38. Sampaio, S.F.M., Sampaio, P.R.F.: Incorporating completeness quality support in internet query systems. In: CAiSE Forum, CEUR-WS.org, pp. 17–20 (2007)

39. Mecella, M., Scannapieco, M., Virgillito, A., Baldoni, R., Catarci, T., Batini, C.: Managing data quality in cooperative information systems. In: On the Move to Meaningful Internet Systems 2002: CoopIS, DOA, and ODBASE. Lecture Notes in Computer Science, vol. 2519, pp. 486–502. Springer, Berlin, Heidelberg (2002)
40. Bentley, R., Horstmann, T., Sikkel, K., Trevor, J.: Supporting collaborative information sharing with the World Wide Web: The bscw shared workspace system. In: Fourth International World Wide Web Conference 1995: The Web revolution, O'Reilly, pp. 63–75 (1995)
41. Naumann, F., Freytag, J., Leser, U.: Completeness of integrated information sources. Inf. Syst. **29**, 583–615 (2004)
42. Fan, W., Geerts, F.: Relative information completeness. ACM Trans. Database Syst. **35**, 1–44 (2010)
43. Emran, N., Embury, S., Missier, P.: Model-driven component generation for families of completeness. In: 6th International Workshop on Quality in Databases and Management of Uncertain Data, Very Large Databases (VLDB) (2008)
44. Emran, N.A., Embury, S.M., Missier, P., Isa, M.N.M., Muda, A.K.: Measuring data completeness for microbial genomics database. In: Intelligent Information and Database Systems, pp. 186–195. Springer, Berlin, Heidelberg (2013)
45. Emran, N.A., Embury, S.M., Missier, P., Ahmad, N.: Reference architectures to measure data completeness across integrated databases. In: Intelligent Information and Database Systems, pp. 216–225. Springer, Berlin, Heidelberg (2013)
46. Emran, N.A., Embury, S., Missier, P.: Measuring population-based completeness for single nucleotide polymorphism (SNP) databases. In: Advanced Approaches to Intelligent Information and Database Systems, pp. 173–182. Springer International Publishing, Berlin (2014)

Cloud Computing: A General User's Perception and Security Awareness in Malaysian Polytechnic

Siti Salmah Md Kassim, Mazleena Salleh and Anazida Zainal

Abstract Cloud computing (CC) is a computing model in which technology resources are delivered over the Internet. Nowadays, it has becoming one of the most popular tools used in educational institutions. The salient features of CC can be exploited for both teaching and administration purposes. This paper aims to look into the acceptance of CC in Malaysian polytechnics (MP) and identify areas that need improvement in terms of awareness. To achieve this aim, related papers in cloud computing were reviewed so as to evaluate the extensiveness of the implementation of CC in MPs. A survey was conducted among polytechnic lecturers. The results of the survey were analyzed, and it revealed that there is positive acceptance of CC in MPs in terms of readiness and perception. However, there is still a lacking on security awareness. Therefore, improvement in terms of creating security awareness among the polytechnic lecturers and strengthen the knowledge on cloud among lecturers are needed.

Keywords Cloud computing · Education · Malaysian polytechnic · Readiness · Security awareness

1 Introduction

In 2010, the Ministry of Higher Education (MOHE) has implemented a transformation plan in the technical education system, which was conducted by Polytechnic Education Department [1]. One of the objectives that MOHE has outlined

S.S. Md Kassim (✉) · M. Salleh (✉) · A. Zainal (✉)
University of Technology Malaysia, 81310, Skudai, Johor, Malaysia
e-mail: sitibuha@gmail.com

M. Salleh
e-mail: mazleena@fc.utm.my

A. Zainal
e-mail: anazida@utm.my

© Springer International Publishing Switzerland 2015   131
A. Abraham et al. (eds.), *Pattern Analysis, Intelligent Security
and the Internet of Things*, Advances in Intelligent Systems and Computing 355,
DOI 10.1007/978-3-319-17398-6_12

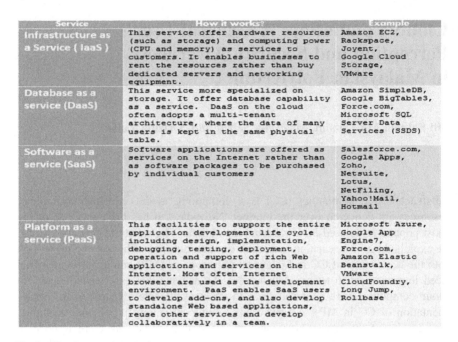

Fig. 1 Cloud computing services

is to enhance the teaching and learning (TnL) using technology. Such enhancement will provide more suitable education for the new generation. In other words, it is essential for MPs to provide high-quality of education and qualify students for the challenges of twenty-first century. In other word, it is essential for MPs to provide high-quality education and equip student for the challenges of twenty-first century learning [2]. Recently, CC is introduced as existing facilities which can supply on demand and availability of data access anywhere at anytime [3]. According to National Institute Standard and Technology (NIST), CC is a model for enabling ubiquitous, convenient, on demand network access to a shared pool of configurable computing resources (e.g., networks, servers, storage, applications, and services) that can be rapidly provisioned and released with minimal management effort or service provider interaction [4]. CC offers services that can be grouped into the following categories as shown in Fig. 1 and provides familiar tools such as email and personal finance to new offerings such as virtual world and social networks [5, 6].

Nowadays, there are few universities in United States and Canada already adopted CC in their institution. As the result of implementing CC in their institutions, they save a lot on software licensing and outsourcing their email services. CC also effectively implements collaborative learning for students at different places and shares resources among its numerous campuses and colleges [7]. However, in Malaysian education CC is sparsely implemented [8]. Malaysian Universities (MU) focusing on the services availability, accessibility and management of the information activities inside MU [9]. The main hindrance for adopting CC is the large number of student's data and information [10]. Compared with MP, student's data, research data, and educational materials are not as much as MU. But how can MU adopt CC? What about the acceptance of CC? These are the questions that will be answered through this paper.

The rest part of this paper that includes literature survey on CC adoption will be described at Part II, and Part III will elaborate on issues on adopting CC in MP, Part IV will reveal the methodology to conduct this study, and then, Part V will focus on questionnaire evaluation, and Part VI will talk over the result and findings and lastly conclude with the future work.

2 Literature Survey on CC Adoption

Moving toward in tandem with the MOHE vision, MU and other higher education institution have to keep invent the TnL method to suit with new students era. For Malaysian case study, as mentioned in paper [9], there is numerous benefits that CC can offer MU and also the challenges that MU might face to realize CC implementation. Security once again addressed as the prior concern in implementing CC.

This claim supports by another paper that also identifies security as one of the main reasons on rejection of CC in university administration [8] as shown in Fig. 2. As the conclusion for this part, there is a positive acceptance of CC in education institution, but security awareness must be the priority concern to make sure that CC can be implemented in Malaysia successfully. Next part will discuss more about how CC can stimulus MP in adopting that technology in TnL.

Resources	Methodology of study	Objective of the study	Findings
Cloud Computing in Malaysia Universities	Comparison of the benefits and challenges of CC in Malaysia University	a) To discuss the benefit CC can give to MU b) To look into the challenges that need to recognized by MU in adapting CC in TnL	**The benefits:** There are numerous benefits that CC can give to MU, such as support with on demand data access, reduce management cost, energy savings and sharing effectiveness. **The Challenges:** Copyright law and patent law. Remotely control by institution (economic, technical and data privacy and audit ability. Secure the Service Level Agreements (SLAs) and backup on storage Service provider selection in terms of integrity and better service offer.
The Policy as Repudiation factors of Adopting Cloud Computing in University Administration	Paper review Explorative case study Interview with administration staff	a) To find out the poor implementation of CC in Malaysian Universities b) To show the main rejection factor on CC in administration activities.	**Four main reason of rejection:** a) security b) accessibility c) policy of acceptance d) applicability

Fig. 2 Malaysian higher education case study

3 Cloud Computing and Malaysian Polytechnic

Even though the education sector continues to receive largest rates allocation when the government provides 21 % of the total budget in 2014, very limited amount allocated among 33 MPs. A total of 789 million ringgit to ensure provision of all government agencies could be spent more effectively [11, 12]. As mentioned before in Part I, MP is not as big as MU. It can be seen from the number of students and also the courses offered by MP compared to MU. Due to this reason, only limited budget allocated for MP. To overcome the limited budget, MPs can adopt CC to reduce the expenditure on hardware and software for teaching and learning and also administration usage. Pay-as-use service will help in reducing the expenses cost to buy more server or personal computer every year and also help in maintenance and upfront cost to run the system on cloud. Parallel with government wish for going green [11], CC allows reducing the consumption of the unused resources; thus, users of CC significantly reduce the carbon footprint [5]. MP can take the advantages of the ready-made applications hosted on robust and dynamic cloud such as email, word processing spreadsheet, collaboration, and media editing. Simplify the method on licensing, installation, and maintenance of individual software also tempted MP to adopt this technology [9, 13]. In order to achieve this goal to adopt CC in MP, this study wants to share the general user acceptance in MP and also the issue on security awareness through survey conducted on the next part.

4 Methodology

This research could be able to envisage the actual situation currently in adapting CC in MP. It focuses for the readiness of all lecturers in MP before CC could be implemented successfully.

4.1 Objectives

There are two main objectives for this study:

(i) to know the general knowledge among MP's lecturer about CC.
(ii) to gain the idea of the level security awareness among MP's lecturer while using CC.

4.2 Data Sampling

This sample was picked randomly. From 150 questionnaires have been distributed, only 45 returned by lecturers and the rest from different categories such as student,

Table 1 Data collection

Polytechnic	State	No. of collection
POLIMAS	Kedah	8
PTSB	Kedah	14
PSP	Penang	5
PBU	Penang	8
PTSS	Perlis	10
Total		45

non-academic staff, and others. Also because of restricted geographical area and time limitation, only five northern polytechnics were successfully collected to complete this initial study. In order to meet the objectives of this paper, that is to know the general knowledge among MP's lecturers about CC and to gain the security awareness level among them, we will only focus on the response from the lecturers. They are the key in implementing this technology in Teaching and Learning (TnL) environment.

4.3 Data Gathering

The study for this work was conducted by using survey question. 45 respondents from five different northern Malaysian polytechnics' (Politeknik Sultan Abdul Halim Muadzam Shah, Jitra-POLIMAS; Politeknik Tuanku Sultanah Bahiyah, Kulim-PTSB; Politeknik Tuanku Syed Sirajuddin, Perlis-PTSS; Politeknik Balik Pulau, Pulau Pinang-PBU; and Politeknik Seberang Perai, Pulau Pinang-PSP) lecturers were successfully collected. Refer to Table 1 for details about the data gathering, and next part will discuss about questionnaire evaluation.

5 Questionnaire Evaluation

The questionnaire has four different components:

(i) Part A: the background of information of the despondence which includes gender, age, role in institution, domain major, and experience in the institution. Table 2 shows the details.
(ii) Part B: general perception or knowledge about CC. For Likert scale score, refer to Table 3.
(iii) Part C: the security awareness using CC. For details refer Table 3.
(iv) Part D: respondents give open comments on security awareness. This part will be used as the suggestion to CC user on security awareness.

From Table 2, the ratio between male and female respondents was 42.3:57.7. About 55.6 % respondents in the age of 22–35 and compare to another group from age 36–58 for 44.4 %. From 45 respondents, 42.4 % are from ICT background and

Table 2 Part A: demography of respondents

Respondence profiles	Frequency	Percentage (%)
Gender		
Female	26	57.7
Male	19	42.3
Age		
≤21	–	–
22–35	25	55.6
36–58	20	44.4
>59	–	–
Major of your expertise at the institution		
Administration	3	6.7
Civil engineering	–	–
Electrical engineering	6	13.3
Mechanical engineering	7	15.6
Information and communication technology	19	42.2
Business faculty	4	8.9
General studied faculty	–	–
Hospitality	4	8.9
Others	2	4.4
Years of the experience at the institution		
<1 year	–	–
2–5 years	6	13.3
6–10 years	22	48.9
>11 years	17	37.8

Table 3 Likert scale

Score	Description
1	Strongly not agree
2	Not really agree
3	Agree
4	Strongly agree

the rest are administration 6.7 %, electrical engineering 13.3 %, mechanical engineering 15.6 %, business faculty 8.9 %, hospitality also 8.9 %, and others like those who currently on study leave with floating status are 4.4 %. Last part of the demography shows related to the experience of respondent regarding their service at respective MP. Most of them have 6–10 years, 48.9 %, followed by 37.8 % who have more than 10 years experience, and only 6 respondents who have the experience of 2–5 years.

The items also can categorize into four groups that is perception, readiness, knowledge, and security awareness. Figure 3 shows details about the category. This category identified based on their response in the questionnaire. These four

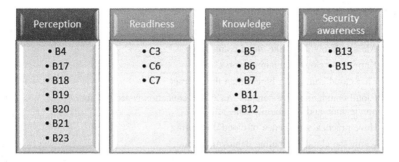

Fig. 3 Category for items/variables that used in questionnaire

Table 4 Reliability of the questionnaire

Cronbach's alpha	Cronbach's alpha based on standardized items	N of items
0.951	0.951	31

components help researcher to identify the extensiveness of the technology implemented in one institution [8–10].

This instrument was tested using Statistical Package for Social Science (SPSS). To see the reliability of the question, all 31 questions were tested using Cronbach's alpha method. According to the value of 0.951 of Cronbach's alpha from Table 4, the questions have been answered and understood well by the respondents. This value also indicates the questions considered reliable. After evaluating the instruments, next session will discuss about the result analysis from this study.

6 Result and Analysis

Table 5 listed the mean value for each items in questionnaire. By using Table 6 to determine the mean value, three indicators have been used. From SPSS calculation for Likert scale, it represents low from mean 1 to 1.33, moderate from 1.33 to 2.66, and high from 2.67 to 4. As shown in Fig. 4, mean value consists of two part, high (>2.67) and moderate (1.34–2.66).

For perception and readiness, the mean value shows high value in range of 2.67 till 2.96. There are seven items for perception and three for readiness. With this value, we can assume there is positive response from MP lecturers in terms of perception and readiness toward accept and adopt CC in their institution.

However, there is still lacking for another part. The moderate mean value revealed something related to the knowledge and security awareness among the MP lecturers. Some of the lecturers need to improve their knowledge regarding CC, and then, they can alert the security awareness that associated with it. The highest mean value shows in perception category. It has a 2.98 score for mean value which shows

Table 5 Result from questionnaire: mean

#	Items	Mean
B1	Already know about the cloud computing	2.44
B2	Know how cloud computing works	2.16
B3	That cloud computing bigger than the Internet	2.23
B4	Cloud computing give benefits Malaysian education especially for polytechnic and community college	2.93
B5	Have practical experience of cloud computing	1.67
B6	Know the cloud computing deployment models	1.77
B7	Have an idea about cloud computing service model	1.67
B8	Ever used Google Docs	2.29
B9	Google Docs can change the sharing setting with others	2.2
B10	That Google Drive have good and secure and sharing services	2.18
B11	Know about Microsoft Azure, Amazon EC2, Rackspase, Google Compute Engine	1.56
B12	Know about AWS Elastic Beanstalk, Win Azure, Heroku, Force.com, GoogleApp Engine	1.51
B13	Used the same password that been used in GoogleDocs for online application	1.56
B14	Already have Google Drive application	2.22
B15	Used the share settings in Google Docs when sharing documents or data with anyone	1.89
B16	Agree cloud computing will help in preparing the effective teaching material by using the application tools provided	2.42
B17	Agree cloud computing driving down the infrastructure for hardware and software cost	2.67
B18	Cloud can increase interoperability between disjoint technologies within and between institutions	2.76
B19	Cloud computing can allowed paperless in workload sharing	2.89
B20	Cloud computing can enabling green computing	2.87
B21	Cloud produce elastic and flexible repository	2.96
B22	Cloud computing will removing the admin burden allows educational facilities to concentrate on their core business and be more productive	2.62
B23	Cloud computing allowing free access applications and other useful tools	2.76
C1	The sharing resources is risky	2.42
C2	The repository of your institution contains confidential data	2.53
C3	Use network/Internet (email or etc.) to share important with someone in your institution	2.8
C4	Put the password to document/data before share it on network or Internet	2.64
C5	Share private and confidential document on network/Internet	2.24
C6	Sharing in network/Internet easier than use the external/USB hard disk	2.93
C7	Cloud computing can make this sharing extra benefits	2.91
C8	Documents or data that keeps in cloud computing is always safe	2.4

Table 6 Indicator of mean value

Mean	Indicator
1–1.33	Low
1.34–2.66	Moderate
2.67–4.00	High

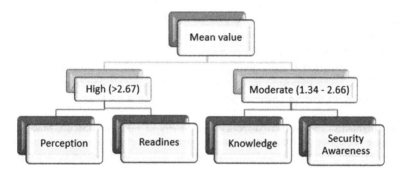

Fig. 4 Mean value for variables

Table 7 Part D: comment from respondents

# Respondent	Comment
R3	Always make backup for the data and set certain level for user
R10	Do not clearly understand the concept of CC
R47	Make extra storage
R63	Frequently change the password and do encryption
R125	Use encryption technique before end of receiving data or document

that respondents agree that CC can support the flexible and elastic repository. For the lowest mean value it shows in knowledge category which is 1.51.

This study also found out some respondents not really familiar with CC. They do not have basic knowledge about CC as shown in Table 7. This part consists of the open comment from respondents. From the respondents' comment, it shows that they more concern about the safety of the data that are sent to cloud in terms of privacy, reliability, and authentication. This is the common issue for any researcher and programmer keep invention the latest idea and method to overcome this issue, but we need to remember, sometimes user is the key to be saved. The user or the people who use the system or application should be educated to practice the most secure procedure in security awareness. Some guidelines or models can be proposed to educate the user. Generally, this study finds out that there are two groups of the respondents. One group knows about the CC and ready to accept this technology in their TnL, but the another group still doubts on security and not so familiar with CC. Both groups need to work together if they wanted to realize the acceptance of CC in MP environment.

7 Conclusion and Future Work

CC can be a tool to enhance the current application that has already implemented in polytechnic transformation plan in future such as e-learning and blended learning. However, this only can be realized if CC is widely accepted in the institution. Therefore, the output from this study can be used as an aid for proposing a new model or guideline to guide lecturers to be more familiar in using CC. It can also help MOHE to take a first step how to adapt CC in the teaching environment and at the same time absorbing all the benefits from CC such as data elasticity, cost-effective, and distributed management system. All of this can substantially reduce their workload. The exposure among lecturers and students can help in expanding the technology in MP and Malaysian education generally.

Acknowledgments This work is supported by Ministry of Education (MOE), Malaysia, and UTM under Vote No. (4L108) and also highly gratitude to all MP's lecturers from northern Malaysia involved in giving an excellent cooperation to make the questionnaire session happened.

References

1. Zain, Mohd: Zuraidah. Universiti Malaysia Perlis, TVET in Malaysia (2008)
2. The National Higher Education Strategic Plan Beyond 2020.: The National Higher Education, Action Plan Phase 2 (2011–2015). http://www.mohe.gov.my/transformasi/fasa2/psptn2-eng.pdf
3. Kim, W., Computing, Cloud: Today and tomorrow. J. Object Technol. **8**(1), 65–72 (2009)
4. Mell, P., Grance, T.: The NIST definition of cloud computing, computer security division, information technology laboratory, National Institute of Standards and Technology (2011)
5. Lamba, H.S., Singh, G.: Cloud computing future framework for e-management of NGO's, arXiv preprint arXiv:1107.3217 (2011)
6. Masud, M.A.H., Xiaodi, H., Jianming, Y.: IEEE 16th international conference on computer supported cooperative work in design (CSCWD), cloud computing for higher education: a roadmap, pp. 552–557 (2012)
7. Okai, S., Uddin, M,. Arshad, A,. Alsaqour, R., Shah, A.: Cloud computing adoption model for universities to increase ICT proficiency, vol. 4(3), SAGE Publications (2014)
8. Badie, N., Che Hussin A.R., Yadegaridehkordi, E.: The policy as repudiation factors of adopting cloud computing in university administration. J. Inf. Syst. Res. Innovation (JISRI), pp. 54–63 (2013)
9. Razak, S.F.A.: Cloud computing in Malaysia universities. Innovative Technol. Intell. Syst. Ind. Appl (CITISIA 2009), pp. 101–106, IEEE (2009)
10. Hashim, A.S., Othman, M.: Cloud computing adoption by universities. Concepts Rev. Int. J. (2014)
11. Amanat Tahun Baharu 2014.: Jabatan Pengajian Politeknik, Kementerian Pendidikan Malaysia. http://politeknik.gov.my/dokumen/files/TEKS%20PENUH%20AMANAT%20TAHUN%20 BAHARU%202014%20KP%20JPP.pdf
12. Malaysia, Perpustakaan Negara, Bajet 2014 (2014)
13. Johnson, L., Levine, A., Smith, R., Smythe, T.: The 2009 horizon report: K-12 edition, Austin. The New Media Consortium, Texas (2009)

The Correlations Between the Big-Five Personality Traits and Social Networking Site Usage of Elementary School Students in Taiwan

Ying-Chun Chou and Chiung-Hui Chiu

Abstract The Big-Five personality traits may influence people's usage of social networking sites (SNSs), and that of children may not be the same as adolescents or adults. This study investigated the relationships between elementary school students' personality traits and their usage of SNSs. Two hundred and forty 6th graders in Taiwan were involved in this work. The results indicated there were no gender differences in students' SNS usage. Extraversion had a significantly positive relationship with SNSs' usage for sharing, branding, monitoring, and learning. Emotional stability had a positive correlation with using SNSs for relaxing. Learning was the most frequent activity carried out on SNSs among the elementary school students examined in this work, and few of them used such sites for expressing, branding, sharing, or organizing.

Keywords Elementary school students · Big-Five-factor model · Social networking site · Personality traits · Online usage

1 Introduction

1.1 Background

In recent years, a number of scholars have worked to facilitate personalized learning experiences by using social networking sites (SNSs), such as Facebook, while many educators have also attempted to integrate such sites into K-12 learning programs. Related research has investigated designing student-centered pedagogies

Y.-C. Chou · C.-H. Chiu (✉)
Graduate Institute of Information and Computer Education,
National Taiwan Normal University, Taipei, Taiwan
e-mail: cchui@ntnu.edu.tw

Y.-C. Chou
e-mail: 80208003E@ntnu.edu.tw

© Springer International Publishing Switzerland 2015
A. Abraham et al. (eds.), *Pattern Analysis, Intelligent Security
and the Internet of Things*, Advances in Intelligent Systems and Computing 355,
DOI 10.1007/978-3-319-17398-6_13

with SNSs [1, 2], examined the use of such sites for academic purposes [3], and analyzed the student-to-student interaction patterns that occur on them [4], or the effects of using SNSs on learning outcomes [5] and friendship [6]. Researchers have also explored various factors that affect SNS usage, such as the user's personality [7], gender [8], age [9], cyberasociality [10], and so on. Many researchers have found that people's personality traits affect their SNS usage, with the so-called Big-Five personality traits being good predictors of this [8, 11]. While these studies have focused on adolescents and adults, not much attention has been paid to elementary school students. Furthermore, we should not assume that children use SNSs in the same way as teens or young adults do [12]. Therefore, the relationships between the personality traits of the elementary school students and their SNS usage behaviors should be investigated in more detail, as the findings could serve as a valuable and reference in the design of more personalized learning courses and websites for such students.

1.2 Purpose and Questions

The primary aim of this study was to investigate the correlation between Taiwanese elementary school students' SNS usage and their personality traits of extraversion, openness to experience, emotional stability, conscientiousness, and agreeableness. The following research questions guided this work:

- Are there any significant correlations between students' SNS usage and their personality traits?
- Are extraversion, openness to experience, emotional stability, conscientiousness, and agreeableness significant antecedent variables of students' SNS usage?

1.3 Theoretical Framework

The current literature defines SNS usage as uploading pictures, sharing information, comments, links or personal statuses, and chatting through Facebook, Twitter, LinkedIn, YouTube, and many other similar sites. Aladwani [13] developed and validated a Facebook use construct, "Gravitating toward Facebook" (GoToFB), which consists of eight dimensions, connecting, sharing, relaxing, organizing, branding, monitoring, expressing, and learning, and can be used to investigate SNS usage.

Previous research suggests that individual students have different preferences in the use of SNSs, which may inhibit or promote the collaborative processes that can occur in online learning environments. Much attention has been devoted to the effects of users' personality, as defined by Big-Five-factor Model [14], which includes the traits of neuroticism, extraversion, openness, agreeableness, and conscientiousness, on the use of SNSs. The 100-question NEO Personality Inventory (NEO-PI) was developed to assess personality traits, and the five personality factors have been shown to relate to the usage of SNSs of teens and adults in different social contexts. Saucier [15] developed the Mini-Markers questionnaire based on the NEO-PI, which consists of 40 items and investigates the following five personality traits: emotional stability, extraversion, openness to experience, conscientiousness, and agreeableness.

2 Method

A survey was conducted in 2014. Two hundred and sixty-six 6th-grade elementary school students (133 boys and 133 girls) in Taiwan participated in this work. Their computer teachers helped to deliver the consent form to parents and to conduct the survey. The survey required the students to go online and fill out the Personality Traits Questionnaire, Sociality on Social networking Sites Questionnaire, and Use of SNSs Questionnaire. The teachers gave students the website address to link to the online self-report questionnaires at the beginning of their computer classes, and they then completed these during regular class hours.

2.1 Measures

The three questionnaires were translated or derived from the following: the "Traditional International English Big-Five Mini-Markers in Chinese" [16], "Cyberasociality Scale" [10], and "Gravitating toward Facebook" [13]. All the questionnaires were written in Chinese. Before the survey, six elementary school students and two elementary school teachers checked the wording of the questionnaires to make sure they were age appropriate, and understandable.

The Personality Traits Questionnaire The Personality Traits Questionnaire was based on the International English Big-Five Mini-Markers instrument, originally developed by Saucier [15]. This was selected to measure the personality traits of the students, because the short form is suitable for children. The questionnaire was translated and validated by Teng et al. in 2011 [16]. The reliability of this Chinese version was established in a study of 370 students in Taiwan, and the Cronbach's alphas of the five dimensions of extraversion, openness to experience, emotional stability, conscientiousness, and agreeableness were reported as 0.91, 0.85, 0.79, 0.86, and 0.89, respectively. In this study, the Cronbach's alpha of the overall instrument was 0.82, and the participants took approximately 5 min to complete the test.

The Sociality on Social Networking Sites Questionnaire The Sociality on SNSs Questionnaire was based on the Cyberasociality Scale developed by Tufekci and Brashears in 2014 [10] for measuring a single dimension of the inability or unwillingness to connect with others through social media. The nine items are measured using a 7-point scale ranging from 1 (strongly disagree) to 7 (strongly agree), with the center point at 4 (neutral). The Cronbach's alpha of this instrument has been reported as 0.75, and it was 0.72 in the current study, with the students taking less than 3 min to complete the test.

Use of Social Networking Sites Questionnaire The Use of SNSs Questionnaire was based on GoToFB developed by Aladwani in 2014 [13]. GoToFB consists of eight dimensions: connecting, sharing, relaxing, organizing, branding, monitoring, expressing, and learning, with thirty-four items in total. The Cronbach's alpha for this scale has been reported as 0.90, and it was 0.93 in this study. It took approximately 6 min for the students to complete the questionnaire.

2.2 Data and Statistical Analysis

IBM SPSS software (version 21) was used to perform Pearson's product–moment correlations between personality traits and the use of SNSs. Multiple regression analysis was also used to analyze the relationships between the personality traits and the use of SNS.

3 Results

Data from 26 participants who failed to complete the online survey were removed from the analysis, leaving a total of 240 participants.

3.1 Demographic Characteristics

Of the two hundred and forty participants, 114 were male (47.5 %) and 126 were female (52.5 %). About 84.2 % of the participants had a SNS account, and 56.7 % participants spent more than 1 h on SNSs per week.

3.2 Students' SNS Usage and Their Personality Traits

The mean scores for the personality characteristics and usage of SNSs of the students are presented in Table 1. Correlation analysis involving all variables was performed, and the correlation coefficients were calculated.

Table 1 Personality characteristics and behaviors

	Mean	SD
Big-Five personality traits		
Extraversion	4.67	0.86
Openness to experience	4.44	0.73
Emotional stability	4.14	0.99
Conscientiousness	4.88	1.03
Agreeableness	5.17	1.03
Use of SNSs		
Connecting	3.10	1.12
Sharing	2.97	1.47
Relaxing	4.20	1.86
Branding	2.55	1.51
Organizing	2.79	1.52
Monitoring	3.08	1.43
Expressing	2.23	1.40
Learning	4.61	1.76
Cyberasociality	4.88	0.88

The results showed that students' extraversion had a significant, positive relationship with their sharing ($r = 0.154$, $p = 0.035$), branding ($r = 0.203$, $p = 0.005$), monitoring ($r = 0.162$, $p = 0.026$), and learning ($r = 0.236$, $p = 0.001$) behaviors. Openness to experience had a significant, positive relationship with branding ($r = 0.188$, $p = 0.010$), organizing ($r = 0.178$, $p = 0.015$), monitoring ($r = 0.156$, $p = 0.033$), and learning ($r = 0.391$, $p = 0.000$). Emotional stability had a significant, positive relationship with relaxing ($r = 0.155$, $p = 0.033$), and conscientiousness had a significant, positive relationship with learning ($r = 0.301$, $p = 0.000$), as did agreeableness ($r = 0.216$, $p = 0.003$).

The results of the multiple linear regression are shown in Table 2. Students' extraversion, openness to experience, emotional stability, conscientiousness, and agreeableness accounted for 5.1 % of their total use of SNSs for branding, $F(5, 182) = 3.014$, $p = 0.012$. Extraversion ($\beta = 0.167$, $p = 0.032$) and openness to experience ($\beta = 0.188$, $p = 0.029$) made significant contributions to the model, while emotional stability, conscientiousness, and agreeableness did not. Extraversion, openness to experience, emotional stability, conscientiousness, and agreeableness accounted for 15.8 % of students' total use of SNSs for learning, $F(5, 182) = 8.034$, $p = 0.000$. Openness to experience ($\beta = 0.298$, $p = 0.000$) made a significant contribution to this model, while the other factors did not. There were no significant antecedent variables for extraversion, openness to experience, emotional stability, conscientiousness, and agreeableness.

Table 2 Multiple regression analysis of the Big-Five personality traits and the use of social networking sites

Predictor	Usage of social networking sites					
	Branding			Learning		
	B	SE B	β	B	SE B	β
Extraversion	0.293	0.136	0.167*	0.229	0.149	0.112
Openness to experience	0.387	0.176	0.188*	0.716	0.193	0.298**
Emotional stability	−0.154	0.115	−0.101	0.091	0.126	0.052
Conscientiousness	−0.060	0.143	−0.041	0.223	0.157	0.131
Agreeableness	−0.078	0.135	−0.053	−0.061	0.148	−0.036
R^2		0.076			0.181	
ΔR		0.051			0.158	
F		3.01*			8.03**	

$*p < 0.05.$ $**p < 0.01$

3.3 Cyberasociality

The data analysis shows that there were no significant correlations between cyberasociality and the Big-Five personality traits. There were no significant correlations between cyberasociality and the usage of SNSs, either.

4 Discussion

More than 80 % of the participants had SNS accounts, and half of them spent more than 1 h per week on such sites. It is interesting to find that learning was the most frequent activity on SNSs, although elementary school students are usually not allowed to use SNSs at school in Taiwan. Relaxing was the second highest use of SNSs reported by the students, and this finding is consistent with Sánchez et al. [3], who also found that most students used SNSs for relaxing. The participants rarely used SNSs for expressing, branding, sharing, and organizing, and this may be due to their cultural backgrounds. Learners from different cultures seem to have different communication patterns in online environments. According to Hofstede et al. [17], Taiwanese people tend to strongly avoid uncertainty, have more feminine traits, and fighting with others make them feel shame and loss of face. Therefore, Taiwanese students are generally reluctant to speak up in class and accustomed to teacher-centered educational practices [18].

The results of this study showed that students' extraversion and openness to experience traits influenced their SNS-based activities, and this supports the findings in Amichai-Hamburger and Vinitzky [11]. The participants with scoring higher on the extraversion and openness to experience traits were more willing to express themselves on SNSs, and this might be because people with such traits like to interact with others and enjoy getting more attention from them. The behavior of branding is helpful in this regard, as it can be used to become more popular on SNSs. In addition, there were no gender differences in personality or behavior on SNSs found in this study, and this is similar to the conclusion of Huang et al. [19].

One limitation of this study is that the participants were all 6th-grade students from one rural elementary school in southern Taiwan, and thus, the results might not be generalizable to other populations in metropolitan schools or other educational levels, and studies of more varied groups could be carried out in future works, in order to compare the results with those obtained in this study.

5 Conclusions

This study investigated the correlations between the personality traits and SNS usage behaviors of 6th-grade elementary school students in Taiwan. Participants scoring higher on the extraversion and openness to experience traits were more likely to engage in branding on SNSs. Moreover, the trait of extraversion had a significant, positive relationship with sharing, branding, monitoring, and learning behaviors. The trait of openness to experience had a significant, positive relationship with branding, organizing, monitoring, and learning. Emotional stability had a significant, positive relationship with relaxing. Finally, conscientiousness and agreeableness both had significant, positive relationships with learning, and most of the elementary school students used SNSs for relaxing and learning. We recommend that further studies can explore the influence of other psychological traits on SNS-related behaviors.

Acknowledgments The authors thank the editor and anonymous reviewers for their remarkably constructive comments. This research was supported by the Ministry of Science and Technology, Taiwan, R.O.C. under Grant No. MOST 103-2511-S-003-024-MY3

References

1. Mills, K.A., Chandra, V.: Microblogging as a literacy practice for educational communities. J. Adolesc. Adult Literacy 55(1), 35–45 (2011)
2. Mihailidis, P., Cohen, J.N.: Exploring curation as a core competency in digital and media literacy education. J. Interact. Media Educ. JIME Spring issue (2013)
3. Sáncheza, R.A., Cortijob, V., Javedc, U.: Students' perceptions of Facebook for academic purposes. Comput. Educ. 70,138–149. doi:10.1016/j.compedu.2013.08.012
4. Koles, B., Nagy, P.: Facebook usage patterns and school attitudes. Multicultural Educ. Technol. J. 6(1), 4–17 (2012). doi:10.1108/17504971211216283
5. Wohn, D.Y., LaRose, R.: Effects of loneliness and differential usage of Facebook on college adjustment of first-year students. Comput. Educ. 76, 158–167 (2014). doi:10.1016/j.compedu.2014.03.018

6. Wang, J.-L., Jackson, L.A., Gaskin, J., Wang, H.-Z.: The effects of social networking site (SNS) use on college students' friendship and well-being. Comput. Hum. Behav. **37**, 229–236 (2014)

7. Ross, C., Orr, E.S., Sisic, M., Arseneault, J.M., Simmering, M.G., Orr, R.R.: Personality and motivations associated with Facebook use. Comput. Hum. Behav. **25**, 578–586 (2009)

8. Correa, T., Hinsley, A.W., Zúñiga, H.Gd: Who interacts on the Web?: the intersection of users' personality and social media use. Comput. Hum. Behav. **26**, 247–253 (2010). doi:10.1016/j.chb.2009.09.003

9. Poellhuber, B., Roy, N., Anderson, T.: Distance students' readiness for social media and collaboration. Int. Rev. Res. Open Distance Learn. **12**(6), 102 (2011)

10. Tufekci, Z., Brashears, M.E.: Are we all equally at home socializing online? Cyberasociality and evidence for an unequal distribution of disdain for digitally-mediated sociality. Inf. Commun. Soc. **17** (4), 486–502 (2014) doi:http://dx.doi.org/10.1080/1369118X.2014.891634

11. Amichai-Hamburger, Y., Vinitzky, G.: Social network use and personality. Comput. Hum. Behav. **26**(6), 1289–1295 (2010)

12. Grimes, S.M., Fields, D.A.: Kids online: a new research agenda for understanding social networking forums. The Joan Ganz Cooney Center at Sesame Workshop, New York (2012)

13. Aladwani, A.M.: Gravitating towards Facebook (GoToFB): what it is? and how can it be measured? Comput. Hum. Behav. **33**, 270–278 (2014)

14. Costa, P.T., McCrae, R.R.: Normal personality assessment in clinical practice: The NEO Personality Inventory. Psychol. Assess. **4**(1), 5 (1992)

15. Saucier, G.: Mini-markers: a brief version of Goldberg's unipolar Big-Five markers. J. Pers. Assess. **63**(3), 506–516 (1994)

16. Teng, C.-I., Tseng, H.-M., Li, I.-C., Yu, C.-S.: International english big-five mini-markers: development of the traditional chinese version. J. Manag. **28**(6), 579–600 (2011)

17. Hofstede, G., Hofstede, G.J., Minkov, M.: Cultures and Organizations: Software of the Mind, 3rd edn. McGraw-Hill USA, NY (2010)

18. Chiu, C.H., Yang, H.Y., Liang, T.H., Chen, H.P.: Elementary students' participation style in synchronous online communication and collaboration. Behav. Inf. Technol. **29**(6), 571–586 (2010). doi:10.1080/01449291003686195

19. Huang, W.-H.D., Hood, D.W., Yoo, S.J.: Gender divide and acceptance of collaborative Web 2.0 applications for learning in higher education. Internet Higher Educ. **16**, 57–65 (2013)

A Cryptographic Encryption Technique of MPEG Digital Video Images Based on RGB Layer Pixel Values

Quist-Aphetsi Kester, Laurent Nana, Anca Christine Pascu, Sophie Gire, Jojo M. Eghan and Nii Narku Quaynor

Abstract With the high increase in the transmission of digital data over secured and unsecured communication channels, security and privacy of such data are critical in this present day of cyberspace and it is a concern to both the transmitter and the receiver. This paper proposes a cryptographic encryption technique of mpeg digital video images based on RGB layer pixel values. The cryptographic encryption technique made use of the Red, Green, and Blue channel in the encryption and securing of the digital images. The programming and implementation were done using MATLAB.

1 Introduction

Many of today's multimedia applications and data transmitted via secured or unsecured network require confidential video transmission. Currently, the advancements in information technology have enabled powerful emerging capabilities, such as

This work was supported by Lab-STICC (UMR CNRS 6285) Research Laboratory, UBO France, AWBC Canada, Ambassade de France-Institut Français-Ghana and the DCSIT-UCC.

Q.-A. Kester (✉)
Faculty of Informatics, Ghana Technology University College, Accra, Ghana
e-mail: kquist-aphetsi@gtuc.edu.gh; kquist@ieee.org

Q.-A. Kester · L. Nana
Lab-STICC (UMR CNRS 6285), European University of Brittany,
University of Brest, Brest, France

A.C. Pascu
Lab-STICC (UMR CNRS 6285), European University of Brittany,
University of Brest, UBO, Brest, France

Q.-A. Kester · S. Gire · J.M. Eghan · N.N. Quaynor
Department of Computer Science and Information Technology, University of Cape Coast,
Cape Coast, Ghana

© Springer International Publishing Switzerland 2015   149
A. Abraham et al. (eds.), *Pattern Analysis, Intelligent Security
and the Internet of Things*, Advances in Intelligent Systems and Computing 355,
DOI 10.1007/978-3-319-17398-6_14

Urban Telepresence, wearable devices, drones. With wearable digital devices constantly emerging and celebrated in the mainstream news media, privacy and security of the public are also becoming a major concern [1]. Urban Telepresence operators can also interact in real time with personnel and sensor assets in that environment and can derive comprehensive shared situational awareness (SA) from a mixed reality (i.e., live-over-virtual-over-time) augmentation of the environment with supporting intelligence, including past/present/forecast information. The deployment of UT capability becomes a force multiplier for military operations as well as civilian safety, security, and emergency response [2]. Also, the advancements in modern-day public key cryptography [3] over the years have provided the bases for securing communications over secured and unsecured communications channels. This makes it easy for keys to be exchanged for secured effective communications over protected and unprotected media of communication [4].

This paper proposes a cryptographic encryption technique of mpeg digital video images based on RGB layer pixel values. The cryptographic encryption technique made use of the Red, Green, and Blue channel in the encryption and securing of the digital images. The paper has the following structure: Sect. 2 Related works, Sect. 3 Methodology, Sect. 4 the explanation of the algorithm, Sect. 5 results and analysis, and Sect. 6 concluded the paper.

2 Literature Review

Asghar, Mamoona Naveed, and Mohammad Ghanbari worked on MIKEY for keys management of H. 264 scalable video coded layers. Their paper investigates the problem of managing multiple encryption keys generation overhead issues in scalable video coding (H.264/SVC) and proposes a hierarchical top down keys generation and distribution system by using a standard key management protocol MIKEY (Multimedia Internet Keying Protocol). Their research goal was in twofold: (1) prevention of information leakage by the selective encryption of network abstraction layer (NAL) units with AES-CTR block cipher algorithm, and (2) reduction of multiple layer encryption keys overhead for scalable video distribution. They combined a MIKEY with the digital rights management (DRM) techniques to derive a mechanism in which every entitled user of each layer has only one encryption key to use [5].

There have been some works done in image cryptography in securing of digital images. Musheer Ahmad and Tanvir Ahmad in their work proposed an efficient encryption method to secure the multimedia color imagery. Complex dynamic responses of multiple high-order chaotic systems were utilized to carry out image pixels shuffling and diffusion processes under the control of secret key [9]. He et al. proposed a stream color image cryptography based on spatiotemporal chaos system. One-way coupled map lattices (OCML) were used to generate pseudorandom sequences and then used to encrypt image pixels one by one [6].

Joint Video Compression and Encryption (JVCE) has gained increased attention in the past couple of years to reduce the computational complexity of video compression, as well as provide encryption of multimedia content for Web services. Pande et al. presented a JVCE framework based on Binary Arithmetic Coding (BAC). They first presented an interpretation of BAC in terms of a skewed binary map and then described 7 other possible chaotic maps which give similar Shannon optimal performance as BAC [7]. Van Wallendael et al., in their paper, described encryption possibilities for the High Efficiency Video Coding (HEVC) standard under development. Bitstream elements, which maintained HEVC compatibility after encryption, were listed and their impact on video adaptation was described [8]. A layered selective encryption scheme for scalable video coding (SVC) was also proposed by Li et al. The main feature of their scheme was making use of the characteristics of SVC [9]. Advanced Video Coding is recently announced and widely used, although the protection means have not been developed thoroughly [10].

3 Methodology

Modern-day cryptography entails complex and advance mathematical algorithm in the encryption of text, and cryptographic techniques for image encryption are usually based on the RGB pixel displacement where pixel of images are shuffled to obtained a cipher image [11]. This paper proposes a cryptographic encryption technique of mpeg digital video images based on RGB layer pixel values. The cryptographic encryption technique made use of the Red, Green, and Blue channel in the encryption and securing of the digital images. A symmetric secret encryption key was generated from the plain image based on features that remained constant for both the ciphered and the plain image before and after the encryption process. The key was then used to encrypt the plain image. The plain image to be encrypted was then encrypted based on pixel displacement algorithm. At the end of the encryption and the decryption process, there was no pixel loss and the quality of the plain image remained unchanged after the decryption process. The proposed technique was implemented on mpeg digital images, and it proved to be very effective at the end.

Figure 1 showed the summary of the image cryptographic approach engaged in the ciphering and the deciphering process of the digital image. Where PI is the plain image and CI is the ciphered image.

From Fig. 1,
where PI is the plain image and CI is the ciphered image.
Salg(PI) = the function Salg() that operates on PI and CI to produce the symmetric secret key SSK.
SSK = the symmetric secret key generated from the image.
ImC = the algorithm for encryption of the plain image.
ImD = the algorithm for the image decryption.

4 The Explanation of the Algorithm

Step1.   Start
Step2.   Extraction of data from a plain image,
Let I= an image=f (R, G, B)
I is a color image of m x n x 3 arrays

$$\begin{pmatrix} R & G & B \\ r_{i1} & g_{i2} & b_{i3} \\ \vdots & \vdots & \vdots \\ \vdots & \vdots & \vdots \\ \vdots & \vdots & \vdots \\ r_{n1} & g_{n2} & b_{n3} \end{pmatrix}$$ (1)

(R, G, B) = m x n
Where R, G, B $\in$ I
(R o G) i j = (R) ij. (G) ij
where r_11 = first value of R
        r= [ri1] (i=1, 2... m) and x $\in$ r_i1 : [a, b]= {x $\in$ I: a $\leq$ x $\geq$ b}
        a=0, b=255 and R= r= I (m, n, 1)
  where g_12 = first value of G
        g= [gi2] (i=1, 2... m) and x $\in$ g_i1: [a, b]= {x $\in$ I: a $\leq$ x $\geq$ b}
        a=0 , b=255 and G= g= I (m, n, 1)
and   b_13 = first value of B
        g= [bi3] (i=1, 2... m) and x $\in$ b_i1 : [a, b]= {x $\in$ I: a $\leq$ x $\geq$ b}
        a=0, b=255 and B=b= I (m, n, 1
  Such that   R= r= I (m, n, 1)
Step3. Extraction of the red component as 'r'
Let size of R be m x n   [row, column] = size (R) = R (m x n)

$$rij= r= I (m, n, 1) = \begin{pmatrix} R \\ r_{11} \\ \vdots \\ \vdots \\ r_{1m} \end{pmatrix}$$

(2)

Step4.   Extraction of the green component as 'g'
Let size of G be m x n   [row, column] = size (G)

$$gij= g= I (m, n, 1) = \begin{pmatrix} G \\ g_{12} \\ \vdots \\ \vdots \\ g_{n2} \end{pmatrix}$$

(3)

Step5.   Extraction of the blue component as 'b'

Let size of B be m x n [row, column] = size (B) = B (m x n)

$$bij = b = I (m, n, 1) = \begin{pmatrix} B \\ b_{i3} \\ \vdots \\ \vdots \\ b_{n3} \end{pmatrix} \tag{4}$$

Step6.   Getting the size of r as [c, p]

Let size of R be [row, column] = size (r) = r (c x p)

Step7.   Engagement of SSK which is the symmetric secret key generated. The key is then engaged to iterate the step 8 to 14.

Step8.   Let r = Transpose of rij

$$r = \begin{pmatrix} R \\ r_{i1} & \cdots & \cdots & \cdots & r_{n1} \end{pmatrix} \tag{5}$$

Step9.   Let g = Transpose of gij

$$g = \begin{pmatrix} G \\ g_{i3} & \cdots & \cdots & \cdots & g_{n3} \end{pmatrix} \tag{6}$$

Step10. Let b = Transpose of bij

$$b = \begin{pmatrix} B \\ b_{i2} & \cdots & \cdots & \cdots & b_{n2} \end{pmatrix} \tag{7}$$

Step11. Reshaping of r into (r, c, p)

$$r = reshape (r, c, p) = \begin{pmatrix} R \\ r_{i1} \\ \vdots \\ \vdots \\ r_{in} \end{pmatrix} \tag{8}$$

Step12. Reshaping of g into (g, c, p)

$$g = reshape (g, c, p) = \begin{pmatrix} G \\ g_{i2} \\ \vdots \\ \vdots \\ g_{n2} \end{pmatrix} \tag{9}$$

Step13. Reshaping of b into (b, c, p)

$$b = reshape (b, c, p) = \begin{pmatrix} B \\ b_{i3} \\ \vdots \\ \vdots \\ b_{n3} \end{pmatrix} \tag{10}$$

Step14. Concatenation of the arrays r, g, b into the same dimension of 'r' or 'g' or 'b' of the original image

$$
= \begin{pmatrix}
R & G & B \\
r_{i1} & g_{i2} & b_{i3} \\
\vdots & \vdots & \vdots \\
\vdots & \vdots & \vdots \\
\vdots & \vdots & \vdots \\
r_{n1} & g_{n2} & b_{n3}
\end{pmatrix}
\tag{11}
$$

Step15. Finally the data will be converted into an image format to get the encrypted image.

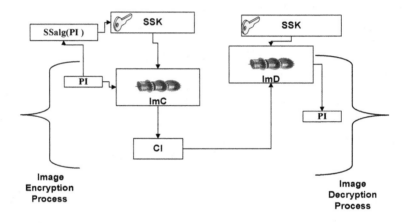

Fig. 1 The summary of the processes engaged

5 The Results and Analysis

The SSK was computed from the image based on features that will remain unchanged for both the ciphered image and plain image.

Let the set of pixel positions in X be x: $x \in X$ and $X \rightarrow x$: $x = x_i = [x_0, x_1, x_2, x_3, \ldots, x_n]$ and $x \in I$ where I is a positive integer.

$$
\Delta = \sum_{k=1}^{n} \Psi_k
\tag{12}
$$

Fig. 2 Three plain frames obtained from an MPEG video. **a** Frame 1. **b** Frame 2. **c** Frame 3

Table 1 Entropy and geometrical mean values of the plain image and the ciphered image

Property	Entropy	Mean
Rp	7.7468	133.6554
Gp	7.4006	110.4604
Bp	7.6244	90.7053
Rc	7.7468	133.6554
Gc	7.4006	110.4604
Bc	7.6244	90.7053

where Ψ_k is the decimal value of x_i

$$\text{SSK} = [(c \times p) + |(\delta \times 103)| + \Delta + |(\text{gm} = (1/n) \cdot \Sigma n_i = 1 x_i)|] \bmod p \quad (13)$$

where $p \in I$, δ = Entropy of image gm is the arithmetic mean for all the pixels in the image.

Figure 2 consists of the plain images used in the experiment, and figure nine is the resulted ciphered image. From Table 1, it can be clearly observed that there was no pixel expansion in the ciphered images and the values before encryption remained unchanged after encryption (Figs. 3, 4, and 5).

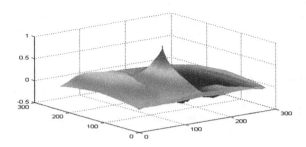

Fig. 3 The graph of the normalized cross-correlation of the plain image in Fig. 2a

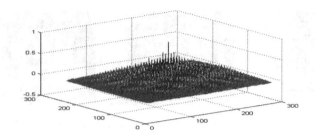

Fig. 4 The graph of the normalized cross-correlation of the matrices of the ciphered image in Fig. 2a

(a)   **(b)**   **(c)**

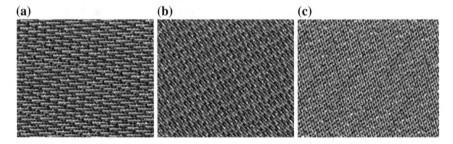

Fig. 5 Three ciphered frames of Fig. 2. **a** Frame 1. **b** Frame 2. **c** Frame 3

6 Conclusion

The cryptographic method used was found to be effective, and there was no pixel loss after the encryption process and this makes it suitable for the encryption and securing of images whose information is needed to be preserved. The implementation of the algorithm in hardware devices or software systems will bring a delay in streaming video images, but with the immergence of fast computers in the future, this will make real-time image encryption processes in off-the-shelf devices possible and faster.

Acknowledgments This work was supported by Lab-STICC (UMR CNRS 6285) at UBO France, AWBC Canada, Ambassade de France-Institut Français-Ghana and the DCSIT-UCC, and also Dominique Sotteau (formerly directeur de recherche, Centre national de la recherche scientifique (CNRS) in France and head of international relations, Institut national de recherche en informatique et automatique, INRIA and currently the Scientific counselor of AWBC).

References

1. Pedersen, I.: Ready to wear (or not): examining the rhetorical impact of proposed wearable devices. In: IEEE International Symposium on Technology and Society (ISTAS), 2013, pp. 201–202, 27–29 June 2013
2. Balfour, R.E., Donnelly, B.P.: The what, why and how of achieving urban telepresence. In: Systems, Applications and Technology Conference (LISAT), IEEE Long Island, pp. 1, 6, 3–3 May 2013. doi:10.1109/LISAT.2013.6578234
3. Huang, Y.J., Liu, F.H., Yang, B.Y.: Public-key cryptography from new multivariate quadratic assumptions. In: Fischlin M., Buchmann J., Manulis M. (eds.) Proceedings of the 15th International Conference on Practice and Theory in Public Key Cryptography (PKC'12), pp. 190–205. Springer, Berlin
4. Applebaum, B., Barak, B., Wigderson, A.: Public-key cryptography from different assumptions. In: Proceedings of the Forty-Second ACM Symposium on Theory of Computing (STOC'10), pp. 171–180. ACM, New York (2010)
5. Asghar, M.N., Ghanbari, M.: MIKEY for keys management of H. 264 scalable video coded layers. J. King Saud Univ. Comput. Inf. Sci. **24**(2), 107–116 (2012)
6. He, J., Zheng, J., Li, Z.B., Qian, H.F.: Color image cryptography using multiple one-dimensional chaotic maps and OCML. In: IEEC'09. International Symposium on Information Engineering and Electronic Commerce, pp. 85, 89, 16–17 May 2009
7. Pande, A., Zambreno, J., Mohapatra, P.: Joint video compression and encryption using arithmetic coding and chaos. In: IEEE 4th International Conference on Internet Multimedia Services Architecture and Application (IMSAA), pp. 1, 6, 15–17 Dec 2010
8. Van Wallendael, G., Boho, A., De Cock, J., Munteanu, A., Van de Walle, R.: Encryption for high efficiency video coding with video adaptation capabilities. In: IEEE International Conference on Consumer Electronics (ICCE), pp. 31, 32, 11–14 Jan 2013
9. Li, C., Yuan, C., Zhong, Y.: Layered encryption for scalable video coding. In: CISP '09. 2nd International Congress on Image and Signal Processing, pp. 1, 4, 17–19 Oct 2009. doi: 10. 1109/CISP.2009.5302934
10. Lian, S., et al.: Selective video encryption based on advanced video coding. In: Advances in Multimedia Information Processing-PCM 2005, pp. 281–290. Springer, Berlin (2005)
11. Bruen, A.A., Forcinito, M.A.: Cryptography, Information Theory, and Error-Correction: A Handbook for the 21st Century, p. 21. Wiley, New York. ISBN 978–1-118-03138-4. http://books.google.com/books?id=fd2LtVgFzoMC& pg = PA21

The Impact of Knowledge Management in Pair Programming on Program Quality

Mazida Ahmad, Ainul Husna Abd Razak, Mazni Omar,
Azman Yasin, Rohaida Romli, Ariffin Abdul Mutalib
and Ana Syafiqah Zahari

Abstract This paper reports on an initiative that determines the most appropriate technique for supporting students' programming ability. The proposed technique combines pair programming (PP) and SECI process that is a knowledge management (KM) model. Combining PP and SECI resulted in the formation of four approaches, which are named as NSNR, NSYR, YSNR, and YSYR. In those four approaches, the subjects who are students of IT-related programs in a higher learning institution complete a set of programming questions. The approaches were then compared based on the subjects' scores in their program codes. Descriptive statistics was used to analyze the gathered data. Generally, the results show that switching the roles (driver and navigator) in PP enhances good quality of coding. Through this study, an initial formation of the KM model and programming technique is contributed in enhancing program quality. Further, future work to be

M. Ahmad (✉) · A.H. Abd Razak · M. Omar · A. Yasin · R. Romli · A.S. Zahari
School of Computing, College of Arts and Sciences, Universiti Utara Malaysia,
06010 Sintok, Kedah, Malaysia
e-mail: mazida@uum.edu.my

A.H. Abd Razak
e-mail: s814049@student.uum.edu.my

M. Omar
e-mail: mazni@uum.edu.my

A. Yasin
e-mail: yazman@uum.edu.my

R. Romli
e-mail: aida@uum.edu.my

A.S. Zahari
e-mail: anasyafiqah91@gmail.com

A. Abdul Mutalib
School of Multimedia Technology and Communication College of Arts and Sciences,
Universiti Utara Malaysia, 06010 Sintok, Kedah, Malaysia
e-mail: am.ariffin@uum.edu.my

© Springer International Publishing Switzerland 2015
A. Abraham et al. (eds.), *Pattern Analysis, Intelligent Security
and the Internet of Things*, Advances in Intelligent Systems and Computing 355,
DOI 10.1007/978-3-319-17398-6_15

159

considered can be a rigorous theoretical formation for constructing other important determinants to enhance program quality because the findings of this research are minimal to SECI model and pair programming technique only.

Keywords Knowledge management · Pair programming · SECI · Code quality

1 Introduction

Pair programming (PP) as one of the key practices in extreme programming has been gaining acceptance among practitioners and the software development community. This success leads to the wide use of PP in educational setting as a computer science or software engineering pedagogical tool especially in programming courses [1].

PP involves two programmers that play different roles, one as a driver and another as a navigator [2]. They work on the same problem from designing to testing phases. In practice, the driver creates and implements the codes, whereas the navigator is responsible for checking errors and suggesting alternatives when necessary [1]. Periodically, the driver and navigator switch their roles, which could induce knowledge sharing among themselves [3]. Indeed, a better structured pair interaction is required by having proper communication within the pair [4]. Although the benefits have been proven, their creativity, brainstorming quality, and speed of programming development task can vary among the pairs, which is hypothesized as partly influenced by gender.

Therefore, this study intends to compare the performance between PP with role rotation and PP without role rotation. In this regard, role rotation refers to a process of changing the role among partners [5]. Programs that are done by female and male students in PP with and without role rotation were analyzed to measure which code is better in terms of their quality. Additionally, the socialization, externalization, combination, and internalization (SECI) model is also injected into this study, encouraging the pairs to apply the knowledge management (SECI model is a knowledge management component) technique in their programming task.

SECI model is known as "knowledge creation theory" which was coined by Nonaka and Takeuchi [6]. It facilitates the understanding of knowledge transformation between tacit and explicit states [2]. In SECI, socialization concerns with sharing mental thinking and experience with others. It is denoted by tacit–tacit. In contrast, externalization concerns in the articulation of tacit knowledge into document form, denoted by tacit–explicit. Meanwhile, combination which is denoted by explicit–explicit refers to supporting explicit knowledge with systematic resources [6]. The model has been decided for consideration because not only as knowledge management component, but it also builds up interaction for knowledge transfer [2]. In conjunction, this study aims at analyzing the impact of knowledge management in PP on program quality.

2 Related Works

PP involves two individual programmers, acting as a team via similar algorithms, design, code, and test, using one computer. In a pair, while the driver types-in programming codes, the navigator constantly and actively navigates the codes and identify whether there is any error or mistake. The navigator also identifies the strategic solution to complete the task or program. When necessary, they switch their roles to improve their work and learn appropriate skills [7]. In software industry, PP has been widely practiced for programming solution, where two programmers working side by side on one computer on the same problem with great success [8]. Winkler et al. [9] highlight the important role of pair programming at increasing coding efficiency and code quality and supports learning of development of team members.

Generally, knowledge management covers the tasks in obtaining, sharing, utilizing, and storing knowledge among individuals in an organization. It is undoubtable that knowledge sharing is an important part of knowledge management and is a crucial task in software development life cycle [3]. It promotes knowledge transmission among individuals in a community or organization and is normally supported by the knowledge sharing mode [10]. The knowledge management modes that enable individuals to exchange knowledge vary from face-to-face communication, conference, knowledge network, and organizational learning. In this study, the face-to-face communication is selected as the knowledge sharing mode in colocated PP practices, which involves two parties, namely contributor and receiver [10] that reflect the navigator and driver, respectively (in PP).

In conjunction, Kavitha and Ahmed [5] have found that PP can be a useful approach to programming course in higher education to facilitate effective knowledge sharing among the students. In PP practice, knowledge sharing involves social interaction, sharing, and constructing knowledge between the partners. In this scenario, the SECI model is applicable to promote sharing and constructing tacit knowledge between partners in generating high-quality programs. Code quality is indicated through number of syntax error and the acceptance level by users in terms of reliability, usability, maintainability, and portability [11].

3 Methodology

This study makes use of survey and experimental design as the method of data collection. The research procedure consists of phase 1 (defining the research context, instrumentation, and selecting variables), phase 2 (experimental process), and phase 3 (analysis and validation). The phases are discussed separately.

3.1 Phase 1

The study began with defining the research context (in which the research procedure is illustrated in Fig. 1), by focusing on a theoretical study that leads to the understanding of the SECI model, program quality, and PP. The instrument was adapted from Mazida [12]. Twenty-two third and fourth semester undergraduate students of College of Art and Sciences (CAS) at Universiti Utara Malaysia (UUM), who enroll in information technology (IT), multimedia, and education in IT programmers, involved in this study. They have learned basic programming course, which is compulsory for first semester students.

3.2 Phase 2

The experiment involved two laboratory experiments (the laboratory experiments were carried out in computer laboratories). Laboratory experiment 1 was carried out without SECI process, while laboratory experiment 2 with SECI process. Identical respondents participated in both laboratory experiments applying PP in solving programming test (the role of driver and navigator apply), in which the test questions were prepared by experts of School of Computing. The aim of the test was to measure the quality of the programs that the subjects produce (in pairs), with the full mark for the program is 40. Hence, the best program will be marked 40. Thus, the closer the mark to 40, the better the program is.

Four approaches have been tested in the laboratory experiments. While laboratory experiment 1 does not apply SECI and laboratory experiment 2 does, the laboratory experiments were carried out with and without role rotations (in the

Fig. 1 Research procedure

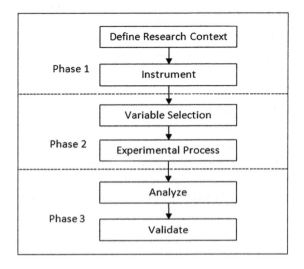

pairs). Therefore, the approaches are named according to their characteristics. The first approach does not apply SECI process and no role rotation/switching; hence, it is named as No SECI No Rotation (NSNR). The second approach also does not apply SECI process but with role rotation/switching. It is named as No SECI Yes Rotation (NSYR). Further, using similar convention, the other two are named as YSNR and YSYR.

3.3 Phase 3

Descriptive statistic consists of frequencies (frequency test) and cross-tabulation (Crosstabs test) on the demographic data was conducted to measure the code quality for each approach. Particularly, frequency provides an overall perspective of the code quality for the four approaches. Meanwhile, Crosstabs measures the code quality for the four approaches categorically [13], in which, in this regard, gender is the variable.

4 Results and Discussions

Based on the frequency and Crosstabs, the findings are discussed in the following subsections.

4.1 Demographic Analysis

Table 1 exhibits the demographic background of the subjects. It is seen that 63.60 % 22 subjects are female and 36.40 % are male. The ratio is common, mapping the real situation, in which most universities in current situation enroll 60 % female versus 40 % male students in every intake.

4.2 Frequency Analysis

Table 2 represents the frequency analysis for data using NSNR approach. It is seen that 50 % of the subjects score between 33 and 40 marks.

Table 3 exhibits the scores in NSYR approach. The scores vary from scales 1 to 5, in which the scores between 33 and 40 are the highest (59.1 %).

Next, Table 4 lists the results of YSNR approach. It is seen that 36.4 % of the subjects score between 33 and 40 marks and between 17 and 24 marks.

Table 1 Respondents' demographic background

	Frequency	Percentage (%)
Gender		
Male	8	36.40
Female	14	63.60
Age		
18–20	5	22.70
21–23	14	63.60
24–26	3	13.60
Programmed		
BSc (IT)	17	77.30
BSc (multimedia)	5	22.70
Current semester		
Semester 1 and 2	3	13.60
Semester 3 and 4	15	68.20
Semester 5 and 6	3	13.60
Semester 7 and 9	1	4.50

Table 2 Frequency of marks in NSNR approach

Scale	Mark	Frequency	Percentage (%)
1	01–08	0	0
2	09–16	0	0
3	17–24	5	22.7
4	25–32	6	27.3
5	33–40	11	50.0
Total		22	100

Table 3 Frequency of marks in NSYR approach

Scale	Mark	Frequency	Percentage (%)
1	01–08	0	0
2	09–16	3	13.6
3	17–24	1	4.5
4	25–32	5	22.7
5	33–40	13	59.1
Total		22	100

Table 4 Frequency of marks in YSNR approach

Scale	Mark	Frequency	Percentage (%)
1	01–08	0	0
2	09–16	0	0
3	17–24	8	36.4
4	25–32	6	27.3
5	33–40	8	36.4
Total		22	100

Table 5 Frequency of marks in YSYR approach

Scale	Mark	Frequency	Percentage (%)
1	01–08	0	0
2	09–16	10	45.5
3	17–24	4	18.2
4	25–32	2	9.1
5	33–40	6	27.3
Total		22	100

Table 6 Percentage of code quality for each approach

Technique	Percentage (%)
NSNR	50.00
NSYR	59.10
YSNR	36.40
YSYR	27.30

Then, Table 5 shows the frequencies for results in YSNR approach. The results are different than other approaches, in which most subjects (45.5 %) score between 9 and 16. It is followed by scores between 33 and 40 marks (27.3 %).

Further, the highest scores for every approach is compared, exhibited in Table 6. The total full mark was converted to 100 % for the percentage values in Table 6. The highest score indicates that the program is good in terms of quality [11]. Referring to the table, NSYR scores the highest (59.10 %).

4.3 Crosstab Analysis

Data were analyzed to measure the scores by the subjects through NSNR, NSYR, YSNR, and YSYR approaches and their gender. It is interesting to measure where there is any different trend between genders.

Table 7 depicts the Crosstabs analysis of NSNR approach with gender details. It is seen that more female subjects (8) score between 33 and 40 marks than male (3). It is understandable that the NSNR is accepted by the subjects.

Table 7 Crosstabs for scores in NSNR and gender

NSNR		Gender		Total
Scale	Mark	Male	Female	
1	01–08	0	0	
2	09–16	0	0	0
3	17–24	2	3	5
4	25–32	3	3	6
5	33–40	3	8	11
Total		8	14	22

Table 8 Crosstabs for scores in NSYR and gender

NSYR		Gender		Total
Scale	Mark	Male	Female	
1	01–08	0	0	
2	09–16	1	2	3
3	17–24	1	0	1
4	25–32	3	2	5
5	33–40	3	10	13
Total		8	14	22

Then, Table 8 depicts the Crosstab analysis for NSYR approach and gender. Similarly, it is found that more female (10) score between 33 and 40 marks than the male (3). The table also indicates that the approach is acceptable by the subjects.

Further, Table 9 exhibits the results of YSNR approach and gender. It is interesting that more female than male for scores between 17 and 24 (female = 5, male = 3) and between 33 and 40 (female = 8, no male). It could be deduced that for male, YSNR is not quite appropriate.

Finally, the results of YSYR were obtained and are displayed in Table 10. Majority of subjects score between 9 and 16 marks, and there are more female (6) than the male (4). Based on the results, in which the subjects score low, YSYR technique might be an inappropriate approach for the subjects.

Further, Table 11 summarizes the Crosstabs analysis of the approaches subjected to their gender. Based on the table, NSYR gives the highest number of subjects

Table 9 Crosstabs for scores in YSNR and gender

YSNR		Gender		Total
Scale	Mark	Male	Female	
1	01–08	0	0	
2	09–16	0	0	0
3	17–24	3	5	8
4	25–32	5	1	6
5	33–40	0	8	8
Total		8	14	22

Table 10 Crosstabs for scores in YSNR and gender

YSYR		Gender		Total
Scale	Mark	Male	Female	
1	01–08	0	0	
2	09–16	4	6	10
3	17–24	4	0	4
4	25–32	0	2	2
5	33–40	0	6	6
Total		8	14	22

Table 11 Frequencies of YSYR respondents' mark

Technique	Percentage (%)
NSNR	50.00
NSYR	59.10
YSNR	36.40
YSYR	27.30

scoring the highest mark. Thus, it shows that NSYR approach encourages subjects to perform and enhance the quality of the program code.

Based on the results in this section, this study reveals that pair rotation contributes significantly toward improving programming quality. This is because knowledge transfer among programmers exists when the rotation between the driver and the navigator takes place. On the other hand, regardless of gender, switching the role will increase the collaboration among programmers in order to achieve the optimal solution, as found in the existing literatures [5, 9, and 14].

In addition, pair rotation acts as peer cross-checking when both programmers share the same experience in problem solving. This can reduce a dominant programmer during the programming tasks. When this happens, knowledge dissemination can be collectively achieved and thus programming quality improved.

5 Conclusion

The purpose of this study is to measure the code quality in programming course by employing an approach that combines knowledge management process and PP technique. The obtained findings can assist educational institutions to determine the best approach in enhancing code quality in programming course. Besides, this study compares each proposed approach to identify the approaches that suit subjects' learning preference. Having compared the approaches, the results suggest that the best approach is NSYR. In addition, the most appropriate approach for female is NSYR, while for male, the most appropriate approaches are NSNR and NSYR. These findings are in line with [9].

Consequently, when designing for programming task, instructors could decide the division of pairs by considering gender effects on the code quality. More importantly, applying SECI process is prospective in ensuring the program quality. Through this study, an initial formation of the KM model and programming technique is contributed to enhance program quality in software engineering education field. This study also contributes to a better understanding of important sharing roles to enhance program quality. Further work to be considered can be a rigorous theoretical formation for constructing other important determinants to enhance program quality because the findings of this research are minimal to SECI model and PP technique only.

Acknowledgments The authors wish to thank the Ministry of Education Malaysia for funding this study under Fundamental Research Grant Scheme (FRGS-SO/CODE:12814).

References

1. Canfora, G., Cimitile, A., Visaggio, C.A.: Lessons learned about distributed pair programming: what are the knowledge needs to address? In: Proceedings of the Twelfth International Workshop on Enabling Technologies: Infrastructure for Collaborative Enterprises, pp. 314–319. IEEE Computer Society, Washington (2003)
2. Md Rejab, M., Omar, M., Ahmad, M.: Knowledge internalization in pair programming practices. J. Inf. Commun. Technol. (JICT). **11**, 163–177 (2013)
3. Chau, T., Maurer, F.: Knowledge sharing in agile software teams. In: Lenski, W. (ed.) Logic Versus Approximation. LNCS, vol. 3075, pp. 173–183. Springer, Heidelberg (2004)
4. Gallis, H., Arisholm, E., Dyba, T.: An initial framework for research on pair programming. In: Empirical Software Engineering, International Symposium, pp. 132–142. IEEE (2003)
5. Kavitha, R.K., Ahmed, M.I.: Knowledge sharing through pair programming in learning environments: an empirical study. In: Education and Information Technologies. pp. 1–15. Springer, US (2013)
6. Nonaka, I., Takeuchi, H.: The knowledge-creating company: how Japanese companies create the dynamics of innovation. Long Range Plan. **29**, 592 (1996)
7. Portel, L., Guzdial, M., McDowell, C., Simon, B.: Success in introductory programming what works? Commun. ACM. **56**, 34–36 (2013)
8. Venkatesan, V., Sankar, A.: Investigation of student's personality on pair programming to enhance the learning activity in the academia. J. Comput. Sci. **10**(10), 2020–2028 (2014)
9. Winkler, D., Kitzler, M., Steindl, C., Biffl, S.: Investigating the impact of experience and solo/ pair programming on coding efficiency: results and experiences from coding Contests. In: Baumeister, H., Weber, B. (eds.) Agile Processes in Software Engineering and Extreme Programming. LNCS, vol. 149, pp. 106–120. Springer, Heidelberg (2013)
10. Fengjie, A., Fei, Q., Xin, C.: Knowledge sharing and web-based knowledge-sharing platform. In: Proceedings of the IEEE International Conference on E-Commerce Technology for Dynamic E-Business, pp. 278–281. IEEE (2004)
11. Omar, M., Romli, R., Hussain, A.: Automated tool to assess pair programming program quality. In: Proceedings of Knowledge Management International Conference Universiti Utara Malaysia, pp. 516–521. (2008)
12. Mazida, A.: An Investigation of Knowledge Creation Processes in LMS-supported Expository and PBL Teaching Methods (Unpublished Doctoral Dissertation): Universiti Sains Malaysia (2010)
13. Wong, W.E., Tingting, W., Qi, Y., Lei, Z.: A crosstab-based statistical method for effective fault localization. In: Proceedings of Software Testing, Verification, and Validation, 2008 1st International Conference, pp. 42–51. IEEE (2008)
14. Srikanth, H., Williams, L., Wiebe, E., Miller, C., Balik, S.: On pair rotation in the computer science course. In: Proceedings of the 17th Conference on Software Engineering Education and Training, pp. 144–149. IEEE (2004)

Social Networks Event Mining: A Systematic Literature Review

Muniba Shaikh, Norsaremah Salleh and Lili Marziana

Abstract Social Networks (SNs) become a major source of reporting new events that happen in real life even before the news channels and other media sources report them nowadays. The objective of this paper is to conduct the systematic literature review (SLR) to identify the most frequently used SN for reporting and analyzing the real-time events worldwide. Furthermore, we recognize the features and techniques used for mining the real-time events from SNs. To determine the literature related to event mining (EM) and SNs, the SLR process has been used. The SLR searching phase resulted 692 total studies from different online databases that went through three phases of screening, and finally 145 papers out of 692 were chosen to include in this SLR as per inclusion criteria and RQs. Based on the data analysis of the selected 145 studies, this paper has concluded that the Twitter micro-blogging SN is the most used SN to repot the events in textual format. The most common features used are n-gram and TF-IDF. Results also showed that support vector machine (SVM) and naive Bayes (NB) are the most frequently used techniques for SNEM. This SLR presents the list of SNs, features, and techniques that are reporting the SN events that can be helpful for other researchers for selection of SNs and techniques for their research.

Keywords Social networks (SNs) · Event mining (EM) · Systematic literature review (SLR) · Social networks event mining (SNEM) · Twitter

M. Shaikh (✉) · N. Salleh
Department of Computer Science, Kulliyah of Information
and Communication Technology, International Islamic University Malaysia,
Kuala Lumpur, Malaysia
e-mail: muniba.shaikh@live.iium.edu.my

N. Salleh
e-mail: norsaremah@iium.edu.my

L. Marziana
Department of Information System, Kulliyah of Information and Communication
Technology, International Islamic University Malaysia, Kuala Lumpur, Malaysia
e-mail: lmarziana@iium.edu.my

© Springer International Publishing Switzerland 2015
A. Abraham et al. (eds.), *Pattern Analysis, Intelligent Security
and the Internet of Things*, Advances in Intelligent Systems and Computing 355,
DOI 10.1007/978-3-319-17398-6_16

1 Introduction

Social Networks (SNs) become a major source of reporting new events that happen in the real life nowadays. No matter the event is related to show business, crisis, emergency, health care, or politics, the SNs report all types of events in no time. SNs have set the new trend of news media. The real-time EM from SNs plays a significant role to provide the event news at first place (as earliest as possible) that is very essential for people from news media, security agencies, emergency services, and other organizations. According to Capurro, D. et al. [1], 65 % people using Internet in USA are also using SNs daily, and Twitter has 500 million users worldwide that use Twitter to update daily events of their life. "Twitter has gained reputation as a way to detect and predict events and user sentiments by observing users posts (tweets) [2]". There is long range of SNs available nowadays that are used for different users and different topics, for example, LinkedIn for professional networking, Sina Weibo micro-blogging for Chinese users mostly, Flicker, PatientsLikeMe for the patients to share the information about their diseases and treatment [1] and many more small-scale SNs. So, the first question arises here is, which SN is used more frequently for reporting the real-time events worldwide and that is also one of the research questions (RQs) of this paper.

The facilities that the SNs provide to users by giving mobile apps, geo-tagging, messengers, mobile uploads of images, hand videos, community sharing, and the common use of smart phones have enhanced the use of SNs in almost every era of research, for instance, health care, surveillance, education, social services, crisis, earthquake, and other natural disasters. SNs are resource heavy therefore require the quick mining of events where the help is required or which possess the national, health threats, or emergency situations [3, 4]. Therefore, the importance of EM of real-time textual data from SNs has been increased recently for different communities.

The aim of the proposed systematic literature review (SLR) is to determine available features of SNs that are essential for further exploration when conducting "event mining (EM)" from SNs. Furthermore, we have investigated the techniques that are used for mining for all type of events reported on SNs and we have also identified SNs that are used mostly by people to report all type of events around the world.

The remainder of the paper focuses on the SLR process to identify relevant studies of SNEM. Section 2 introduces the overview of SLR process. Section 3 presents the results of SLR and the answers of research questions. Finally, Sect. 4 provides the conclusions.

2 The Systematic Literature Review (SLR) Process

SLR is a research process that defines the sequence of activities needed to carry out the literature review in a systematic way. SLR is a tool that aims to answer all RQs based on the analysis of research evidences collected through the systematic search process and analysis [5]. So, this section presents the SLR process.

2.1 Research Questions

One of the core steps of the SLR is to formulate the RQs that are going to be addressed in the review [6]. Table 1 shows attributes of the RQs that are structured according to PICOC. PICOC is the criteria for structuring the RQs that was specified by Petticrew and Roberts [6]. PICOC is based on five attributes: population, intervention, comparison, outcome, and context. In this SLR, the comparison attribute is not applicable since our review will be focusing on the social network event mining (SNEM) (see Table 1). In SLR, all studies that explore EM of SN textual or image meta-data are included. A SLR is needed to find out which SN is used at most by people around the world to report all kinds of events. Furthermore, we demonstrate the techniques and features that have been used in literature for SN data analysis. This section presents systematic review process in SNEM for textual data only.

The SLR aimed to answer the following primary RQ:

1. Primary question:

   - RQ1: Which SN is widely used for EM?
   - RQ2: What techniques and features are used for EM from SNs?

Table 1 PICOC (population, intervention, comparison, outcome, and context)

S. No.	Term	Content or definition
1	Population	SN's information (examples, Tweets, FB posts)
2	Intervention	Event mining (EM)
3	Comparison	Null
4	Outcome	Features, algorithm/technique/matrices/approach, experiment
5	Context	Within the domain of event mining, topic mining, opinion mining, and sentiment analysis

2.2 Identification of Relevant Literature

The strategy used to construct the search terms is as follows [7–9]:

1. Derive *main terms* used in the review questions (i.e., based on the PICOC) (see Table 1).
2. List the keywords mentioned in the articles (based on pilot study).
3. Search for synonyms and alternative words.
4. Join synonyms and *main terms* using Boolean *OR*.
5. Combine *main terms* from PICOC using Boolean *AND*.

So, the result of 5 strategies defined above for constructing the searching string (SS) is presented in Table 2. The SS presented in Table 2 is the initial SS used for searching the literature from different databases such as IEEEXplorer and Springer, but this format (complex query) is not supported by all the databases. Therefore, the simpler SS has been used for literature retrieval instead of the SS defined in Table 2. So, the final SS used for searching literature is (("event mining" OR "event detection") AND ("social network" OR "social media" OR Facebook OR Twitter OR micro-blogging)). Once the SS string is finalized, the next step is to start the *search process*. The search process starts by the selection of online databases. So, in this research, the selected 4 online databases are as follows: IEEEXplorer, ACM Digital library, ScienceDirect, and Springer. The selection of online databases has been a result of the pilot study that index EM and SN studies. The online databases have been chosen based on pilot study done. Furthermore, online search engines such as Google and "Google Scholar" have also been used for searching the relevant literature.

2.3 Study Selection Criteria

Inclusion Criteria's

Only studies that contain the PICOC terms or are related to at least one of the RQs will be considered for inclusion in SLR. This research includes studies that investigate the use of SN's information for mining useful data (events or patterns).

Table 2 Concatenation of all possible terms by using Boolean *OR* and *AND*

(SN's information OR social networks OR social networking OR social networking sites OR Facebook OR Twitter OR social media OR social network analysis) AND (event mining OR event detection OR mining OR opinion mining OR sentiment analysis OR topic mining)
AND (features, algorithm OR technique OR matrices OR approach OR experiment)

Furthermore, the concern of this SLR is to identify, which SN is used at most to report daily events and important features, techniques used in EM from SNs.

Exclusion Criteria's

ExC1. Studies that does not match the RQ.
ExC2. Studies written in a language other than English.
ExC3. Publications on video surveillance or video as data source will not be included.
ExC4. Abstract papers, tutorials, posters, presentations, proceedings, and program of scientific events, or editorial letters.
ExC5. Studies that do not contain the data analysis and results.

2.4 Study Selection Process

Preliminary Selection Process During the initial selection process, we first perform screening of the titles and abstracts to see the relevance of the sources. The titles and abstracts are extracted and compiled into a list (for example in an Mendeley), and for those that are found relevant, the full paper is retrieved. Full papers will only be obtained if they meet the minimum requirement of the inclusion criteria [10].

Final Selection Process In the final selection process, the researcher will review paper details. During this phase, hard copies of the selected paper will be obtained and thorough reading will be performed for each study. During this time, any paper which does not fulfill the inclusion criteria mentioned above will be discarded, and the reference in the Mendeley library will be updated accordingly.

2.5 Study Quality Assessment Checklist

In assessing the quality of studies, we developed a checklist which consists of six questions pertaining to the quality aspect of an article. The following checklist (Table 3) was adapted from [5, 7] to be used when evaluating the research. In the quality checklists, the "yes" answer is given only if the criterion of the question is being met (Score = 1). If the criterion is not met, then the "no" answer will be given (Score = 0). "Partially" answer will be given if the study answered the question partly (Score = 0.5). However, if the paper does not give clear or enough information for us to decide, then it will be remarked as "not reported." In this assessment, the higher the points obtained by a paper, the better will be its quality. The resulting total quality score for each study ranges between 0 (very poor) and 10 (very good). However, the studies with lower score will not be excluded. The quality score is only an indicator of whether a study is highly reliable or not since the information is useful during the evidence synthesis.

Table 3 Study quality checklist

#	Items	Quality score
1	Is it about event mining using SNs?	Yes/no/partially
2	Does article has described that particularly which SN has been chosen for reporting events?	Yes/no/partially
3	Are the aim(s) of study clearly stated? [5]	Yes/no/partially
4	Does the article have stated a valid research question? [5]	Yes/no/partially
5	Does the article describe the techniques and features used for SNEM?	Yes/no/partially
6	Was the formulation of the data analysis well conveyed? [5]	Yes/no/partially

2.6 Data Extraction Strategy

The data extraction form is designed in order to carry out the "data extraction process" [5]. After the final selection of papers, data extraction will be carried out on all papers that passed the screening process. During this stage, all important information based on RQs was extracted and placed in the data extraction form. The extracted data filled in data extraction form is further analyzed for answering the RQs.

3 Results

3.1 Introduction or Synthesis of Data

This section presents the synthesis of evidence found as result of searching process of SLR. The initial phase of the search process identified 692 studies using the final SS. Among 692 studies, 444 studies were found from Springer database by running the "final SS," 140 studies were retrieved from ScienceDirect, ACM resulted 41 studies, and 67 studies found from IEEEXplorer by running the same SS. Now, the study selection process defined in Sect. 2.4 has been followed. As mentioned earlier, in various cases, the same publication can be extracted from more than a single database [10]. Therefore, automatic (using Mendeley duplication removal option) and manual removal of publications with the same title was carried out. Afterward, the exclusion criteria described in Sect. 2.5 were applied, resulting in a dramatic reduction of the number of publications. As a result of "primary and final selection process" defined in Sect. 2.4, total of 145 papers were retrieved. The resulting 145 papers have been included in this SLR, and the next step is to check the quality score. Only two studies were found to achieve very low score as they were survey reports and book chapter on "text mining techniques," the remaining studies (90 %) achieved above average quality scores, and a lesser number (10 %)

Table 4 List of SNs that reports all textual EM

Name of SNs	Total studies	List of studies
Twitter	101	S1, S2, S3, S5, S10, S8, S9, S11, S13, S14, S15, S16, S18, S19, S20, S39, S21, S23, S27, S28, S29, S30, S31, S32, S33, S34, S35, S36, S37, S38, S40, S41, S43, S68, S69, S70, S47, S48, S49, S54, S58, S60, S61, S63, S64, S66, S67, S71, S72, S74, S75, S76, S77, S79, S80, S82, S83, S86, S87, S88, S89, S91, S93, S95, S96, S98, S100, S102, S103, S104, S105, S107, S108, S109, S112, S114, S118, S119, S120, S121, S122, S123, S125, S126, S127, S128, S129, S130, S131, S132, S133, S134, S135, S138, S139, S140, S141, S143, S144, S145
Flickr	10	S6, S17, S42, S50, S51, S116, S114, S136, S137, S138
Sina Weibo	9	S20, S25, S46, S78, S81, S90, S99, S106, S117
YouTube	6	S92, S114, S116, S136, S137, S138
Facebook	6	S1, S19, S86, S100, S112, S113
Meetup	2	S85, S111
Instgram	1	S5
Plurk	1	S45
LinkedIn	1	S19

of studies were ranked good and very good. Finally, data extraction process has been started as per data extraction form and filled in it for further data analysis.

3.2 Answers of Research Questions

This paper described a SLR process including the protocols of SLR that investigated studies of EM from SN content. The SLR identified the answers of RQs that are as follows:

RQ1: Which SN is widely used for EM?

According to our SLR, micro-blogging SN has been used as major source of reporting all real-time (text) events and hot topics as compared to other SNs because of its format and popularity. Furthermore, the micro-blogging data are easily accessible as Twitter API allows us to extract the real-time data and so analysis can be performed to generate early warnings of events. Among micro-blogging SNs, Twitter is most popular, as it is reported by maximum number (101 out of 145) of studies. SW is the Chinese micro-blogging SN that is mostly used in china. However, Plurk is the most popular micro-blogging SN in Taiwan.[1] So as mentioned

---

[1]http://www.alexa.com/siteinfo/plurk.com.

in Table 4, 101 studies out of 145 use Twitter as the main source of data collection. Apart from micro-blogging, other SNs such Flickr, Meetup, Facebook, and YouTube are also used for textual EM as shown in Table 4. Table 4 also represents that Flickr is used in 10 studies, Sina Weibo in 9 studies, YouTube in 6 studies, Facebook in 6 studies, Meetup in 2 studies, and Instgram, LinkedIn, and Plurk in 1 study each.

This paper focuses only on English language so not many papers are found that use SW and Plurk because they are written in Chinese language (language barrier). The few studies on SW and Plurk that are included in our SLR are because the studies were written in English. Furthermore, Flickr is used for image data, YouTube is used for videos, and Meetup is a mobile application used for meetings and office appointments. The reason why we have included few papers on Flickr and YouTube is the use of the meta-data (textual information of images and videos, e.g., titles, tags, and spatio-temporal features). Furthermore, a part of SNs news Web sites, blogs, RSS feeds, Wikipedia are the major source of for EM data. But our focus is on SNs, textual data in English language. Therefore, the studies using images (Flickr), Videos (YouTube), blogs, and news Web sites have been not included.

RQ2: What techniques and features are used for EM from SNs?

As per the findings of our research, the techniques that are mostly used in this era are naive Bayes (NB) and support vector machine (SVM) classifiers that are used in 34 studies. The features used for SNEM include n-gram or co-occurrence of words (reported in 33 studies), term frequency–inverse document frequency (TF-IDF) (reported in 23 studies) are used mostly. Besides that "tokenization" is reported in one study and is also "hash tags" has been reported in one study. However, **Jaccard** similarity measure is reported by two studies, bag of word in four and part of speech in ten studies.

4 Conclusions

This paper has described the use of the SLR process and protocols of SLR in detail, which enabled us to find studies reporting for EM of SN textual content only. The studies on EM of SN textual content and the analysis of 145 papers found as a result of SLR were useful to answers the RQs. This SLR determined that 101 studies out of 145 use Twitter (SN) to mine and analyze the daily life events, so Twitter is a SN that is used more frequently for reporting, mining, and analyzing textual events as compared to other SNs. According to this SLR, the other SNs that have been used for SNEM are Flickr, Sina Weibo, YouTube, Facebook, Meetup, Instgram, Plurk, and LinkedIn. It also been concluded by this SLR study that Sina Weibo is the highest used SN in China but mostly in Chinese Language, and this SLR included only 9 studies using Sina Weibo which are written in English. The aim of this paper was to identify the useful features and techniques used for SNEM and analysis.

Based on our data analysis, NB and SVM are the most used techniques. However, the highly used features for SNEM are n-gram and TF-IDF.

So, as a result, Twitter is used at most for English textual SNEM as compared to other SNs, and for Chinese language textual SNEM, Sina Weibo is highly used. Furthermore, it can be concluded that techniques used for SNEM are SVM and NB, and features are n-gram and TF-IDF.

Acknowledgments This research was partially funded by the Ministry of Higher Education Malaysia under RAGS research grant (RAGS12-001-0001).

References

1. Capurro, D., Cole, K., Echavarria, MI., Joe, J., Neogi, T., Turner, A.M.: The use of social networking sites for public health practice and research: a systematic review. J. Med. Internet Res. (JMIR) **16**(3) (2014)
2. Sakaki, T., Okazaki, M., Matsuo, Y.: Earthquake shakes twitter users: real-time event detection by social sensors. In: Proceedings of the 19th International Conference on World Wide Web, WWW '10, pp. 851–860. ACM, New York (2010)
3. Lamb, J., KR, Puskar, Tusaie-Mumford, K.: Adolescent research recruitment issues and strategies: application in a rural school setting. J. Pediatr. Nurs. **16**(1), 43–52 (2001)
4. Ginsberg, J., Mohebbi, M.H., Patel, R.S., Brammer, L., Smolinski, M.S., Brilliant, L.: Detecting influenza epidemics using search engine query data. Nature **457**, 1012–1014 (2009). doi:10.1038/nature07634
5. Ibrahim, M.A., Salim, N.: Opinion analysis for twitter and Arabic tweets: a systematic literature review. J. Theor. Appl. Inf. Technol. **56**(3) (2013)
6. Petticrew, M., Roberts, H.: Systematic Reviews in the Social Sciences: A Practical Guide. October
7. Kitchenham, B., Charters, S.: Guidelines for performing systematic literature reviews in software engineering. Technical Report EBSE 2007–001, Keele University and Durham University Joint Report (2007)
8. Salleh, N.: Protocol for systematic review of pair programming. Available on https://www.cs.auckland.ac.nz/norsaremah/research.htm (2008)
9. Salleh, N., Mendes, E., Grundy, J.: Empirical studies of pair programming for cs/se teaching in higher education: a systematic literature review. IEEE Trans. Soft. Eng. **37**(4), 509–525 (2011)
10. Spolaôr, N., Cherman, E.A., Metz, J., Monard, M.C.: A systematic review on experimental multi-label learning. Technical Report 392, ICMC-USP, São Carlos—SP (2013)

Personalized Learning Environment (PLE) Experience in the Twenty-First Century: Review of the Literature

Che Ku Nuraini Che Ku Mohd and Faaizah Shahbodin

Abstract In the fast-changing world of the early twenty-first century, education is also changing. The use of ICT in education lends itself to more student-centered learning settings. Given this changing landscape of teacher education, the purpose of this chapter is to explore new educational approaches to enhance teachers' ICT capabilities in the twenty-first-century learning environment. The literature indicates a brief explanation of twenty-first-century education about the roles of (i) student, (ii) teacher, (iii) curriculum, (iv) classroom, and (v) information and communication of technology (ICT). The new approach in education nowadays is introduced, which is personalized learning environment (PLE). PLE enables learners to organize their learning, provides the freedom to choose content, and allows communication and collaboration with others easily. In conclusion, the chapter concludes with recommendations for continued improvements in twenty-first-century education in order to ensure the opportunities of higher education remain open to as many students as possible.

Keywords Education · Information literacy · Twenty-first century · Technology · ICT

1 Introduction

New technologies have significantly entered our lives, and online services offer the opportunity for sustainable regional development. Electronic services offered by the new information and communication technologies (ICT's) have proved as an

C.K.N.C.K. Mohd (✉) · F. Shahbodin
Faculty of Information and Communication Technology, Department of Interactive Media, Universiti Teknikal Malaysia Melaka, Hang Tuah Jaya, 76100 Durian Tunggal, Melaka, Malaysia
e-mail: cknuraini@email.com

F. Shahbodin
e-mail: faaizah@utem.edu.my

© Springer International Publishing Switzerland 2015   179
A. Abraham et al. (eds.), *Pattern Analysis, Intelligent Security and the Internet of Things*, Advances in Intelligent Systems and Computing 355, DOI 10.1007/978-3-319-17398-6_17

important tool in efforts to disseminate e-learning in modern education [1]. Given advancements in Web technology, global agencies, organizations, and publishers began proposing and promoting the use of standards for representing e-learning content associated with e-learning systems or educational content [2]. E-learners can metaphorically be considered as "organisms" in a virtual learning environment. Their navigational behavior can be construed as movement directed by some factors to enable them to achieve the learning goals [3]. The Internet had great impact on e-learning due to the fact that it is an effective and economical medium for making information available to dispread individuals [4]. The general aim of e-learning platforms is to provide information and practical opportunities for students in order to help them to increase their knowledge and skills on a particular topic [5]. The world is changing rapidly in a lot of ways, but the dominant change is in ICT. Knowledge of ICT is very important, especially for teachers and students because without a good knowledge, it will be a constraint in implementing information literacy in teaching and learning [6].

ICT can help deepen students' content knowledge, engage them in constructing their own knowledge, and support the development of complex thinking skills [7–9]. The various kinds of ICT products available and having relevance to education, such as teleconferencing, email, audio conferencing, television lessons, radio broadcasts, interactive radio counseling, interactive voice response system, audio cassettes, and CD ROMs, have been used in education for different purposes [10–12]. The ability to manage and deliver online courses has become an important aspect of the learning models, and this importance has created a tremendous dependency upon e-learning systems as educators strive to deliver quality education to their learners [4].

According to McLoughlin and Lee [13], digital-age students want an active learning experience that is social, participatory, and supported by rich media. Despite attempts by institutions of higher education to harness technology to facilitate learning through online courses, college students more frequently drop out of online courses than they do traditional, face-to-face courses [14]. The concept and application of e-learning has become progressively more prevalent in educational settings ranging from modern post-secondary institutions to the smallest and the most remote rural schools; in addition, e-learning systems now have an integral role in many educational organizations [4].

The emergence of ICT has made it possible for teachers and students to collaborate with each other in diverse ways [15]. Today, the dimension emphasized in the definition of the concept of education is the process of assisting students in acquiring the skills to access and use information more than conveying the knowledge from teacher to student. This traditional method not only fails to meet the needs of modern society, but also excludes or at least neglects adult education which is emphasized in the informal training process, but excluded at the definition level should be expanded to include lifelong and unlimited education [16]. Today, it is vital to design different materials in different teaching environments and to use them for different purposes.

2 Personalized Learning Environment (PLE)

Many techniques have been applied in e-learning application in order to predict the learning route based on the learning knowledge behavior [17]. This technology-based learning technique means, among several things, accessing through the Internet short training modules that cover specific topics available as self-study [18]. PLEs appeared as a new construct in the e-learning literature which finds its support of social media and steadily gains ground in the e-learning field as an effective platform for student learning [19]. PLE is not only a social landscape, but also a personal space which belongs and controlled by the learner [20]. According to the concept of a "social hub," the social learning management system of the study program focuses on connecting students' PLEs [21]. As instructors and instructional designers move toward personalized learning with the hope of increasing learner motivation and ultimately learner achieving, research on best practices for using technology to successfully accomplish this must be explored because in its current state, the research in these areas is limited [22].

They are not only to seek information but also to share information by taking advantage of digital and networked technologies [23]. In opposition to obsolete learning theories and concepts, modern- and learner-centered concepts and approaches such as personalized learning environments (PLE) and connectivism have emerged [24]. The increase of personal computing technologies, primarily Web 2.0 technologies, has made it easier for learners to create their own learning systems [25]. However, if students are not clear with their learning goals and are uncertain how to appropriate, relevant technologies to achieve these goals, an effective PLE would not occur at all [23]. PLE uses many content sources, applications, and tools for qualified learning. In fact, PLE is often used in our online lives unintentionally. People may use PLEs for formal and informal learning, sharing, communicating, and collaborating with others. Social networks, bookmarks, start pages, blogs, etc., all can be considered components of PLE. Furthermore, PLE is useful for:

- Socializing with other learners;
- Customizable content; and
- Different, easy, and interactive way for learning.

The personalized e-learning system helps teacher to save a lot of time for learning and helps the teacher to examine the learning progress of students [17]. PLEs need on the one hand to focus on technical issues, regarding information exchange between services and user interface problems [26]. The online environment is one application that has been important for the development of connectivism. PLE is based on a connectivism and design with connectivist principles. PLEs and connectivism share some common traits. Common principles of connectivist learning and PLEs are as follows:

- Diversity;
- Autonomy;
- Interaction/Connectedness; and
- Openness.

2.1 SymbalooEDU—the Personal Learning Environment Platform

SymbalooEDU is created in 2010. It is an educational version of the original Symbaloo application founded in Holland in March 2007. It is a software application that enables learners to organize, integrate, and share the online content in one setting or personal learning environment (PLE). The platform also allows educators to create mixes of tailored resources and share these mixes with students. Once the Web sites are shared, students can integrate them into their own SymbalooEDU PLE, where they are free to use, add and share content with their peers and tutors. SymbalooEDU works by enabling users to simply construct customizable tiles which are linked to URLs of online resources. Once the user has created a grid of tiles or Web mix, it can be shared with others via email. Figure 1 shows the SymbalooEDU environment.

2.2 Liferay Portal Community Edition, (http://www.liferay.com)

Liferay portal was created in 2000 and boasts a rich open-source heritage that offers organizations a level of innovation and flexibility unrivaled in the industry. The main objective of the portal is to enable live discussions with native speakers and personal practice at transition between jobs, thanks to the integration in the PLE of tools that can exchange data, such as online dictionaries, pronunciation, microblogging, video conferencing, multimedia, and discussion tools. In a political context where social sites are often blocked, a PLE in which services can easily be replaced by equivalent non-blocked ones is essential. The integration with mobile phones is also important as part of the activities is carried out at a distance. Figure 2 shows the screenshot of the PLE used in a French as a second language class at the Shanghai Jiao Tong University.

Fig. 1 SymbalooEDU environment

Fig. 2 Screenshot of the PLE used in a French as a second language class at the Shanghai Jiao Tong University

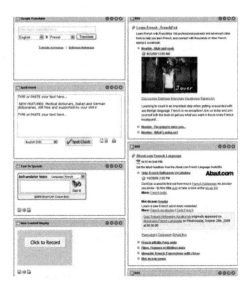

2.3 The Hybrid Institutional-Personal Learning Environment (HIPLE)

HIPLE context is an online course on e-Government. It is introduced by Peña-López [27]. There are three characters in HIPLE. There is a character (ONcampus) which is a student that, for unspecified reasons, just wants to access the virtual campus to study and that everything that happens on the campus remains unknown for the outer world. There is a second character (ictlogist) that is also a student and uses several Web 2.0 tools for learning (call it a Personal Learning Environment or PLE), among them Twitter, and just does not want to use *two* nanoblogging tools, one on-campus and another one off-campus. A third character (OFFcampus) is a professional working on e-Government such as use Twitter to interact with other people on the field. Figure 3 shows the screenshot of the hybrid institutional-personal learning environment (HIPLE). Basically, there are two conversations:

- Inside the campus, a closed conversation that neither benefits from "outside" conversations nor contributes to them. Including the student remaining unknown to other people on the field.
- Outside campus, an open, but not-permeating-the-campus conversation and that forces some people attend *two* conversations on the same field, mostly with different people but similar purposes.

Fig. 3 Screenshot of the
hybrid institutional-personal
learning environment
(HIPLE)

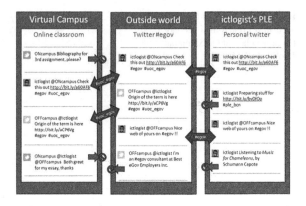

2.4 The PLExus Prototype

PLExus prototype is created by Kolås and Staupe [28]. PLExus provides a student
interface allowing customized views of learning objects and learning activities
based on pedagogical method, media type, learning objective type, proficiency
stage, etc. In a pedagogical-based PLE-like PLExus, the student is able to customize
the learning environment. This requires that learning objects (LO) and learning
activities (LA) are saved and retrieved in such a manner that one student could
reach the learning objective through a presentation, while other students reach the
same learning objective through example discovery, demonstration, or collaboration [29]. Figure 4 shows the conceptual model of PLExus. The conceptual model is
built around the use of topic maps, since the topic maps are suitable as the core of a
powerful PLE with information administration, search, and navigation as important
components.

Fig. 4 The conceptual model
of PLExus

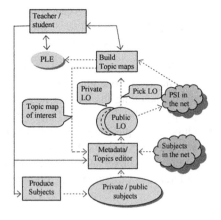

3 Twenty-First Century of Education in Malaysia Context

The term "twenty-first" has become an integral part of educational thinking and planning for the future. However, despite learning about the skills that students will need to develop to become successful in the twenty-first century, as well as what beliefs about education may be worth hanging onto or throwing away, schools and teachers are left trying to figure out what their role needs to be in the education of their twenty-first-century students. In the twenty-first century, teachers have to develop their systematic thinking by coordinating the various components for creating new learning environments which are included curriculum, content, students, teaching and assessment methods, and technology [30]. In this context, education systems, besides constantly developing to meet the needs of the current era, have been obliged to focus on the future and to go beyond the needs of the current era.

Malaysian students are below par when compared to their contemporaries in other countries, acknowledged Education Minister II Datuk Seri Idris Jusoh. Although literacy rates were rising in Malaysia, it was vital to assess and compare the Malaysian education system against international standards. During the 18th Malaysian Education Summit, literacy rates are raised in Malaysia, and it is vital to assess and compare our education system against the international standards. Out of 74 countries, Malaysia ranked in the bottom third in the Program for International Student Assessment (PISA) 2009+. This is below the international and OECD (Organization for Economic Co-operation and Development) [31].

Idris stated that the need for the Education Blueprint is justified in the context of raising international standards, the government aspiration of better preparing Malaysian children for the needs of the twenty-first century, increased public and parental expectations of education policy. The Higher Education Blueprint will also be introduced in order to ensure consistency with the primary and secondary education system and allow for seamless progression in terms of educational offerings, opportunities, and advancement. The Higher Education Blueprint will address challenges such as empowering university governance, democratizing to higher education, and improving graduate employability. Malaysia Education Blueprint (MEB) has offered a vision of the education system and students' aspirations that Malaysia both needed and deserved and outlined eleven strategic and operation shifts that would be required to achieve that vision.

Education is a key area that is crucial toward achieving the country's aspirations of becoming a high-income, knowledge-based nation by 2020. There is no doubt that new educational technologies are always charged with exciting pedagogical properties and there is an understanding of the type of knowledge learners ideally need to develop in the twenty-first century. Realizing the future that we want for our national education, the Ministry has introduced 11 shifts to transform our education system. These shifts include introducing new initiatives and strengthening existing ones. Each shift complements and supports the 5 aspirations of the Education Blueprint, which are access, quality, equity, unity, and efficiency.

3.1 Student

The focus of student learning in this classroom is different. The focus is no longer on learning by memorizing and recalling information, but on learning how to learn. Now, students use the information they have learned and demonstrate their mastery of the content in the projects they work on. Students learn how to ask the right questions, how to conduct an appropriate investigation, how to find answers, and how to use information. The emphasis in this classroom is on creating lifelong learners. With this goal in mind, students move beyond the student role to learn through real-world experiences.

Teachers will use a variety of performance-based assessments to evaluate student learning. Tests that measure a student's ability to memorize and to recall facts are no longer the sole means of assessing student learning. Instead, teachers use student projects, presentations, and other performance-based assessments to determine students' achievement and their individual needs. The goal of the twenty-first century is to prepare students to become productive. As lifelong learners, they are active participants in their own learning. They seek out professional development that helps them to improve both student learning and their own performance.

3.2 Teacher

Teaching in the twenty-first century has to require an emphasis on understanding how to use information technologies. Teachers need to instruct students on computer usage, to legitimate methods of Internet research, and how to identify useful information. Additionally, this focus on technology can open up a world of new resources to support traditional teaching methods, such as the incorporation of software programs in the classroom. The new approaches, such as focusing on thinking skills rather than technical skills and providing various contexts different from ordinary classroom lessons, help teachers to develop adaptive expertise [30].

Teachers are no longer teaching in isolation. The teacher, the school, and the textbook companies can individualize instruction for the different types of learners [23]. The teachers also have been equipped to face the challenges and complexities of the teaching and learning in the twenty-first century; and what directions should be taken to better prepare the new generation of teachers [30]. They now co-teach, team teach, and collaborate with other department members. Teachers know that they must engage their students in learning and provide effective instruction using a variety of instructional methods as well as technologies. To do this, teachers keep abreast of what is happening in the field. As lifelong learners, they are active participants in their own learning. The new direction of ICT education for teachers lies primarily in the development of a set of adaptive and transferable knowledge and skills, so that teachers are better able to adapt to the challenging and complex nature of future learning environments [30]. However, effective school reform begins by taking existing practice as a way of tapping into what motivates teachers as a starting point for change [32].

3.3 Curriculum

The curriculum must become more relevant to what students will experience in the twenty-first-century workplace. To develop intentional learners, the curriculum must go beyond helping students gain knowledge for knowledge's sake to engaging students in the construction of knowledge for the sake of addressing the challenges faced by a complex and global society. Teachers today are stressed by the current state of affairs in education, and many feel that they do not have the time to design and deliver a twenty-first-century curriculum. They feel pressured to teach to the test, putting their students and themselves through a regimen of memorizing huge amounts of facts so that they can pass their standardized tests.

According to the Ulriksen's concept, "implied student" to be a useful one because it allows us to acknowledge that we make many assumptions about students, what they will be like, what they will know, how they will learn, and how they will interact [33]. Different program structures and modes of study are associated with different understandings of the implied student and we base important decisions about curriculum upon those assumptions. Students are expected to draw on various knowledge bases, integrate them, conduct increasingly more sophisticated analyses as they progress through college, and use their integrated knowledge to solve complex problems.

3.4 Classroom

In the twenty-first-century classroom, teachers are facilitators of student learning and creators of productive classroom environments, in which students can develop the skills they might need at present or in future. An interactive teacher is by definition one that is fully aware of the group dynamics of a classroom. As Dörnyei and Murphey [34] explained, the success of classroom learning is very much dependent on:

- How students relate to each other and their teacher;
- What the classroom environment is;
- How effectively students cooperate and communicate with each other; and
- The roles not only the teacher plays, but also the learners engage in.

According to Harmer [35], the term "facilitator" is used by many authors to describe a particular kind of teacher, one who is democratic (where the teacher shares some of the leadership with the students) rather than autocratic (where the teacher is in control of everything that goes on in the classroom), and one who fosters learner autonomy (where students not only learn on their own, but also take responsibility for that learning) through the use of group and pair work and by acting as more of a resource than a transmitter of knowledge.

3.5 Information, Media, and Information Communication and Technology (ICT) Literacy

E-learning may therefore be a tool for direct transmission of knowledge, without spatial limitations, knowledge that is needed to formulate the philosophy toward all crises which follow one another in the early twenty-first century [1]. Information technology is undergoing a technological revolution that is very fast. This is because the technology has become a media medium to deliver information and communication, especially in teaching and learning in this cyber era. Programs that are largely ICT skill-based are unlikely to prepare pre-service teachers to learn how to deal with the problem of complexity-making intimate connections among content, pedagogy, and technology [36]. If the student refuses and did not follow any course on information technology, they are not likely to know how to use the latest information technology tools [6]. The nature of technologies for teaching and learning has become increasingly social, collective, and multi-modal since the emergence and rapid adoption of Web 2.0 and cloud technologies [30]. At the same time, technology transfers some responsibility for learning for students [20]. Al-Khasawneh et al. [37] reported that the use of the Internet has contributed to education, such as providing the opportunity to improve quality and to study in a broader context. Peters [38] noted that the progress of the Internet has brought positive changes to the way teachers' teach, students' learn and communicate. Internet revolution does not only find information globally, it even forges closer ties between human to communicate.

However, any expectation that teachers would or could change to constructivist practices is problematic, because it was based much less on evidence than on wishful thinking and speculation [39, 40]. Many factors simultaneously influence teaching practice, which means that predicting change in this practice can never be a completely certain affair. ICTs have been in schools for a number of years, and teachers' lack of constructivist practices with ICT can now also be interpreted as a disconnection between the theoretical conceptualizations of how ICT should be used in schools and the day-to-day reality of teaching with ICT [41]. Livingstone [42], states in the application of ICT in teaching and learning, knowledge and skills are of key importance. This is because without the knowledge and techniques in the search of information resources, information literacy cannot be applied in teaching and learning. Educators and students have to turn to ICT, particularly the Internet to enable them to become independent thinkers and effective decision makers [43].

4 Impact on Education in Malaysia

Primary and secondary school education standards in Malaysia need to improve, particularly so in bridging the gap between urban and rural areas. This is to ensure that access comes together with quality education of international standards. The challenge was producing knowledgeable, competent, and globally competitive human capital. The solution to this is the Malaysia Education Blueprint (MEB) 2012–2025, which was launched as well as the soon to be released National Education Blueprint for Higher Education 2015–2025 (Higher Education Blueprint).

Malaysia has consistently demonstrates high levels of expenditure on education. This has resulted in almost universal access to primary education and significant improvements in access to secondary education. However, the current review has shown that we need to invest on factors that have the highest impact on student outcomes. The rapid penetration of increasingly sophisticated technologies into every facet of society is causing significant shifts in how, when, and where we work, how individuals, companies, and even nations understand and organize themselves, and how educational systems should be structured to prepare students effectively for life in the twenty-first century [44].

5 Conclusion

The new role of the teacher in the twenty-first-century classroom requires changes in teachers' knowledge and classroom behaviors. If students are to be productive members of the twenty-first-century workplace, they must move beyond the skills of the twentieth century and master those of the twenty-first century. Teachers are entrusted with mastering these skills as well and with modeling these skills in the classroom. The characteristics of the twenty-first-century classroom will be very different from those in the classrooms of the past because the focus is on producing students who are highly productive, effective communicators, inventive thinkers, and masters of technology.

It would most likely require a shift in emphasis within the syllabus, from specific content knowledge toward non-cognitive outcomes, values, and citizenship education, as well as a strong emphasis on informal learning [45]. Professional development in support of these constructivist practices would also need to allow for the depth and complexity of teachers' commitment to their current approaches to teaching.

Acknowledgments I would like to take this opportunity to thank the anonymous reviewers for their thoughtful and careful feedback. This work was supported by short grant PJP/2013/FTMK (23D)/S01265, Universiti Teknikal Malaysia Melaka (UTeM).

References

1. Ioannis V. Kirkenidis and Zacharoula S. Andreopoulou, "e-Learning and the aspect of students in forestry and environmental studies", *Journal of Agricultural Informatics*, 2015 Vol. 6(1) pp. 80–87. Hungarian Association of Agricultural Informatics European Federation for Information Technology in Agriculture, Food and the Environment, 2015
2. Al-Yahya, M., George, R., Alfaries, A.: Ontologies in E-learning: review of the literature. Int. J. Softw. Eng. Appl. **9**(2), 67–84 (2015). ISSN: 1738-9984 IJSEIA. Copyright © 2015 SERSC
3. Bagarinao, R.T.: Students' Navigational pattern and performance in an e-learning environment: a case from UP Open University, Philippines. Turk. Online J. Dist. Educ. (TOJDE) **16**(1), Article 7 (2015). ISSN 1302-6488
4. Palanivel, K.. Kuppuswami, S.: Towards service-oriented reference model and architecture to e-learning systems. Int. J. Emerg. Trends Technol. Comput. Sci. (IJETTCS) **3**(4), 146–155 (2014). ISSN 2278-6856
5. Karadimas, N.V., Karamanoli, V.: E-Psychology in research level using asynchronous e-learning platform in military environment. In: Recent Advances in Electrical Engineering and Educational Technologies (2014). ISBN: 978-1-61804-254-5
6. Ruhizan, M.Y., Norazah, M.N., Mohd Bekri, R., Faizal, A.N.Y., Jamil, A.B.: Vocational education readiness in Malaysia on the use of E-portfolios. J. Tech. Educ. Train. (JTET) **6**(1), 57–71 (2014). ISSN 2229-8932
7. Kozma, R.: National policies that connect ICT-based education reform to economic and social development. Human Technol. **1**(2), 117–156 (2005)
8. Kulik, J.: Effects of Using Instructional Technology in Elementary and Secondary Schools: What Controlled Evaluation Studies Say (Final Report No. P10446.001). SRI International, Arlington, VA (2003)
9. Webb, M., Cox, M.: A review of pedagogy related to information and communications technology. Technol. Pedagogy Educ. **13**(3), 235–286 (2004)
10. Bhattacharya, I., Sharma, K.: India in the knowledge economy—an electronic paradigm. Int. J. Educ. Manage. **21**(6), 543–568 (2007)
11. Sanyal, B.C.: New functions of higher education and ICT to achieve education for all. Paper prepared for the Expert Roundtable on University and Technology-for Literacy and Education Partnership in Developing Countries, International Institute for Educational Planning, UNESCO, Paris, Sept 10–12 2001
12. Sharma, R.: "Barriers in using technology for education in developing countries. Singap. Schools' Comput. Educ. **41**(1), 49–63 (2003)
13. McLoughlin, C., Lee, M.: Personalised and self regulated learning in the Web 2.0 era: International exemplars of innovative pedagogy using social software. Australas. J. Educ. Technol. **26**(1), 28–43 (2010)
14. Hart, C.: Factors associated with student persistence in an online program of study: a review of the literature. J. Inter. Online Learn. **11**(1), 1541–4914 (2012)
15. Yunus, M.M., Salehi, H., Chenzi, C.: Integrating social networking tools into ESL writing classroom: strengths and weaknesses. Eng. Lang. Teach. **5**(8), 42–48 (2012)
16. Arkün, S., Akkoyunlu, B.: A study on the development process of a multimedia learning environment according to the ADDIE model and students' opinions of the multimedia learning environment. Int. Educ. Multimedia **17**, 1–19 (2008)
17. Sivakumar, N., Praveena, R., Saranya, S.: Improving content personalization through ant optimization in E-learning. Adv. Nat. Appl. Sci. **9**(6), 581–586 (2015). © 2015 AENSI Publisher
18. Lopez, E.O., Morales. G.E., Hedlefs. I., Gonzalez, C.J.: New empirical directions to evaluate online learning. Int. J. Adv. Psychol. (IJAP) **3**(2)

19. Dabbagh, N., Kitsantas, A.: Personal learning environments, social media, and self-regulated learning: a natural formula for connecting formal and informal learning. In: Internet and Higher Education (2012)

20. Mohd, C.K.N.C.K., Shahbodin, F., Pee, N.C.: Exploring the potential technology in personalized learning environment (PLE). J. Appl. Sci. Agric. 9(18), 61–65 (2014)

21. Hölterhof, T., Nattland, A., Kerres, M.: Drupal as a social hub for personal learning. In: Proceedings of the PLE Conference 2012, Aveiro, Portugal

22. Davis, M.R.: Researchers tackle personalized learning. In: Education Week, pp. 30–38 (2011)

23. Mohd, C.K.N.C.K., Shahbodin, F., Pee, N.C., Hanapi, C.: Mapping of personalized learning environment (PLE) among Malaysian s secondary school. In: Proceeding of the International Conference on Advances In Computing, Communication and Information Technology—CCIT 2014. Copyright © Institute of Research Engineers and Doctors. All rights reserved. ISBN: 978-1-63248-010-1

24. Kesim, M., Altınpulluk, H.: The future of LMS and personal learning environments. J. Procedia Soc. Behav (2013). Sci. CY-ICER 2013. ELSEVIER. 2013

25. Johnson, M., Liber, O.: The personal learning environment and the human conditions: From theory to teaching practice. Interactive Learning Environments 16(1), 3–15 (2008)

26. Ullrich, C., Shen, R., Gillet, D.: Not yet ready for everyone: an experience report about a personal learning environment for language learning (2010)

27. Peña-López, I.: Interview: introducing the HIPLE: hybrid institutional-personal learning environment. In: ICTlogy, p. 81, June 2010

28. Kolas, L., Staupe, A.: The PLExus prototype: a PLE realized as topic maps. In: Spector, J.M., Sampson, D.G., Okamoto, T., Kinshuk, Cerri, S.A., Ueno, M., Kashihara. A. (eds.) Proceedings of the 7th IEEE International Conference on Advanced Learning Technologies (ICALT'07), pp. 750–752. IEEE Computer Society Press, Washington, D.C. (2007)

29. Heinich, R., Molenda, M., Russell, J.D., Smaldino, S.E.: Instructional Media and Technologies for Learning, 7th edn. Merrill Prentice Hall, Upper Saddle River (2002)

30. Kim, H., Choi, H., Han, J., So, H-J.: Enhancing teachers' ICT capacity for the 21st century learning environment: three cases of teacher education in Korea. Australasian Journal of Educational Technology 2012, Vol 28 (6), pp. 965–982, 2012

31. The Star Online, Nation.: Malaysian students are below par, says Idris. Published on Friday May 16, 2014. Accessed on March 12, 2015. Retrieved from http://www.thestar.com.my/News/Nation/2014/05/16/Malaysian-students-are-below-par-says-Idris/

32. Fullan, M.: The Six Secrets of Change: What Leaders the Best Do to Help Their Organisations Survive and Thrive. Jossey-Bass, California (2008)

33. Ulriksen, L.: "The implied student. Stud. High. Educ. 34(5), 517–532 (2009)

34. Dörnyei, Z., Murphey, T.: Group Dynamics in the Language Classroom. Cambridge University Press, Cambridge (2003)

35. Harmer, J.: The Practice of English Language Teaching, 4th edn. Longman, Harlow (2007)

36. So, H.J., Kim, B.: Learning about problem-based learning: Student teachers integrating technology, pedagogy, and content knowledge. Australas. J. Educ. Technol. 25(1), 101–116 (2009)

37. Al-Khasawneh, A., Khasawneh, M., Bsoul, M., Idwan, S., Turan, A.H.: Models for using internet technology to support flexible e–learning. Int. J. Manage. Educ. 7(1), 61–70 (2013)

38. Peters, L.: Global Education: Using Technology to Bring The World to Your Student. International Society for Technology in Education, Washington, D. C (2009)

39. Selwyn, N.: From state-of-the-art to state-of-the-actual. Technol. Pedagogy Educ. 17, 83–87 (Introduction to a special issue) (2008)

40. Underwood, J.: Research into information and communications technologies: where now? Technol. Pedagogy Educ. 13, 135–143 (2004)

41. Convery, A.: The pedagogy of the impressed: how teachers became victims of technological vision. Teachers Teach. 15(1), 25–41 (2009)

42. Livingstone, S.: Critical reflections on the benefits of ICT in education. Oxf. Rev. Educ. 38(1), 9–24 (2012)

43. Quigley, M.: ICT ethics and security in the 21st century: new developments and applications. Inf. Sci. Ref., Hershey (2011)
44. McClarty, K.L., Orr, A., Frey, P.M., Dolan, R.P., Vassileva, V., McVay, A.: A Literature Review of Gaming in Education. Pearson's Research Reports (2012)
45. Orlando J.: ICT, constructivist teaching and 21st century learning. CL Curriculum Leadersh. J. **9**(11) (2011). ISSN: 1448-0743
46. Mohd, C.K.N.C.K., Shahbodin, F., Pee, N.C.: Personalized learning environment (PLE) experience in the 21st century: review of literature. In: 4th World Congress on Information and Communication Technologies (WICT14). Paper published by IEEE (2014)
47. Quigley, M.: ICT Ethics and Security in the 21st Century: New Developments and Applications. Information Science Reference, Hershey (2011)

Author Biographies

Che Ku Nuraini Che Ku Mohd

was born on June 7, 1986, at Terengganu, Malaysia. She is a PhD student attached at Universiti Teknikal Malaysia Melaka (UTeM), Melaka, Malaysia. She received her Degree in Computer Science (Interactive Media) in 2008 and a Master of Science in Information and Communication Technology (Multimedia) in 2011. Her research interest is on the multimedia applications, problem-based learning, user interface design, and personalized learning environment.

Faaizah Shahbodin

was the main supervisor for author 1. She was born on September 6, 1970, at Perak, Malaysia. She received her Degree in Computer Science in 1994 from Universiti Utara Malaysia (UUM), and Master in Computer Science in 1997 from Queensland University of Technology (QUT), Brisbane, Australia. She was a researcher, and a project supervisor for several interests is primarily on computers in education projects in UNIMAS, Kolej Latihan Telekom (Kolej Multimedia), and UTeM for 15 years. She completed her PhD in multimedia education systems at University Kebangsaan Malaysia (UKM). Her research interest is on the problem-based learning, multimedia applications, creative contents, and user interface design.

Social Networks Content Analysis for Peacebuilding Application

Muniba Shaikh, Norsaremah Salleh and Lili Marziana

Abstract Peace provides the freedom to express our views, to relate with others people and create cooperation, and social networks (SNs) provide that platform. SNs can play a very important role to improve peacebuilding (Pb) applications as current peace-related studies witness that violence- and Pb-related reports are communicated through different SNs applications. People and victims of the conflicts make use of SNs and its applications to cast their concerns. However, the major setback of these SNs is to manage the huge amount of SNs data and to extract the topic specific (Pb related) information. There is lack of research done on SNCA by Pb perspective. Therefore, the objective of this research is to perform CA, means to identify which (*SN*) what (*data*) how (*to extract*)? Furthermore, what features and techniques should be used for content analysis (CA) of Pb-related data? This research proposes framework for automatic SNs data extraction (DE) and CA to achieve our objective. The proposed framework shows that Twitter is most popular SN for Pb CA purpose and proposed framework presents the searching criteria and custom filters to extract the topic specific data. Moreover, the research proposes to use lexical analysis (LA) method to extract the SNs features, first-order context representation (CR) technique to represent the context of the extracted features, DBSCAN clustering algorithm for data management by making different clusters, ranking algorithm, Log-likelihood ratio, and SVM techniques for CA and classification. The proposed framework aims to help in conducting SNCA to support Pb application in order to take important information from the sea of SNs data to predict violence-related information or incidents that will help peacekeepers for

M. Shaikh (✉) · N. Salleh
Department of Computer Science, Kulliyyah of Information and Communication
Technology—International Islamic University, Selangor, Malaysia
e-mail: muniba.shaikh@live.iium.edu.my

N. Salleh
e-mail: norsaremah@iium.edu.my

L. Marziana
Department of Information System, Kulliyyah of Information and Communication
Technology—International Islamic University, Selangor, Malaysia
e-mail: lmarziana@iium.edu.my

© Springer International Publishing Switzerland 2015
A. Abraham et al. (eds.), *Pattern Analysis, Intelligent Security
and the Internet of Things*, Advances in Intelligent Systems and Computing 355,
DOI 10.1007/978-3-319-17398-6_18

communicating and maintaining peace-related news (may it be natural disaster or manmade terrorism activities) around the world.

Keywords Peacebuilding (Pb) · Social networks (SNs) · Content analysis (CA) · Data extraction (DE)

1 Introduction

Peace is the spirit of mankind's existence; it is a foundation of our survival and achievement as a species. Peace provides us the freedom to express our views, to relate with others people and create cooperation.[1] At the same time, social networks (SNs) provide the platform to connect, meet, share ideas, knowledge, and experience [1]. SNs become one of the major communication channels now a days that millions of people around the world are using SNs in their daily life. According to Dambach [2], 'the ultimate objective of Pb is to reduce and eliminate the frequency and severity of violent conflict'. However, since more than past six decades, many countries of the world are facing conflicts, violence, wars, and other type of these problems. Still there are no specific answers of the questions that what tasks, tools, and techniques peacebuilding (Pb) entails [3]. Many profit and non-profit organizations are working on Pb strategies to overcome these problems. However, there is a lack of research on CA of SNs data to support Pb application. Content analysis (CA) is a method to carry out the research to inspect artifact of social communication. CA is a bundle of procedures that draw a valid inference from the text [4]. Therefore, this research proposes the SNs data extraction (DE) and CA framework to extract Pb-related data reported through SNs and to conduct analysis of the data to predict important patterns. The overall purpose of CA is to extract context from the content that is being analyzed by knowing the fact that '*Who (says) What (to) Whom (in) what Channel and (with) What Effect?*' [5].

Therefore, this research proposes the DE and CA framework that aims to answer the RQs; How to extract SNs topic specific data?, What are the 'searching criteria,' pre-processing methods, what are the important SNs attributes or features?, how to extract context from SN content and finally what techniques should be used for clustering and classification of SN content? The proposed DE and CA framework represents the DE procedure and clustering, context analysis, and representation techniques. These techniques can help in giving context to content in order to take important information to predict Pb- or violence-related information and incidents and, at the same time, can be used by the peacekeepers for communicating early updates (warnings) about incidents may it be natural disaster or manmade terrorism activities to support Pb applications.

---

[1]http://worldunderstandingandpeace.com/2007/11/19/peace-some-thoughts-on-peace-what-is-peace-and-how-important-is-peace-in-our-life/.

The remainder of the paper is organized as follows: Sect. 2 summarizes the literature review; Sect. 3 describes the problem addressed by this research; Sect. 4 presents the proposed DE and CA framework and answers the RQs; and finally, Sect. 5 provides conclusion of this research.

2 Literature Review

2.1 The Role of SNs in Non-profit Organizations

The papers [6, 7] describe that how SNs can be so helpful or useful for non-profit organizations (such as NGOs, social workers, charity organizations) for supporting people. Furthermore, SNs and social media (SM) can be used for large enterprises to connect professionals. Now a days, these technologies are revolutionising company intranets, so these papers conclude that SNs can provide great help for non-profit organizations and for work. Therefore, in this research, the power of SNs is intended to use for Pb purpose besides with other machine learning and ICT tools.[2]

2.2 SN Content Analysis Tool

Ushahidi[3] is a Swahili word that means *testimony* or *witness*; it is non-profit company that develops open source softwares for information collection, visualization, and interactive mapping.[4] Ushahidi is a platform for people to share any information, stories, or evidences that they have got related to crisis, through SMS or email. Ushahidi is most relevant software to our research but the focus of our research and Ushahidi is different, because Ushahidi project focuses moreover on crisis mapping and visualization, as per our focus is on conflict resolution and peacebuilding. However, Ushshidi is a very big and open source project that has many applications and APIs, so we will try to use some of its APIs that are relevant to our research.

3 Problem Statement

As SNs become today's fastest media of communication, they play a vital role in spreading news and information throughout the globe in no time. However, the major setback of these SNs is to manage the huge amount of SNs data and to extract the

---

[2]http://www.simplysmile2012.com/. Last checked on Dec'12.

[3]http://ushahidi.com/about-us. Last checked on Dec'12.

[4]http://en.wikipedia.org/wiki/Ushahidi.

topic-specific (Pb-related) information. There is lack of research done on SNCA by Pb perspective. Therefore, the objective of this research is to perform CA, means to identify which (*SN*) what (*data*) how (*to extract*)? Furthermore, what features and techniques should be used for CA of Pb-related data? Thus the ultimate aim of proposing DE and CA framework is to manage SNs content, cluster, and classify data, as per the topic, nature (positivity, negativity, threat, etc.) that will help in Pb applications and organizations to provide early warnings to people in case of any threat.

4 Proposed Social Networks Content Analysis and Data Extraction Framework

The proposed framework describes CA and DE framework to extract the Pb-relevant data from SN such as Twitter and presents the CA process and techniques as shown in Fig. 1. Data play very important role for CA, if data are not relevant then

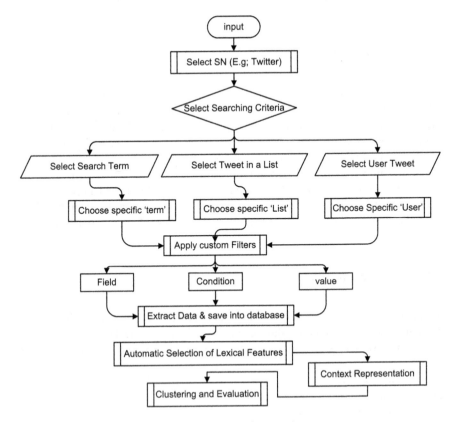

Fig. 1 Twitter data extraction and content analysis framework

whole process can produce unexpected results (corrected but not relevant to that context). Therefore, DE framework needs to be proposed in order to extract topic-specific data [8]. CA framework shows how to perform the analysis on extracted data and what will be results.

1. **Step-01**: **Select SN**: The first step is to choose SN based on research era. So, in this research, Twitter has been selected as a main SN because it is most frequently used SNs in Pb incidents [9–11].
2. **Step-02**: **Select Searching Criteria**: Twitter has too much data, so it needs to be chosen what kind of data is required. So, for Twitter data, searching criteria is defined as follows: As we are interested in only tweets related to Pb era, we can choose among three searching criteria listed above to narrow down the number of tweets and extract the tweets that are related to Pb era only.

   (a) **Select Search Term**: This criteria can be applied to extract tweets based on specified 'search term (ST)'. For example, if someone wants to extract tweets that only contain ST such as 'Pb' or 'peace'. So, by applying this searching criteria, the only tweets will be extracted that contains the specified Pb ST.
   (b) **Select Tweet in a List**: This criteria can be applied to extract tweets only from a specific list (group of users).
   (c) **Select User Tweet**: This criteria can be applied to extract tweets only from a specific 'users' that are chosen.

   As a sample data, the 'Tweet in List' option is selected. The purpose of selecting this searching option is to extract data only from a specific list (group of users). We created a separate list and named as 'Prayforpeace'. 'Prayforpeace' list only contains users that are tweeting about 'peace activities' or 'peace application'. 'Tweet in List' searching option allows to choose and save only tweets from that specific list (Prayforpeace) and whenever there is a new tweet of any user from that list, it will be automatically saved in database as new record or in excel as new row. One can choose any searching criteria as per one's need. The next step is to apply custom filters.
3. **Step-03**: **Apply Custom Filters**: It is very important to apply filters in order to remove noise and unwanted data or tweeter attributes, as a single tweet can contain 70–200 attributes (fields). In step1, 'searching criteria,' some sort of filters are already applied by selecting 'specific list of tweets' to select specific tweets. However, there is still need to apply customize filters in order to delete unwanted attributes or fields from each tweet. Custom Filters consist of three pieces as shown in Table 1 that need to be filled out in one custom filter, which are the following:

Table 1 Twitter custom filter samples

#	Field	Condition	Value
1	Text	Text contains	Peace
2	User name	Text—exactly matches	Ken Banks

- **Field**: This tells what specific field or tweeter attribute needs to be checked for data to filter off of. This research identifies some most important features for Pb-related data based on our pilot study [11–13] that are the following: 'text (tweet),' 'time,' 'date,' 'location,' 'user id,' 'user's full name,' 'user screen name,' 'followers count,' 'following count,' 'url,' 'image,' etc.
- **Condition**: The logic applied by your filter. The conditions are actually the criteria of data selection. The list of conditions is the following:

  - Text—Exact Matches,
  - Text—Not Exact Matches,
  - Text—Contains,
  - Text—Not contains,
  - Text—Starts with,
  - Text—Not Starts with,
  - Text—Ends with,
  - Text—Not Ends with,
  - Number—Greater than, and
  - Number—Less than.

- **Value**: This is the value your filter checks against the condition and attribute, as shown in Table 1.

  More than one filters can be created as per need with 'AND' and 'OR' conditions. For example, Twitter attribute called 'text', which is the actual tweet content, can be chosen in the 'field' and the 'condition' can be 'text contains' (any one condition from the list), and 'Value' can be 'peace', as shown in Table 1 (row-1). Once the data are extracted from Twitter and filtered, then the filtered data will be stored directly to the excel file or in Mysql database and also forwarded to lexical analyzer.

4. **Step-04: Lexical Analysis**: Lexical analysis (LA) is based in extracting lexical features that are single words or word pairs [14, 15]. Lexical features are the textual words found in any data set without and any grammatical information. The list of Lexical feature is given below:

   (a) **Unigrams**: Unigrams are single words that occur n times or most frequently occurring words [14, 16].
   (b) **Bigrams**: Bigrams are two single words (2 unigrams) occur together in same sequence, may contain intervening words between them [17]. For example: 'social networks' and 'networks social' are two separate bigrams because the sequence of words is different. In 'social contacts can be easily found by online networks', in that sentence the two words (social & networks) are same (1 bigram) as intervening words between the two words (social & networks) will be ignored.
   (c) **Co-occurrences**: Co-occurrences are simply unordered bigrams. For example, 'social network' and 'network social' will be two occurrence of the same word and can also ignore the actual intervening in the actual data.
   (d) **Target co-occurrences**: Target co-occurrences are those unordered bigrams that contain the specific target word.

5. **Step-05: Context Representation (CR)**: In this step, the processed data will go under transformation process to provide the context to content. The processed data had been broken into smaller chunks of data (semantic or lexical features) by giving different tags.

- **First-Order context Representation** is one kind of CR that shows the direct relationship between 'words' or direct connection between the 'nodes' (entities, words). In first-order representation, each 'context' is represented by a 'row' and 'lexical features' as 'columns' in the matrix. The frequencies of each feature should be calculated and placed in the matrix. The context is represented by the matrix values.

6. **Step-06: Clustering and Evaluation**: This research uses density-based spatial clustering application with noise (DBSCAN) for clustering the data. Based on the results of 'lexical analysis' and 'context representation' steps, the data are divided into different clusters by DBSCAN. After finding the relevant clusters of documents, the next step is *evaluation of data*. The evaluation can be done by finding the associations (links) between the lexical features and contexts on which the clusters are made (distinguished).Therefore, this research uses *Ranking Algorithm* and *Log-likelihood ratio* [18] to find associations or links between the lexical features. Furthermore, this research use *SVM* algorithm as it is one of the top 10 classification algorithms [19]. Classification algorithm is used to train the classifier on small knowledge base or training data. The ranking algorithms can be used to find the most important features or for rating the features that helps in information verification and comparison. Finally, the data are absolutely structured with much more context that facilitates users to prioritize the information. The input data have been completely transformed into useful knowledge and that knowledge will be saved in knowledge base for future use and will be displayed to user interface.

5 Conclusion

This research proposed the framework for SNs CA and DE for extraction and analysis of the Pb (topic specific) data. Based on our literature review, the given framework has been proposed which results that Twitter is most used SN for CA. However, the most used SN features are time, location, user id, and text. Furthermore, CA framework proposes to use SVM, ranking algorithm, Log-likelihood ratio, and DBSCAN clustering techniques for SNCA. Clustering helps data management and easy retrieval of information. The ultimate aim of this proposed framework is to assist the Pb applications by organizing and analyzing SNs content. The proposed framework can be used to extract data and perform CA for any topic other than Pb as well.

Acknowledgment This research was partially funded by the Ministry of Higher Education Malaysia under RAGS research grant (RAGS12-001-0001).

References

1. Boyd, D.M., Ellison, N.B.: Social network sites: definition, history, and scholarship. J. Comput. Mediated Commun. (JCMC) **13**(1):article 11 (2007)
2. Dambach, C.: What is peacebuilding? Alliance for peacebuilding. http://www. allianceforpeacebuilding.org/?aboutpeacebuilding. Accessed 14 Sept 2012
3. Karki, S., Bennett, R., Nepal, N.: Youth and peacebuilding in Nepal: the current context and recommendations. In: sfcg.org, Jan 2012
4. Weber, R.P.: Basic Content Analysis, vol. 49. Sage Publications, Incorporated, London (1990)
5. Lasswell, H.D.: The structure and function of communication in society. Commun. Ideas **37** (1948)
6. DiMicco, J., Millen, D.R., Geyer, W., Dugan, C., Brownholtz, B., Muller, M.: Motivations for social networking at work. In: Proceedings of the 2008 ACM Conference on Computer Supported Cooperative Work, CSCW'08, New York, NY, USA, pp. 711–720. ACM (2008)
7. Waters, R.D., Burnett, E., Lamm, A., Lucas, J.: Engaging stakeholders through social networking: how nonprofit organizations are using Facebook. Public Relat. Rev. **35**(2), 102–106 (2009)
8. Chakrabarti, S., van den Berg, M., Dom, B.: Focused crawling: a new approach to topic-specific web resource discovery. Comput. Netw. **31**(1116), 1623–1640 (1999)
9. Goolsby, R.: Lifting elephants: twitter and blogging in global perspective. In: Social Computing and Behavioral Modeling, pp. 1–6. Springer, New York (2009)
10. Goolsby, R.: Social media as crisis platform: the future of community maps/crisis maps. ACM Trans. Intell. Syst. Technol. **1**(1), 7:1–7:11 (2010)
11. Shaikh, M., Salleh, N., Marziana, L.: Social networks peacebuilding event mining: a systematic literature review. Published in proceedings of 3rd International Conference on Advanced Computer Science Applications and Technologies (ACSAT), pp.119–124, (2014)
12. Hirata, N., Sano, H., Shiramatsu, S., Ozono, T., Swezey, R.M.E., Shintani, T.: A web agent based on exploratory event mining in social media. In: International Conference on Advanced Applied Informatics, IIAI'12, pp. 236–241. IEEE (2012)
13. Kitaguchi, S., Miyanishi, T., Seki, K., Uehara, K.: Interactive disaster information search system for microblog by minimal user feedback (2013)
14. Chen, H.: Dark Web: Exploring and Data Mining the Dark Side of the Web, vol. 30. Springer, New York (2011)
15. Moens, M.F.: Information Extraction: Algorithms and Prospects in a Retrieval Context, vol. 21. Springer, New York (2006)
16. Niu, Y., Zhu, X., Li, J., Hirst, G.: Analysis of polarity information in medical text. In: AMIA Annual Symposium Proceedings, vol. 2005, p. 570. American Medical Informatics Association (2005)
17. Ted Pedersen: A decision tree of bigrams is an accurate predictor of word sense. In: Proceedings of the Second Annual Meeting of the North American Chapter of the Association for Computational Linguistics, pp. 79–86 (2001)
18. Java, A., Song, X., Finin, T., Tseng, B.: Why we twitter: understanding microblogging usage and communities. In: Proceedings of the 9th WebKDD and 1st SNA-KDD 2007 Workshop on Web Mining and Social Network Analysis, WebKDD/SNA-KDD'07, New York, NY, USA, pp. 56–65. ACM (2007)
19. Wu, X., Kumar, V., Ross Quinlan, J., Ghosh, J., Yang, Q., Motoda, H., McLachlan, G.J., Ng, A., Liu, B., Yu, P.S., Zhou, Z., Steinbach, M., Hand, D.J., Steinberg, D.: Top 10 algorithms in data mining. Knowl. Inf. Syst. **14**(1), 1–37 (2008)

Tree-base Structure for Feature Selection in Writer Identification

Nooraziera Akmal Sukor, Azah Kamilah Muda, Noor Azilah Muda, Yun-Huoy Choo and Ong Sing Goh

Abstract Handwriting is individualistic where it presents various types of features represent the writer's characteristics. Not all the features are relevant for Writer Identification (WI) process and some are irrelevant. Removing these irrelevant features called as feature selection process. Feature selection select only the importance features and can improve the classification accuracy. This chapter investigated feature selection process using tree-base structure method in WI domain. Tree-base structure method able to generate a compact subset of non-redundant features and hence improves interpretability and generalization. Random forest (RF) of tree-base structure method is used for feature selection method in WI. An experiment is carried out using image dataset from IAM Hand-writing Database. The results show that RF tree successively selects the most significant features and gives good classification performance as well.

Keywords Feature selection · Writer identification · Tree-base structure · Random forest tree

N.A. Sukor · A.K. Muda (✉) · N.A. Muda · Y.-H. Choo · O.S. Goh
Faculty of Information and Communication, Universiti Teknikal Malaysia Melaka, Melaka, Malaysia
e-mail: azah@utem.edu.my

N.A. Sukor
e-mail: mai_sukor@yahoo.com

N.A. Muda
e-mail: azilah@utem.edu.my

Y.-H. Choo
e-mail: huoy@utem.edu.my

O.S. Goh
e-mail: goh@utem.edu.my

© Springer International Publishing Switzerland 2015
A. Abraham et al. (eds.), *Pattern Analysis, Intelligent Security and the Internet of Things*, Advances in Intelligent Systems and Computing 355,
DOI 10.1007/978-3-319-17398-6_19

1 Introduction

Handwriting analysis is divided into two: handwriting identification (HI) and handwriting recognition (HR). HI is used to identify the writer of the given handwritten document while HR deals with the content and meaning of handwritten text. There are two models in HI: writer identification (WI) and writer verification (WV). WI is a process to determine the writer of a given handwriting sample and WV determines whether the samples are belong to the same writer or different writer [1, 2]. This study is focused on the WI domain.

Handwriting sample consisted of various types of features. These features are unique due to the writer's characteristics to individuality, thus causing challenges in the identification process. Some features are irrelevant, not provided useful information and decrease the performance of a classifier. To solve the main issue in WI which is to acquire the significant features reflect the author, feature selection process was applied. Feature selection is a process to identify and select the most significant features from presented features in handwriting documents and eliminate the irrelevant features. The selected features give major impact to the identification process and hence, increase the classification accuracy.

The structure of this chapter is as follows: Section 2 describes the feature selection process in WI followed by Sect. 3 that explains the random forest tree method for feature selection process. A simulation experiments and the results are presented in Sect. 4. Section 5 provides the experimental discussion and finally conclusion is elaborated in Sect. 6.

2 Feature Selection

Feature selection is a process to select the most significant features and eliminate the useless one from the raw data. The main objective of the feature selection process is to obtain the most minimal size subset of features as long as the classification accuracy does not significantly decreased and the result of the selected features class distribution is as close as possible to original class distribution [3]. Feature selection method does not alter the original representation of variable, but merely selects a subset of them. Practical experience has shown that presented of too much irrelevant and redundant information will degraded performance of a classifier [4].

There are three types of feature selection method: filter method, wrapper method, and embedded method [5]. Filter method scores the relevance features and removes the low-scored features, while wrapper method used an induction algorithm by exploring the space of features subsets to estimate the merit of feature subsets. Simultaneously, the selection process in embedded method is done inside the induction algorithm, being far less computationally intensive. Embedded method is used to explore on feature selection process to identify the significant features, and

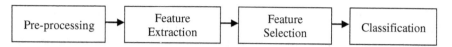

Fig. 1 Framework of study

tree structure-based method is decided to use due to its unique structure growing un-prune splitting branches into a proper tree. An algorithm and framework of random forest (RF) of tree structure-based method are applied for feature selection process. RF tree is one of the embedded methods where it is involved classification and regression tree (CART) during the development of tree.

Implementation of feature selection process after feature extraction task has significantly improved the classification accuracy [4]. This study adopted traditional framework of WI with an additional task; feature selection process takes place between feature extraction task and classification task. Figure 1 shows the framework used in the experiment.

During feature extraction phase, the handwriting attributes that are obtained would enable to differentiate the writing style of different writers. Word images are extracted by using the United Moment Invariant (UMI) [6] that represents the word features. UMI is used because it can be applied in all conditions with a good set of discriminate shapes features [7]. UMI has a capability in the description of image shape where it can find the similar unique features for the same class (writer). Good features are those that satisfy the two requirements which are the small intra-class (same writer) invariance and the large inter-class (different writer) invariance [8].

The classification process is used to find the best class that is closest to the classified pattern. This process is important in order to verify how effective the proposed method is used as the feature selection method. Random forest (RF) tree is selected as the classifier in this study because it is proven that RF tree is an effective classifier in WI in previous studies. The performance measurements for evaluating the performance of these feature selection methods depend on the number of selected features and the classification accuracy.

3 Random Forest Tree

Random forest (RF) tree is introduced by Breiman in 1983. The RF tree is an ensemble learning method consisted of a bagging of un-pruned decision tree method for CART. Each tree is constructed using a different bootstrap sample from the original data. The bootstrap sample is one-third of the cases left out of the training data and not used in the construction of the tree. This is called as out-of-bag (OOB) data.

In the classification trees, input feature vector is classified in every tree in the forest, and output is referred to the class that received the majority of "votes." While regression trees, feature vector is classified according to the average of the responses over all the trees in the forest. This study uses the regression tree of RF tree because of the nature of the handwriting data which is obtained from the measuring process. Regression calculates relationship between predictors and response variables;

$$\hat{y} = T(x) = \sum_{m=1}^{M} \hat{c}_m I(x \in \widehat{R}_m) \tag{1}$$

where

$\hat{y}$ = training set
$T(x)$ = tree
M = bagging
$\hat{c}_m$ = prediction error
$\widehat{R}_m$ = impurity measurement

To gain prediction error, the algorithm is as follows:

$$\hat{c}_m = \frac{1}{N_m} \sum_{x_i \in R_m}^{M} y_i \tag{2}$$

where

N_m = number of samples
y = testing set

The factor used to estimate the importance of variable is the OOB error to estimate the prediction error and measured the variable importance (VI) through permutation. OOB is used to get a running unbiased estimate of the classification error as trees are added to the forest. The importance of features will be measured by the factor of VI and the mean square error (MSE).

The VI analyzed and revealed the important features by predicting correct classification of RF tree based on the OOB data while MSE is the squared error for OOB vectors, averaged over all ensemble trees and is divided by the standard deviation taken over the trees, for each variable. During constructing the trees, each branch is split according to the VI value. The OOB error is computed as well as the squared error for OOB (MSE). Then, the average of the value of each predictor (feature) is computed. To select the important features, the VI value of each predictor is compared with the MSE value. The predictors that contained high values of the VI and the MSE are the significant features. When the MSE value is high, the invariance of the same writer is smaller than the different writer. To identify the writer, the similarity error for intra-class must be lower than the inter-class [9] because of having the small intra-class invariance indicates closeness to the real author. Table 1 shows the comparison of the traditional RF algorithm with the proposed RF algorithm.

Table 1 Comparison between traditional RF and proposed RF

Traditional RF algorithm	Proposed RF algorithm
1. Draw *ntree* bootstrap samples from the original data	1. Draw *ntree* bootstrap samples from the original data
2. From bootstrap samples, grow an un-pruned classification or regression tree	2. From bootstrap samples, grow an un-pruned classification or regression tree
2.1. At each internal node, randomly select *mtry* predictors and determine the best split using only these predictors (bagging can be thought of as the special case of the random forests obtained when *mtry* = *p*, the number of predictors)	2.1. At each internal node, randomly select *mtry* predictors and determine the best split using only these predictors where each of the split is based on the VI value
3. Predict new data by aggregating the predictions of the *ntree* trees (majority votes for classification, average for regression)	3. Compute the MSE value by averaging over all trees for each variable
	4. Final prediction generates important features containing the VI ≥ MSE value

4 Experiment and Result

This section provides explanation about experiment that performs the random forest tree as feature selection method.

4.1 Dataset

The experiment is executed using the dataset from the IAM Handwriting Database which contained samples of English word handwritings. Only 60 classes out of 657 classes were used to be experimented in this work. From these classes, 4400 instances are collected. Refer to Fig. 2, 4400 instances were divided randomly into five groups to form the training and testing data in the classification task. Each set of data will be divided to 80 % for training data and 20 % for testing data.

This experiment uses discretization data as it was proven that it can lead to the better accuracy in classification phase compared with the undiscretized data [7]. Discretization is a process of dividing or classifying the continuous attributes into one of the regions or intervals. The values that are lying in each interval are mapping to the same value for converting the numerical value that can be treated as being symbolic [10]. The discretization process took place between feature extraction process and feature selection process in the WI. The method of equal width binning was used in the discretization process that is performed on the feature vectors extracted before passing them to the feature selection process.

The experiment was conducted using MATLAB® Software, Version 7.9 (R2009b), for feature selection process and Waikato Environment for Knowledge Analysis (WEKA) 3.7.4 for classification process.

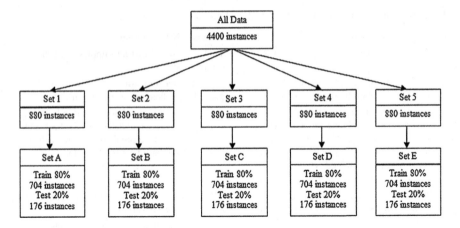

Fig. 2 Data collection for training and testing

4.2 Experimental Result

See Tables 2 and 3.

4.3 Comparison with Other Method

See Tables 4 and 5.

Table 2 Experiment result of discretized dataset

Random forest tree on discretized dataset							
	Execution	#1	#2	#3	#4	#5	Intersection
Set A	Feature	f1, f3, f4	f1, f3	f1, f3	f3	f3	f3
	Classification (%)	99.10	99.10	99.10	99.32	99.32	99.19
Set B	Feature	f3	f1, f3, f4	f3	f1, f3, f7	f1, f3	f3
	Classification (%)	99.53	99.30	99.53	99.30	99.30	99.39
Set C	Feature	f1, f3	f1, f3, f4	f3	f1, f3, f4	f1, f3	f3
	Classification (%)	99.09	99.09	99.89	99.09	99.09	99.25
Set D	Feature	f1, f3	f3	f1, f3	f3	f3, f4	f3
	Classification (%)	98.79	99.56	98.79	99.56	99.12	99.16
Set E	Feature	f1, f3	f1, f3	f1, f3	f1, f3	f3, f7	f3
	Classification (%)	98.98	98.98	98.98	98.98	98.98	98.98

Table 3 Experiment result of undiscretized dataset

Random forest tree on undiscretized dataset

	Execution	#1	#2	#3	#4	#5	Intersection
Set A	Feature	f1, f3, f4, f5	f1, f2, f3, f8	f1, f3, f4	f1, f2, f3, f5	f1, f2, f3, f5, f8	f1, f3
	Classification (%)	45.65	45.31	45.76	46.55	46.33	45.92
Set B	Feature	f1, f3, f8	f1, f3, f5	f1, f3, f5	f1, f3, f6	f1, f3, f4	f1, f3
	Classification (%)	47.94	47.94	47.94	48.06	48.30	48.04
Set C	Feature	f1, f2, f3, f5	f1, f2, f3	f1, f2, f3	f1, f3, f4	f1, f3, f5	f1, f3
	Classification (%)	40.71	40.59	40.59	40.82	40.02	40.55
Set D	Feature	f1, f2, f3, f4	f1, f2, f3	f1, f2, f3	f1, f2, f3, f5	f1, f2, f3	f1, f2, f3
	Classification (%)	30.98	30.43	30.43	30.21	30.43	30.50
Set E	Feature	f1, f2, f3	f1, f3, f5	f1, f3, f4	f1, f3, f5	f1, f3, f5	f1, f3
	Classification (%)	39.34	38.32	40.36	38.32	38.32	38.93

5 Experiment Discussion

The experiments of each dataset are conducted in five executions in order to gain the most significant results. There are two measurements used to verify the performances of the RF tree during the feature selection process; the number of features selected and the classification accuracy are discussed below:

(a) Number of Feature

The number of features is the primary consideration of this study. Table 2 shows the number of features selected by the RF tree using the discretized dataset, and Table 3 shows the number of features selected by the RF tree using the undiscretized dataset. Based on the feature selection results, it is shown that this feature selection method succeed to select the most significant features and eliminate the useless ones for both type of datasets.

Based on the result, it is shown that this feature selection method yields different subsets of importance features on both type of dataset. According to the Table 2, all the sets (A, B, C, D, and E) yield the third feature (f3) as the most importance feature. It is however a different situation with the results shown in Table 4. All the sets produced first feature (f1) and third feature (f3) as the unique features except the set D which produced the first feature (f1), the second feature (f2), and the third feature (f3).

It is worth mentioned that although this feature selection method yielded different features in every execution of each dataset, they seem to always include the third feature (f3) in their results. Therefore, it can be concluded that the third feature (f3) is the most significant feature, and it is chosen as the significant unique feature in order to proof the individuality of the handwriting.

Table 4 Comparisons with other techniques using discretized dataset

Discretized dataset							
Method	Criteria	Set A	Set B	Set C	Set D	Set E	Intersection
CFS	Number of selected features	6	6	6	5	6	4
	List of selected features	f1, f2, f3, f5, f7, f8	f1, f3, f4, f5, f6, f7	f1, f3, f4, f5, f6, f7	f1, f3, f5, f7, f8	f1, f3, f4, f5, f6, f7	f1, f3, f5, f7
	Classification accuracy (%)	94.24	97.18	97.18	94.01	97.18	95.95
LVF	Number of selected features	4	4	4	4	4	4
	List of selected features	f2, f3, f4, f6	f2, f3, f4, f6	f2, f3, f4, f6	f2, f3, f4, f6	f2, f3, f4, f6	f2, f3, f4, f6
	Classification accuracy (%)	97.40	97.40	97.40	97.40	97.40	97.40
FCBF	Number of selected features	8	8	8	8	8	8
	List of selected features	f1, f2, f3, f4, f5, f6, f7, f8	f1, f2, f3, f4, f5, f6, f7, f8	f1, f2, f3, f4, f5, f6, f7, f8	f1, f2, f3, f4, f5, f6, f7, f8	f1,f2, f3, f4, f5, f6, f7, f8	f1, f2, f3, f4, f5, f6, f7, f8
	Classification accuracy (%)	98.08	98.00	98.06	97.57	97.62	97.87
SFS	Number of selected features	3	2	2	2	2	2
	List of selected features	f3, f6, f8	f3, f6	f3, f6	f3, f6	f3, f6	f3, f6
	Classification accuracy (%)	97.45	97.32	96.67	96.47	96.42	96.87
SFFS	Number of selected features	2	2	3	3	2	2
	List of selected features	f3, f6	f3, f6	f3, f6, f8	f3, f6, f8	f3, f6	f3, f6
	Classification accuracy (%)	96.66	97.02	96.85	96.32	96.51	96.67

(continued)

Table 4 (continued)

Discretized dataset							
Method	Criteria	Set A	Set B	Set C	Set D	Set E	Intersection
CI + SFFS	Number of selected features	3	3	2	2	2	2
	List of selected features	f1, f3, f6	f1, f3, f6	f3, f6	f3, f6	f3, f6	f3, f6
	Classification accuracy (%)	97.92	97.46	97.19	96.98	97.03	97.32
SFS + MIC	Number of selected features	f3, f6, f8	f3, f6	f3, f6	f3, f6	f3, f6	f3, f6
	List of selected features	3	2	2	2	2	2
	Classification accuracy (%)	97.45	97.32	96.67	96.47	96.42	96.87
SFFS + MIC	Number of selected features	f3, f6	f3, f6	f3, f6, f8	f3, f6, f8	f3, f6	f3, f6
	List of selected features	2	2	3	3	2	2
	Classification accuracy (%)	96.66	97.02	96.85	96.32	96.51	96.67
RF tree (proposed technique)	Number of selected features	1	1	1	1	2	1
	List of selected features	f3	f3	f3	f3	f1, f3	f3
	Classification accuracy (%)	99.19	99.39	99.25	99.16	98.97	99.19

Third feature (f3) is the most significant feature because it contained high value of VI. At the same time, the VI value of this feature is more than the MSE value which means that f3 is the most important feature compared with others. The high value of VI and MSE means low similarity error of intra-class. On the other hand, high MSE value shows low similarity error of the intra-class and low MSE value shows high similarity error for inter-class. Lower similarity error indicates closeness to the real author. Good features are those that satisfy two requirements which are small the intra-class invariance and the large inter-class invariance [8].

Table 5 Comparisons with other techniques using undiscretized dataset

Undiscretized dataset							
Method	Criteria	Set A	Set B	Set C	Set D	Set E	Mean
ReliefF (benchmark)	Number of selected features	8	8	8	8	8	8
	Classification accuracy (%)	45.99	45.99	45.99	45.99	45.99	45.99
CFS (benchmark)	Number of selected features	1	1	1	1	1	1
	Classification accuracy (%)	4.29	4.29	4.29	4.29	4.29	4.29
LVF (benchmark)	Number of selected features	4	4	4	4	4	4
	Classification accuracy (%)	45.65	45.65	45.65	45.65	45.65	45.65
FCBF (benchmark)	Number of selected features	1	1	1	1	1	1
	Classification accuracy (%)	4.29	4.29	4.29	4.29	4.29	4.29
SMFS (benchmark)	Number of selected features	1	1	1	1	1	1
	Classification accuracy (%)	4.29	4.29	4.29	4.29	4.29	4.29
SFS + NBaves (benchmark)	Number of selected features	8	7	8	6	8	7.4
	Classification accuracy (%)	45.99	48.19	45.99	30.76	45.99	43.38
SFS + IB1 (benchmark)	Number of selected features	2	2	5	2	2	2.6
	Classification accuracy (%)	30.85	30.85	40.93	30.85	30.85	32.87
SFS + 1R (benchmark)	Number of selected features	2	1	1	1	1	1.2
	Classification accuracy (%)	38.75	4.29	4.29	4.29	4.29	11.18
SFS + RFgrest (benchmark)	Number of selected features	8	8	8	8	8	8
	Classification accuracy (%)	45.99	45.99	45.99	45.99	45.99	45.99

(continued)

Table 5 (continued)

Undiscretized dataset							
Method	Criteria	Set A	Set B	Set C	Set D	Set E	Mean
SFS − NSA (benchmark)	Number of selected features	5	4	5	5	4	4.8
	Classification accuracy (%)	45.76	48.06	40.02	30.21	40.25	40.86
RF tree (proposed technique)	Number of selected features	2	2	2	3	2	2.2
	Classification accuracy (%)	45.92	48.04	40.55	30.50	38.93	40.79

(b) Classification Accuracy

Number of features selected is not always an indicator of a successful feature selection process. Therefore, further validation must be justified and validated through the identification performance, which is the classification accuracy. The feature subsets are tested against the classification, which uses the RF tree as the classifier.

According to Table 2, all the set of data produced high classification accuracy. Set B produced the best accuracy (99.39 %), followed by set C (99.25 %). The next best accuracy is set A (99.19 %), set D (99.16 %), and finally set E (98.98 %). The results of the RF tree are shown to be stable because of the nature of the data that is consistently allowed the RF tree to perform well. Besides, due to the behavior of these methods which can specifically identify the unique features in the dataset, therefore it produced the highest performance result.

According to Table 3, there is no feature selection method that yields classification accuracy more than 50 %. Even though the feature selection is performed, the lower classification results have already been expected, due to the various shapes of words that have been used to represent the writers. Based on the classification results of undiscretized dataset, set B (48.04 %) produced the highest classification accuracy, followed by set A (45.92 %). The next highest classification accuracy is presented by set C (40.55 %) and set E (38.93 %). Finally, the worst classification accuracy is hold by set D (30.50 %).

The failure of the undiscretized dataset produced better classification accuracy is because the RF tree is more suitable when handling high-dimensional data, these method analyzed the correlation between features, which are the feature relevance and the feature redundancy. These methods performed poorly when they failed to find the correlation between features. Even though there are no feature selection methods capable to increase the classification accuracy more than 50 %, the number of features is still reduced, and thus reduced the workload of the classification task.

Based on the experimental results, third feature produced high classification accuracy because it succeeded to find the correlation between the features. Meaning that, during the classification process, the presented of the third feature (f3) gives more impact to the handwriting of authors compared with other features. The correlation between the features are depended on the proximity measurement where it measures the frequency of the features in training sample (in-bag and OOB data) ended up in the same terminal node. Frequently presented of third features(f3) means contained high value of VI where it shows that third feature is important feature compared to other.

For discretized dataset as shown in Table 4, RF tree is the best feature selection method where it produced highest classification accuracy about 99.19 % with one feature selected. It is followed by FCBF, LVF, SFFS (classifier CI), SFS, SFS (classifier MIC), SFFS, SFFS (classifier MIC), and finally CFS. RF tree succeed to select the most significant feature and produced high classification accuracy because of its unique structure developing the tree by using VI value in splitting the branches.

According to the result shown in Table 5, highest classification accuracy is about 45.99 % produced by ReliefF and SFS (classifer RForest). Then, it is followed by the performance of CFS, SFS (classifier NBayes), SFS (classifier NSA), RF tree, SFS (classifier IB1), SFS (classifier 1R), LVF, FCBF, and finally SMFS. Those methods performed better than RF tree is because they had ability to handle the various shape of words which have been used to represent the writer.

6 Conclusion

This chapter presented the performance of the RF tree as the feature selection method in the WI. For the discretized dataset, the third feature (f3) is declared as the most significant features with the classification accuracy of 99.19 %. Meanwhile for the undiscretized dataset, the first feature (f1) and the third feature (f3) are declared as the most significant features with the classification accuracy of 40.79 %. As a conclusion, the proposed RF tree is an effective tree structure-based method when it succeeded to select the only significant features during the feature selection process and gives high percentage of the classification accuracy. The RF tree performances are measured based on the VI and the MSE in identifying the valuable features. The larger these values show, the more valuable the features to the work. For future works, some modification in the RF tree algorithm needs to be implemented in order to reduce the over fitting problem during the selection of the significant features.

Acknowledgments This work was funded by the Ministry of Higher Education Malaysia and Universiti Teknikal Malaysia Melaka (UTeM) through the Fundamental Research Grant Scheme—FRGS/2/2013/ICT02/FTMK/02/4/F00187.

References

1. Muda, A.K., Shamsuddin, S.M., Darus, M.: Invariants discretization for individuality representation in handwritten authorship. In: 2nd International Workshop on Computational Forensic, Springer, Washington, DC (2008)
2. Muda, A.K.: Authorship Invarianceness for Writer Identification Using Invariant Discretization and Modified Immune Classifier. Universiti Teknologi Malaysia, Johor (2009)
3. Dash, M., Liu, H.: Feature selection methods for classification. Intelligent Data Analysis: An Internat. J. 1(3) (1997)
4. Yu, L., Liu, H.: Efficiently handling feature redundancy in high-dimensional data. In: Proceedings of the Ninth ACM SIGKDD International Conference on Knowledge Discovery and Data Mining (KDD-03), pp. 685–690. Washington, DC, August, 2004
5. Xu, J., Yang, G.: Hong Man and Haibo He, L1 Graph based on sparse coding for Feature Selection, Lecture Notes in Computer Science, Vol. 7951, pp. 594–601 (2013)
6. Yinan, S., Yuechao, W., Weijun, L.: United moment invariants for shape discrimination. In: Proceedings on IEEE International Conference Robotics, Intelligent Systems and Signal Processing. IEEE (2003)
7. Pratama, S.F., Muda, A.K., Choo, Y.H., Muda, N.: Computationally inexpensive sequential forward floating selection for acquiring significant features for authorship invarianceness in writer identification. International Journal on New Computer Architectures and Their Applications (IJNCAA) 1(3), 581–598. ISSN: 2220-9085 (2011) (The Society of Digital Information and Wireless Communications)
8. Khotanzad, A.: Invariant image recognition by Zernike moments. IEEE Trans. Pattern Anal. Mach. Intell. 12(5), 489–497 (1990)
9. Srihari, S.N., Cha S.-H., Lee, S.: Establishing handwriting individuality using pattern recognition techniques. In: Document Analysis and Recognition, 2001. Proceedings. Sixth International Conference on, pp. 1195–1204, 10–13 Sept 2001
10. Pratama, S.F., Muda, A.K., Choo, Y.H.: Feature selection methods for writer identification: a comparative study. In: International Conference on Computer and Computational Intelligence (ICCCI 2010)

Factors Affecting the Effective Online Collaborative Learning Environment

Sharifah Nadiyah Razali, Faaizah Shahbodin, Hanipah Hussin
and Norasiken Bakar

Abstract Interest in collaboration is a natural outgrowth of the trend in education towards active learning. Many researchers have found advantages of collaborative learning; it improves academic performance, promotes soft skills development (communication, collaboration, problem solving and critical thinking skills) and increases satisfaction in the learning experience. However, several studies have reported the opposite. Therefore, this paper aims to determine the factors to be considered in creating an effective online collaborative learning environment. In order to achieve the aims, this study was conducted qualitatively in the form of a document review. The results indicate three main factors that affect the effectiveness of online collaborative learning environments such as learning environment, learning design and learning interaction. An online learning interaction model is also proposed according to the results. This study will continue to determine the elements that can clarify all the factors which have been identified in this study.

Keywords Collaborative learning · Online collaborative learning · Learning interaction

S.N. Razali (✉) · F. Shahbodin · N. Bakar
Faculty of Information and Communication Technology,
University of Technical Malaysia Melaka, Melaka, Malaysia
e-mail: shnadiyah@yahoo.com

F. Shahbodin
e-mail: faaizah@utem.edu.my

N. Bakar
e-mail: norasiken@utem.edu

H. Hussin
Centre for Languages and Human, University of Technical Malaysia Melaka,
Melaka, Malaysia
e-mail: hanipah@utem.edu.my

© Springer International Publishing Switzerland 2015   215
A. Abraham et al. (eds.), *Pattern Analysis, Intelligent Security
and the Internet of Things*, Advances in Intelligent Systems and Computing 355,
DOI 10.1007/978-3-319-17398-6_20

1 Introduction

The benefits of collaboration in learning have been proven by social constructivism [1]. According to the study by Johnson & Johnson [2], learning tends to be most effective when students are in the position to work collaboratively in expressing their thoughts, discussing and challenging ideas with others, and working together towards a group solution to the given problem. Zhu [3] defines collaborative learning as a social interaction involving the acquisition and sharing of experience or knowledge amongst learners and teachers. Collaborative learning, which in an online environment is typically referred to as online teams or online groups, refers to instructional activities for getting students to work together online to achieve common educational goals.

Interest in collaboration is a natural outgrowth of the trend in education towards active learning, whereby students become involved in constructing their own knowledge through discovery, discussion and expert guidance. Many published reports have outlined the advantages of collaborative learning, suggesting that it improves academic performance, promotes soft skills development (communication, collaboration, problem solving and critical thinking skills) and increases satisfaction in the learning experience (refer Table 1).

MA and 馬慧穎 [4] tried to identify the interaction patterns and discourse quality of a CSCL environment. She found a positive relationship between the quality of the collaborative process and the quality of cognitive skills fostered. Besides that, she also found that effective collaborative learning can contribute to the establishment of a learning community and that it fosters high-order thinking through knowledge processes. Because of the tedious and time-consuming coding process, she suggested other researchers to computerize the coding process.

Research done by Kabilan et al. [5] reported on pre-service teachers' meaningful experiences in collaborative projects and how they had enriched their professional development. The results showed their professional development engagements were enriched by envisioning professional development, gaining and enhancing in

Table 1 Summary of collaborative learning benefits

References	Performance	Soft skills				Satisfaction
		Communication	Collaboration	Problem solving	Critical thinking	
MA and 馬慧穎 [4]	X	X	X		X	
Kabilan et al. [5]		X	X	X		
Chen [6]	X					X
Lee and Lim [7]	X	X				
Zhu [3]	X					X

five skills (planning and researching, problem solving, the fundamental notion of learning, language skills and computing skills), sharing and exchanging information, knowledge ideas, views and opinions related to the tasks given and also teachers socializing within and between groups. For future research, they suggested that other researchers should also focus on additional popular online platforms such as Facebook, Academia.edu and LinkedIn as tools for their online professional development projects.

With the growth of web 2.0 technology, Chen [6] investigated the differences between students' learning outcomes and satisfaction in a class using an online social networking tool (Facebook) amongst different learning styles. There were four learning styles: diverger, assimilator, converger and accommodator. He found that the converger group performed better and showed a more positive attitude towards Facebook compared to other learning style groups. In the converger group's perception, Facebook facilitated their interaction with others and improved content understanding in the class. For the future, he suggested examining the effects on different levels of learners to link the relationship of learning styles and the online social networking tool (Facebook).

Lee and Lim [7] investigated the important issues when it comes to students evaluating their peers in team project-based learning by analysing each message and comparing them with peer evaluation results. They classified the messages into four types: managerial, procedural, social and academic messages. The findings showed that all message types, except academic messages, predicted peer evaluation results. They concluded that students find social contribution to be more important compared to cognitive contribution when they evaluate peers. They suggested other research be done to compare the relationship between learning outcome by instructor's evaluation, peer evaluation and interaction message types.

Zhu [3] found that online collaborative learning can enhance students' knowledge construction. He examined satisfaction with the online learning environment, their online performance and knowledge construction via online group discussions of students in two different cultural contexts (Flemish and Chinese). The results showed there was a relationship between student satisfaction and academic achievement in an innovative e-learning environment. It also showed that online learning systems can enrich students' collaborative learning activities as well as their knowledge construction via group interaction. However, it was found that instructors evaluate the quality of the final product without knowledge of the teamwork process. Therefore, it was suggested that, in the future, researchers may want to not only study cognitive learning outcomes, but also social skills in collaborative learning outcomes.

The benefits of Collaborative Learning have been summarized in Table 1.

Contrary to this, other research has shown evidence that online learning can pose an even greater challenge for collaborative work than face-to-face (F2F) learning. According to the study by Chiong and Jovanovic [8], establishing and maintaining an active collaboration is a challenging task due to the lack of active participation by group members in their group work. Results from the interview session on collaborative learning experience in the research by Zhang and Han [9] showed that there exists group tension towards the fairness of being given the same mark. Educators are not able to assume that every student makes an equal contribution to the group work and then allocate the same marks to all members [10]. Therefore, educators must allocate marks based on a student's contribution to encourage students to participate actively in their group work activity [11].

Lee and Lim [7] found that instructors may not observe all the processes occurring within student groups and the evaluation is done only on the quality of the final product, ignoring the teamwork process. They suggested, instructors should closely monitor group interaction messages and do peer evaluations. Wang [12] also suggested that educators, including teachers and lecturers, should closely monitor how their students work together in a collaborative learning process for effective learning to take place. By monitoring the collaborative learning process, it can help educators keep track of students' ongoing performance. Therefore, this study aims to determine the factors to be considered in creating an effective online collaborative learning environment.

2 Materials and Methods

In order to achieve the aim, the study was conducted qualitatively in the form of a document review. According to the study by Sallabas [13, 14] and Best and Kahn [13, 14], the document review method is the most appropriate tool to collect information in a qualitative study. Stewart [15] defines materials and resources that can be used as documents to carry out the analysis and interpretation of which are (i) journals and books, (ii) research literature and (iii) reports from scholarly research papers and materials. Several previous studies including reports, conference proceedings and journals were referred to as a literature review. The collected data were then analysed using a matrix table [16].

3 Results and Discussion

Based on a review of documents, those factors affecting the effectiveness of online collaborative learning environments are summarized in Table 2.

A matrix table has been drawn to determine the main factors affecting the effectiveness of online collaborative learning environments using Strauss and Corbin's model. The results are illustrated in Table 3.

Based on the analysis shown in Table 3, the researchers determined three factors that affect the effectiveness of online collaborative learning: learning interaction, learning design and learning environment.

In previous research done by Moore [23], he proposed three types of interaction in his interaction theory using the three constructs of instructor–student–content

Table 2 Factors that affect the effectiveness of online collaborative learning environments

References	Factors
Vygotsky [1]	• Tenor/personal (learners' relationships)
	• Mode/behaviour (language/textual)
	• Fields/environment (social activity)
Tu and Corry [17]	• Social context/constructed from the CMC users' characteristics and their perception of the CMC environment (social form, informal and casual communication, personal and sensitive means of communication, the recipients, social relationships, access/location and perceptions on media)
	• Online communication/attributes of the language used online and the applications of online language (stimulating, expressive, conveying feelings and emotions, meaningful, easily understood keyboarding skills, expressiveness, characteristics of discussion and language skills)
	• Interactivity/activities in which CMC users engage and the communication styles they use (CMC as pleasant, immediate, responsive and comfortable with familiar topics, response time, communication styles/skills and the size of discussion groups)
Gerbic [18]	• CMC environment (easy access, familiarity, group size, technical problems, lack of participation, spontaneous exchanges, a lot of information, express thoughts in text rather than speech, written messages, posting message anxiety)
	• Curriculum (interesting discussion topic, link online discussions with assessment, voluntary, integrates online discussions into a course, interaction satisfaction, course workload and programme culture)
	• Student (subject familiarity, confidence level, reading preferences, lack of time, motivation, time management, extra workload, commitment to online discussion and online discussion role and value)
Sun et al. [19]	• Learner (computer attitude, computer anxiety, Internet competence)
	• Instructor (response time, e-learning attitude)
	• Course (flexibility, quality)
	• Technology (technology quality, Internet quality)
	• Design (perceived usefulness, perceived ease of use)
	• Environment (assessment, interaction)
Ali [20]	• Learner
	• Learning process
	• Content (subject matter)
	• Learning environment
	• Time constraints for learning
	• Lecturer
Kaur et al. [21]	• People (dynamic, patience, subject knowledge, clear instruction, fellow students and support staff)
	• Structure (clear delineation and comprehensive activities)
	• Environment (accessibility, navigation and support)
	• Resources (varied, well selected and learning style)
Filigree [22]	• Technology (integrates learning spaces and flexible learning environment)
	• People (training, guidance and support)
	• Process (high-quality content, content relevance to subject and adapt pedagogical tools and models)

Table 3 Matrix table

Construct	Construct		
	Learning interaction	Learning design	Learning environment
Vygotsky [1]			
Personal factors (tenor)	✓		
Behaviour (mode)		✓	
Environment (field)			✓
Tu and Corry [17]			
Social context	✓		
Interactivity		✓	
Online communication			✓
Gerbic [18]			
CMC environment			✓
Curriculum		✓	
Student	✓		
Sun et al. [19]			
Learner	✓		
Instructor	✓		
Course		✓	
Technology			✓
Design		✓	
Environment			✓
Ali [20]			
Interaction	✓		
Process		✓	
Learning environment			✓
Kaur et al. [21]			
People	✓		
Structure		✓	
Resource		✓	
Environment			✓
Filigree [22]			
People	✓		
Process		✓	
Technology			✓

(refer Fig. 1). In the model, the three types of interaction are identified as learner–content interaction, learner–instructor interaction and learner–learner interaction.

In the early stages of a collaborative learning environment, a number of studies have defined interaction involves only the relationship between learner [1, 17, 18]. However, recent studies define interactivity not only involves learners with learners, but also involves the relationship between learners and teachers [19–22].

Fig. 1 Moore (1989)
interaction model

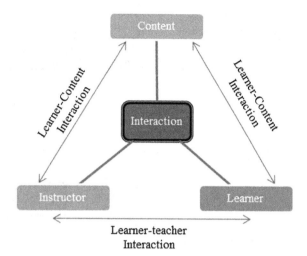

Previous researchers used different terms to define the relationship learner–learner and learner–teacher relationship such tenor, social context, student, learner and instructor, learner and teacher and people. Therefore, in this study learning inter-activity will be used to define the interaction between learner–learner relationship and leaner–teacher relationship.

In learner–teacher interaction, teacher has to encourage students actively par-ticipate in online discussion using provided platform. Providing a suitable platform can facilitate and increase interaction and collaboration between learners. It also helps teachers monitor student engagement. In previous study, Sharifah Nadiyah and Faaizah [24] suggested integrated current LMS with Facebook to enable stu-dents and lecturers communicate on Moodle through Facebook and also to facilitate online collaborative learning [25]. Yeo and Quek [26] found technology mediation has supported interaction. Previous researchers used different terms to define the learning platform such as: technology, field, CMC environment, environment and online environment, but this study will use the term learning environment to define the platform using in learning.

Teacher has responsible to provide guidelines for all tasks. To promote inter-action between learner and task, teacher also needs to develop strategies and technique. In this study, learning design will be used to define the activity or process or structure of learning. There were a few different terms used by previous researchers such as resources, content, curriculum and mode. Therefore, in this study, the model will be developed using the following three constructs: interaction, design and environment. All the construct will be used to develop proposed pro-totype in order to enhance student soft skills: communication, collaboration, problem solving and critical thinking skills [27].

The researchers proposed four interactions, which are learner–learner interac-tion, learner–teacher interaction, design interaction and environment interaction (refer Fig. 2). There are two types of interactions in learner interaction:

Fig. 2 Proposed online
learning interaction model

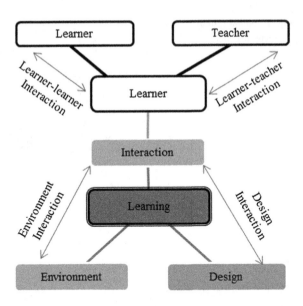

learner–learner interaction and learner–teacher interaction. In an online environment, the learner interaction can happen in either a synchronous or an asynchronous way. In a synchronous way, learners interact at the same time; whilst in an asynchronous way, the learners are not required to interact at the same time. Design interaction is an interaction between learners and a given task. The task has the ability to enrich learners' behaviour. The interaction between learners and the environment is called environment interaction.

4 Conclusion

Technology can be used to encourage learning process, support communication setting, assess learning activities, manage resources and create learning materials [28]. Technology is also seen as an important enabler for improving student–learning outcomes, but to get the greatest value from technology, best practices are required. There are five levels of collaboration maturity proposed by Filigree [22]: basic, partially implemented, integrated, collaborative and transformative. The report emphasized that collaborative learning is heavily rooted in the idea that learning is inherently social and can be facilitated with technology and proper practices. Collaborative learning not only promotes social skills, but also facilitates retention, improves the experience and enhances creativity. With higher levels of collaboration, greater results will be delivered.

Previous section has determined the factors affecting the effective online collaborative learning. In the next stage, this research will determine the elements that can clarify all the factors which have been identified in the previous section. Currently, the model is only in a conceptual phase and requires significant development before it could be used to gather data.

Acknowledgments Sharifah Nadiyah Razali would like to gratefully acknowledge the financial support of Universiti Teknikal Malaysia Melaka (UTeM) and the Ministry of Higher Education of Malaysia for her PhD study.

References

1. Vygotsky, L.S.: Mind in Society. Harvard University Press, Cambridge (1978)
2. Johnson, R.T., Johnson, D.W.: Cooperation and Competition: Theory and Research. Interaction Book Company, Edina (1989)
3. Zhu, C.: Student satisfaction, performance, and knowledge construction in online. Educ. Technol. Soc. **15**(1), 127–136 (2012)
4. MA, Wai Wing Ada, and 馬慧穎.: Computer supported collaborative learning and higher order thinking skills: A case study of textile studies (2009)
5. Kabilan, M.K., Adlina, W.F.W., Embi, M.A.: Online collaboration of English language teachers for meaningful professional development experiences. Engl. Teach. Pract. Critique **10** (4), 94–115 (2011)
6. Chen Y.: Learning styles and adopting Facebook technology. In: Technology Management in the Energy Smart World (PICMET), pp. 1–9 (2011)
7. Lee, H.-J., Lim, C.: Peer evaluation in blended team project-based learning; what do students find important? Educ. Technol. Soc. **15**(4), 214–224 (2012)
8. Chiong, R., Jovanovic, J.: Collaborative learning in online study groups: an evolutionary game theory perspective. J. Inf. Technol. Educ. Res. **11**, 81–101 (2012)
9. Zhang, Z., Han, Z.: A phenomenographic study into conceptions of social relations in online collaborative learning—case study of China higher education learners. In: IEEE International Conference on Computer Science and Software Engineering (ICCSSE), pp. 146–149 (2008)
10. Wang, Q.: Using online shared workspaces to support group collaborative learning. Comput. Educ. **55**(3), 1270–1276 (2010)
11. Swan, K., Hiltz, S.R.S., Shen, J.: Assessment and collaboration in online learning. J. Asynchronous Learn. Netw. **10**(1), 45–62 (2006)
12. Wang, Q.: Design and evaluation of a collaborative learning environment. Comput. Educ. **53** (4), 1138–1146 (2009)
13. Sallabas, M.E.: Analysis of narrative texts in secondary school textbooks in terms of values education. Afr. J. Bus. Educ. **1**(3), 59–63 (2013)
14. Best, J.W., Kahn, J.V.: Research in Eduction. A Viacom Company, USA (1998)
15. Stewart, A.M.: Research Guide for a Students and Teachers. State University of New York, New York (2009)
16. Strauss, A., Corbin, J.: Basics of Qualitative Research: Grounded Theory Procedures and Techniques. Sage, Newburry Park (1990)
17. Tu, C.-H, Corry, M.: Social presence and critical thinking for online learning. In Annual Conference of American Educational Research Association (AERA) (2002)
18. Gerbic, P.: To post or not to post: undergraduate student perceptions about participating in online discussions. In: Proceedings of the 23rd Annual Ascilite Conference: Who's Learning? Whose Technology? No. 1995, pp. 271–281 (2006)
19. Sun, P., Tsai, R., Finger, G., Chen, Y., Yeh, D.: What drives a successful e-learning? An empirical investigation of the critical factors influencing learner satisfaction. Comput. Educ. **50**, 1183–1202 (2008)

20. Ali, H.: A comparison of cooperative learning and traditional lecture methods in the project management department of a tertiary level institution in Trinidad and Tobago. Caribb. Teach. Sch. **1**(1), 49–64 (2011)
21. Kaur, A., Shriram, R., Ravichandran, P.: A framework for online teaching and learning: the S-CARE pedagogical model. In 25th AAOU Annual Conference, pp. 1–12 (2011)
22. Filigree, C.: Instructional Technology and Collaborative Learning Best Pratices :Global Report and recommendations. SMART Technologies (2012)
23. Moore, M.G.: Three types of interaction. Am J Distance Educ. **3**(2), 1–6 (1989)
24. Sharifah Nadiyah, R., Faaizah, S.: The usage of CIDOS and social network sites in teaching and learning processes at Malaysian polytechnics. Int. J. Comput. Technol. **13**(4), 4354–4359 (2014)
25. Sharifah Nadiyah, R., Faaizah, S., Norasiken, B., Hanipah, H., Mohd Hafiez, A.: The need of incorporating Cidos with Facebook to facilitate online collaborative learning. WIT Trans. Inf. Commun. Technol. **58**, 1089–1097 (2014)
26. Yeo, T., Quek, C.: Investigating design and technology students' peer interactions in a technology-mediated learning environment: a case study. Australas. J. Educ. Technol. **27**(4), 751–764 (2011)
27. Sharifah Nadiyah, R., Hanipah, H., Faaizah, S.: 21st century core soft skills research focus for integrated online project based collaborative learning model. J. Appl. Sci. Agric. **9**(11), 63–68 (2014)
28. Che Ku Nuraini M.C.K., Faaizah S., Ahmad Naim C.P.: Mapping of personalized learning environment (PLE) among Malaysian's secondary school. In: International Conference on Advances in Computing, Communication and Information Technology, pp. 13–16 (2014)

Comparing Features Extraction Methods for Person Authentication Using EEG Signals

Siaw-Hong Liew, Yun-Huoy Choo, Yin Fen Low,
Zeratul Izzah Mohd Yusoh, Tian-Bee Yap and Azah Kamilah Muda

Abstract This chapter presents a comparison and analysis of six feature extraction methods which were often cited in the literature, namely wavelet packet decomposition (WPD), Hjorth parameter, mean, coherence, cross-correlation and mutual information for the purpose of person authentication using EEG signals. The experimental dataset consists of a selection of 5 lateral and 5 midline EEG channels extracted from the raw data published in UCI repository. The experiments were designed to assess the capability of the feature extraction methods in authenticating different users. Besides, the correlation-based feature selection (CFS) method was also proposed to identify the significant feature subset and enhance the authentication performance of the features vector. The performance measurement was based on the accuracy and area under ROC curve (AUC) values using the fuzzy-rough nearest neighbour (FRNN) classifier proposed previously in our earlier work. The results show that all the six feature extraction methods are promising. However, WPD will induce large vector set when the selected EEG channels increases. Thus, the feature selection process is important to reduce the features set before combining the significant features with the other small feature vectors set.

S.-H. Liew · Y.-H. Choo (✉) · Z.I.M. Yusoh · T.-B. Yap · A.K. Muda
Faculty of Information and Communication Technology, Universiti Teknikal Malaysia
Melaka (UTeM), 76100 Durian Tunggal, Melaka, Malaysia
e-mail: huoy@utem.edu.my

S.-H. Liew
e-mail: siawhong.liew@gmail.com

Z.I.M. Yusoh
e-mail: zeratul@utem.edu.my

A.K. Muda
e-mail: azah@utem.edu.my

Y.F. Low
Faculty of Electronics and Computer Engineering, Universiti Teknikal Malaysia Melaka
(UTeM), 76100 Durian Tunggal, Melaka, Malaysia
e-mail: yinfen@utem.edu.my

© Springer International Publishing Switzerland 2015
A. Abraham et al. (eds.), *Pattern Analysis, Intelligent Security
and the Internet of Things*, Advances in Intelligent Systems and Computing 355,
DOI 10.1007/978-3-319-17398-6_21

Keywords Electroencephalograms · Feature extraction · Person authentication · Feature selection

1 Introduction

Person authentication using brainwaves particularly aimed to differentiate client from imposter based on the distinctive features hidden in the electroencephalograms (EEG) signals. EEG signals are unique but highly noisy, weak and difficult to process. Therefore, feature extraction plays an important role in extracting more relevant and meaningful information to facilitate better analysis. It is important to represent noisy, weak and non-stationary raw data such as EEG signals in a better manner. Features extraction stage involves the transformation of the raw data signal into a relevant data structure which is known as feature vector. A good feature vector tends to suppress noise, disclose important information and eliminate redundant data [1].

In the study of signal processing, feature extraction methods such as Fast Fourier Transform (FFT), autoregressive (AR) model and wavelet transform (WT) are widely used in many signal processing studies. Nevertheless, they are prone to different shortcomings, thus has jeopardized the performance in signal analysis. The FFT method provides useful information but only from frequency domain. Features with the combination of time domain and frequency information can improve the classification performance of EEG signals [2]. On the other hand, the AR model cannot capture transient features from the EEG signals [3]. The feature vectors of WT are rather complex and it will increase the difficulty to get accurate transcendent information.

Feature extraction methods such as FFT, AR and WT are not promising in non-stationary signals, i.e. the EEG signal. In our earlier work [4], some of the feature extraction methods were identified from the literature and were used in analysing EEG signals for person authentication purposes. No comparison was done on the selected feature extraction methods towards the classification performance. In recent studies, WPD, Hjorth parameter and mutual information are claimed to be good and appropriate for EEG signals analysis [5, 6]. Therefore, this study aimed to compare the feature extraction methods used in [4] and other methods recommended in the literature for person authentication analysis using the non-stationary EEG signals.

The rest of this chapter is organized as follows: Section 2 presents the proposed feature extraction methods in this study. Section 3 discusses the dataset, the experimental design, feature selection technique and the fuzzy-rough nearest neighbour (FRNN) classification technique. Section 4 depicts the results and discussion while Sect. 5 draws the conclusions and the direction of the future work.

2 The Proposed Feature Extraction Methods

Raw EEG data are non-stationary, noisy, complex and difficult to analyse. Therefore, feature extraction is needed to extract the relevant information or characteristics from the EEG signals. Features extracted from EEG signals are unique between subjects and sufficient for person authentication [7]. Different features provide different discriminative power for different subjects. Most of the authentication systems will make use of features combination architecture. The results were able to demonstrate the significant improvement in the system performance [8]. The feature extraction methods used in this study are as follows:

2.1 Wavelet Packet Decomposition (WPD) [3, 5]

In [5], the authors have demonstrated that WPD is an excellent feature extraction method for non-stationary signals such as EEG signals, and it is very appropriate for EEG signal analysis. WPD provides a multi-level time-frequency decomposition of signals, and it is able to provide more significant features. The wavelet decomposition splits the original signal into detail and approximation. After that, the approximation is split itself into next level approximation and detail. This process will be repeated until n-level. On the other hand, the detail also split itself into the next level to yield more than different ways to encode the signal. Figure 1 shows a complete decomposition tree of a signal.

Research work in [3] has proven that Daubechies with order 4 (DB4) wavelet and sixth level of wavelet packet decomposition is appropriate parameter in order to analyse the EEG signals with 256-Hz sampling rate. Since the frequency of useful EEG signals is lower than 50-Hz, therefore, we use 25 sub-bands in each channel. The combination with the time domain and frequency domain can provide more significant features; we characterized the time–frequency distribution of EEG signals by combining the features below:

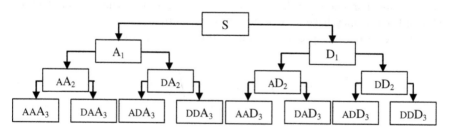

Fig. 1 Level 3 of wavelet packet decomposition tree [9]

Average coefficients in sixth sub-band. A total of 9 channels were selected from 64 available channels in the dataset and the sampling rate for each channel is 2^8, the sub-band means (Mj) at jth level is defined as in Eq. (1):

$$M_j = \frac{1}{2^j} \sum_{k}^{2^8} d_j(k) \tag{1}$$

where $d_j(k)$ represents the coefficient of WPD at jth level and kth sample. 25 sub-bands were used for WPD as the frequency of useful EEG signal is lower than 50 Hz. Therefore, the dimensions of feature vector for average coefficients are 225.

Wavelet packet energy in each sub-band. In the perspective of wavelet packet energy, WPD decomposes signal energy on different time–frequency plain; the integration of square amplitude of WPD is proportional to signal power. The sub-band entropy is defined as in Eq. (2):

$$E(j,n) = \int |S(t)|^2 dt = \sum_{k} \left(d_j^n(k) \right)^2 \tag{2}$$

where $n = 0, 1, 2, \ldots, 2^j$. Since we were selected 9 channels in this research, the dimension of feature vector for wavelet packet energy is 225.

2.2 Hjorth Parameter [6, 10]

Hjorth parameter is essential to analyse EEG signals in both time and frequency domains. It can extract the property of EEG signals efficiently [6]. Hjorth parameters are used to compute the quadratic mean and the dominant frequency of EEG signals on each side of the brain, we used the first two Hjorth descriptors in 1970 and 1973, namely activity and mobility. From the activity and mobility in the EEG signals, it reflects the global trend of a signal, for visual analysis. Hjorth parameters were used in various online EEG analyses, such as in sleep staging in order to compute the amplitude and the main frequency of a signal. These descriptors are chosen because they have a low calculation cost [10].

Let us consider the spectral moment of order zero and two

$$m_0 = \int_{-\pi}^{\pi} S(w)dw = \frac{1}{T} \int_{t-T}^{t} f^2(t)dt \tag{3}$$

$$m_2 = \int_{-\pi}^{\pi} w^2 S(w) dw = \frac{1}{T} \int_{t-T}^{t} \left(\frac{df}{dt}\right)^2 dt \tag{4}$$

where $S(w)$ represents the power density spectrum and $f(t)$ represents the EEG signal within an epoch of duration T. The first two of Hjorth parameters is given by

$$\text{Activity} : h_0 = m_0 \tag{5}$$

$$\text{Mobility} : h_1 = \sqrt{\frac{m_2}{m_0}} \tag{6}$$

h_0 is the square of the quartic mean and h_1 reflect that frequency of dominant. These quantities that are in discrete forms, where $h_0(k)$ and $h_1(k)$ at a sampled time of k, are calculated within a sliding window of 1 s length using the open-source software library BioSig.

Besides that, Hjorth parameter has also used the fourth-order spectral moment m_4 to define a measure of the bandwidth of the signal, called complexity.

$$\text{Complexity} : h_2 = \sqrt{\frac{m_4}{m_2} - \frac{m_2}{m_0}} \tag{7}$$

The first parameter is activity which represents the signal power, mobility represents the mean frequency, and complexity represents the change in frequency.

2.3 Mean [4]

Mean, also known as average, can be obtained by summing up of all EEG potential value and divides by the number of samples. The expression of the mean is given in Eq. (2) as follows:

$$\bar{x} = \frac{1}{n} \cdot \sum_{i=1}^{n} x_i \tag{8}$$

where, n is the number of data and x_i is the value of the data.

2.4 Coherence [11]

Coherence is used to measure the degree of linear correlation between two signals. The correlation between two time series at different frequencies can be uncovered by coherence. The range value for the magnitude of the squared coherence estimate is between 0 and 1. The value of 0 for the coherence function means the independence between two signals while a value of 1 for the coherence function means the complete linear dependence. The formula of coherence is given as follows:

$$C_{xy}(f) = \frac{|P_{xy}(f)|^2}{P_{xx}(f)P_{yy}(f)} \tag{9}$$

where $C_{xy}(f)$ is a function of the power spectral density, (P_{xx} and P_{yy}) of x and y and the cross-power spectral density (P_{xy}) of x and y.

2.5 Cross-correlation [12]

The main purpose of the cross-correlation is to measure the similarity between two channels. Cross-correlation is also known as sliding dot product, which is used to find occurrences of a known signal in unknown one. Furthermore, it is a function of the relative delay between the signals which can be applied in pattern recognition and cryptanalysis. Two input signals will be used to compute the cross-correlation:

- Channel 1 with itself: ρ_X;
- Channel 2 with itself: ρ_Y;
- Channel 1 with channel 2: ρ_{XY}.

The correlation ρ_{XY} between two random variables x and y with expected values, μ_X and μ_Y, and standard deviation, σ_X and σ_Y is given as:

$$\rho_{XY} = \frac{\text{cov}(X, Y)}{\sigma_X \sigma_Y} = \frac{E((X - \mu_X)(Y - \mu_Y))}{\sigma_X \sigma_Y} \tag{10}$$

where $E(\cdot)$ is the expectation operator, and $\text{cov}(\cdot)$ is the covariance operator.

2.6 Mutual Information (MI) [7, 8]

Mutual information theory represents the quantity that can measure the mutual dependence of the two signals. It is defined as the difference between the sum of entropies within the time series of two channels and their mutual entropy. Logarithms of base 2 were used in the experiment to measure MI in bit.

3 Materials and Methods

3.1 Data Description and Data Preparation

In this study, an EEG dataset from UCI Machine Learning Repository were used in the experiments. The full dataset consists of three versions of data with different subject size, i.e. small (1 subject), large (10 subjects) and full dataset (122 subjects). Large dataset were used in this research and it consists of 10 subjects with 64 channels electrode placement. Each individual is completed with a total number of 60 trials and sampled at 256 Hz (3.9-ms epoch). Due to many redundant trials in one of the subjects, it was replaced by another subject from the full dataset. The swap was performed to ensure that the prediction ability is not biased due to the redundant data in both training and testing phase in a particular subject.

Instead of treating the classification as a ten-class problem, the classifier was trained with only two outputs, i.e. the client and the imposter. The data were split into 80 % of training and 20 % of testing. For training data, 16 trials of S1 object and 32 trials of S2, both match and not match will be selected. On the other hand, there are 4 trials of S1 object and 8 trials of S2 object, both match and not match cases were selected for testing data. It is because the amplitude of the EEG signals will be different when the subject performed different task. The signal data in S2, both match and not match, involve analysis of the picture whether it is match or not match with the previous picture. This is different from the EEG signals as S1 object does not involve analysis as such.

Only the lateral and midline electrodes were used in this study because they have been proven good and able to provide stronger signals in response to visual stimuli [13]. The O1, OZ, O2, PO7 and PO8 are the selected lateral active electrodes, while the FPZ, FZ, OZ, CZ and PZ are the selected midline active electrodes.

3.2 Feature Selection

The WPD method tends to induce large vector set especially when the selected EEG channels increases. Thus, the feature selection process is important to reduce the features set before combining the significant features with the other small feature vectors set. Three common models for feature selection are the filter model, the wrapper model and the embedded model. Correlation-based feature selection (CFS) is a good feature selection method which is able to reduce dimensionality without affecting accuracy [14]. It is a fast and correlated-based filter algorithm that is applicable in discrete and continuous problems [14]. The CFS algorithm evaluates the feature subset according to the correlation-based heuristic merit. A good feature subset contains high correlation between features and the class [15]. In this study, the experiments were designed in two levels, i.e. select the attributes from WPD feature vectors using CFS algorithm before combining with other feature vectors;

and to apply the CFS feature selection across all feature vectors at the same time. The results in Sect. 4 show the influence of different feature vectors on the classification performance.

3.3 Fuzzy-Rough Nearest Neighbour (FRNN) Classification

The performance quantification in this study was based on the accuracy and area under ROC curve (AUC) measurements using the FRNN classifier proposed previously in our earlier work. FRNN classifier introduced by Jensen and Cornelis [16] is an algorithm which combined the strength of fuzzy sets, rough sets and nearest neighbour classification approach motivated by human decision making. The implementation of FRNN algorithm was carried out using the fuzzy-rough version of WEKA data mining tools. In FRNN algorithm, the nearest neighbours are used to construct the fuzzy lower and upper approximations to quantify the membership value of a test object to determine its decision class, and test instances are classified based on their membership to these approximations. FRNN classification approach outperformed various nearest neighbours' approaches such as support vector machine and Naïve Bayes prediction models. FRNN was used in [4] and has gained good results for person authentication using EEG signals. The accuracy and AUC were recorded at 90.17 % and 0.904, respectively.

4 Result and Discussion

The experimental data were preprocessed in the same way as mentioned in Sect. 3. The same classification method and performance measures were used to ensure a fair comparison on different feature extraction methods. The extracted data were also further tested in the two experiment settings to investigate the effect of feature selection process. Table 1 shows the comparison of classification performance using 3 features extraction methods as reported in [4] and 6 features extraction methods as proposed in this study. The 3 feature extraction methods are mean, cross-correlation and coherence while the 6 feature extraction methods are mean, cross-correlation, coherence, Hjorth parameter, mutual information and WPD. The mutual information feature values were normalized to the interval of [0, 1] since the extracted values were relatively small compared with other features.

The average classification accuracy for 3 features set is slightly lower than the result using 6 features set while the AUC shows contrary result where the 3 features set is slightly outperform the 6 features set. It might due to the large number of feature vectors (567 features) which have affected the classification performance. Feature selection was proposed and implemented in the second stage of the experiments. Apart from that, it is obvious that person 7 has obtained the highest accuracy and AUC while person 4 has gained the lowest accuracy and AUC for

Table 1 Comparison of classification performance using 3 feature extraction methods versus 6 features extraction methods

Person	3 feature extraction methods		6 feature extraction methods	
	Accuracy (%)	AUC	Accuracy (%)	AUC
Person 1	87.50	0.924	96.67	0.981
Person 2	86.67	0.788	89.17	0.843
Person 3	88.33	0.922	85.83	0.909
Person 4	80.83	0.704	78.33	0.681
Person 5	93.33	0.954	91.67	0.993
Person 6	88.33	0.924	88.33	0.854
Person 7	99.17	1.000	97.50	1.000
Person 8	90.83	0.895	88.33	0.826
Person 9	90.00	0.936	92.50	0.934
Person 10	96.67	0.990	95.00	0.992
Average	90.17	0.904	90.33	0.901

both 3 features and 6 features. The results are in line with the earlier work reported in [4] where most of the VEP signals for person 7 are more consistent and hence the results are the best compared with others. In contrast, the EEG data of person 4 are incomplete and thus have influenced the classification performance.

In the second stage of experiment, feature selection method, i.e. CFS, was used to select the important attributes before the classification process. From a total of 567 attributes, only 21 attributes are selected. The performance of the selected attributes demonstrated using FRNN classifier is as shown in Table 2.

Commonly, the results are generally more promising when implementing feature selection process. However, in this experiment, both the classification accuracy and AUC values were worse than the results without feature selection. The average accuracy was recorded at 87.00 % while the AUC was recorded at 0.760. The dominating feature, i.e. WPD, has causing the bias in feature selection process, thus resulting in lower accuracy and AUC readings.

In order to avoid bias on the large feature vector, i.e. the WPD features, feature selection was applied to identify significant attributes among the WPD features before combining with the Hjorth parameter, mean, coherence, cross-correlation and mutual information. The classification results are as shown in Fig. 2. From the results, it is proven that a better way of avoiding bias by large feature vectors is to apply separate feature selection process on individual large feature vector before combining with other small feature vectors. Refer to Table 2 for the comparison of feature selection on WPD feature vector.

The classification accuracy and AUC with the feature selection applied on WPD only are both better than applying feature selection across all feature vectors at the same time. The average accuracy and AUC were reported an increase of 4.67 % and

Table 2 Comparison of classification performance without feature selection (FS), with feature selection (FS) and feature selection (FS) on WPD only for 6 feature extraction methods

Person	Without FS		With FS		FS on WPD only	
	Accuracy (%)	AUC	Accuracy (%)	AUC	Accuracy (%)	AUC
Person 1	96.67	0.981	85.83	0.827	98.33	0.991
Person 2	89.17	0.843	85.83	0.618	93.33	0.902
Person 3	85.83	0.909	84.17	0.715	94.17	0.968
Person 4	78.33	0.681	86.67	0.690	92.50	0.937
Person 5	91.67	0.993	90.83	0.693	99.17	0.999
Person 6	88.33	0.854	80.00	0.678	91.67	0.944
Person 7	97.50	1.000	92.50	0.959	98.33	1.000
Person 8	88.33	0.826	87.50	0.725	94.17	0.975
Person 9	92.50	0.934	86.67	0.734	90.00	0.949
Person 10	95.00	0.992	90.00	0.962	98.33	0.999
Average	90.33	0.901	87.00	0.760	95.00	0.966

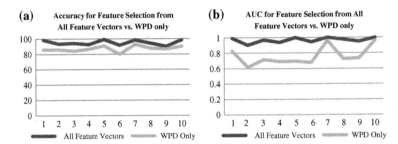

Fig. 2 The comparison of classification accuracy (**a**) and AUC (**b**) for implementing feature selection on all feature vectors versus on only the WPD feature vector

0.065, respectively. Therefore, appropriate feature selection process has proven to be good in person authentication analysis especially when many EEG channels are used and the extracted feature vectors are large in quantity.

5 Conclusion

In this chapter, we have compared and discussed on few well-known feature extraction methods. It can be concluded that WPD cannot perform well in person authentication classification. It must combine with other feature extraction methods in order to get higher accuracy and AUC. Apart from that, feature selection also plays important roles. Last but not least, WPD, Hjorth parameter, mean, coherence, cross-correlation and mutual information are good to extract important attributes for person authentication, and feature selection is needed if the size of feature vectors is large.

Acknowledgments This work was funded by the Ministry of Higher Education Malaysia and Universiti Teknikal Malaysia Melaka (UTeM) through the Fundamental Research Grant Scheme, FRGS/2/2013/ICT02/FTMK/02/6/F00189.

References

1. DeryaUbeyli, E.: Statistics over features: EEG signals analysis. Comput. Biol. Med. **39**, 733–741 (2009)
2. Mensh, B.D., Werfel, J., Seung, H.S.: BCI competition 2003–data set Ia: combining gamma-band power with slow cortical potentials to improve single-trial classification of electroencephalographic signals. IEEE Trans. Biomed. Eng. **51**, 1052–1056 (2004)
3. Dingyin, H., Wei, L., Xi, C.: Feature extraction of motor imagery EEG signals based on wavelet packet decomposition. In: IEEE/ICME International Conference on Complex Medical Engineering (CME), pp. 694–697, IEEE Press, New Jersey (2011)
4. Liew, S.H., Choo, Y.H., Low, Y.F.: Fuzzy-rough nearest neighbour classifier for person authentication using EEG signals. In: International Conference on Fuzzy Theory and Its Applications (iFUZZY), pp. 316–321. IEEE Press, Taipei (2013)
5. Ting, W., Guo-zheng, Y., Bang-hua, Y., Hong, S.: EEG feature extraction based on wavelet packet decomposition for brain computer interface. Measurement **41**, 618–625 (2008)
6. Oh, S.H., Lee, Y.R., Kim, H.N.: A novel EEG feature extraction method using Hjorth parameter. Int. J. Electron. Electr. Eng. **2**, 106–110 (2014)
7. Jian-feng, H.: Multifeature biometric system based on EEG signals. In: Proceedings of the 2nd International Conference on Interaction Sciences, pp. 1341–1345. ACM, New York (2009)
8. Jian-feng, H.: Biometric system based on EEG signals by feature combination. In: International Conference on Measuring Technology and Mechatronics Automation (ICMTMA), pp. 752–755. IEEE Press, Changsha (2010)
9. Gokhale, M.Y., Khanduja, D.K.: Time domain signal analysis using wavelet packet decomposition approach. Netw. Syst. Sci. **3**, 321–329 (2010)
10. Cecchin, T., Ranta, R., Koessler, L., Caspary, O., Vespignani, H., Maillard, L.: Seizure lateralization in scalp EEG using Hjorth parameters. Clinical Neurophysiol. **121**, 290–300 (2010)
11. Safont, G., Salazar, A., Soriano, A., Vergara, L.: Combination of multiple detectors for EEG based biometric identification/authentication. In: IEEE International Carnahan Conference on Security Technology (ICCST), pp. 230–236. IEEE, Boston (2012)
12. Riera, A., Soria-Frisch, A., Caparrini, M., Grau, C., Ruffini, G.: Unobtrusive biometric system based on electroencephalogram analysis. EURASIP J. Adv. Signal Process. **2008**, 1–8 (2008)
13. Odom, J.V., Bach, M., Brigell, M., Holder, G.E., McCulloch, D.L., Tormene, A.P.: ISCEV standard for clinical visual evoked potentials (2009 update). Doc. Ophthalmol. **120**, 111–119 (2009)
14. Hall, M.A.: Correlation-based feature selection for discrete and numeric class machine learning. In: ICML'00 Proceedings of the 17th International Conference on Machine learning, pp. 359–366, USA (2000)
15. Lu, X., Peng, X., Liu, P., Deng, Y., Feng, B., Liao, B.: A novel feature selection method based on CFS in cancer recognition. In: IEEE 6th International Conference on Systems Biology (ISB), pp. 226–231, China (2012)
16. Jensen, R., Cornelis, C.: Fuzzy-rough nearest neighbour classification and prediction. Theoret. Comput. Sci. **412**, 5871–5884 (2011)

A Comparative Study of 2D UMI and 3D Zernike Shape Descriptor for ATS Drugs Identification

Satrya Fajri Pratama, Azah Kamilah Muda, Yun-Huoy Choo and Ajith Abraham

Abstract Drug abuse is a threat to national development. Generally, drugs can be identified based on the structure of its molecular components. This procedure is becoming more unreliable with the introduction of new amphetamine-type stimulants (ATS) molecular structures which are increasingly complex and sophisticated. An in-depth study is crucial to accurately identify the unique characteristics of molecular structure in ATS drug. Therefore, this chapter is meant for exploring the usage of shape descriptors (SD) to represent the drug molecular structure. Two-dimensional (2D) united moment invariant (UMI) and three-dimensional (3D) Zernike are selected and their performances are analyzed using drug chemical structures obtained from United Nations Office of Drugs and Crime (UNODC) and various sources. The evaluation identifies the most interesting method to be further explored and adapted in the future work to fully compatible with ATS drug identification domain.

Keywords United moment invariant · 3D Zernike · Moment invariants function · ATS drugs · Drugs identification · Molecular structure

S.F. Pratama · A.K. Muda · Y.-H. Choo · A. Abraham (✉)
Computational Intelligence and Technologies (CIT) Research Group, Center of Advanced Computing and Technologies, Faculty of Information and Communication Technology, Universiti Teknikal Malaysia Melaka, Hang Tuah Jaya, 76100 Durian Tunggal, Melaka, Malaysia
e-mail: ajith.abraham@ieee.org

S.F. Pratama
e-mail: satrya@student.utem.edu.my

A.K. Muda
e-mail: azah@utem.edu.my

Y.-H. Choo
e-mail: huoy@utem.edu.my

A. Abraham
Machine Intelligence Research Labs (MIR Labs), Scientific Network for Innovation and Research Excellence, Auburn, WA, USA

© Springer International Publishing Switzerland 2015
A. Abraham et al. (eds.), *Pattern Analysis, Intelligent Security and the Internet of Things*, Advances in Intelligent Systems and Computing 355, DOI 10.1007/978-3-319-17398-6_22

237

1 Introduction

Abuse of amphetamine-type stimulants (ATS) drugs has become a global, harrowing social problem. Law enforcement authorities are still struggling to find a concrete solution to prevent drugs abuse due to the existence of new brand or unfamiliar ATS drug substances. However, less attention is given to the shape similarity search that can lead to identification of unknown substances in the field of chemoinformatics.

Generally, drugs can be identified based on the structure of its molecular components. This procedure is becoming more unreliable with the introduction of new ATS molecular structures which are increasingly complex and sophisticated. However, due to the limitations of the current test kit to detect new brand or unfamiliar ATS drug, it presents a challenge to both national law enforcement authorities and scientific staff of forensic laboratories. In addition, existing drug test kits are sometimes prone to false positive detection.

Both of the two-dimensional (2D) and three-dimensional (3D) chemical structures are basically represented as a shape. Two-dimensional shape descriptors (SD) have been developed which can be generally divided into boundary and area based. Meanwhile, 3D SD focus on volume-based and surface-based descriptors. 3D SD also have been described as more powerful and accurately represent a component structure's shape. Thus, 3D SD are believed to be used to identify molecular structure of ATS drug's chemical components, even for a new brand of ATS drug due to their similar ring substitutes.

This paper aimed to explore the distinguishing strength of 2D united moment invariant (UMI) [1] SD compared with 3D Zernike [2] SD in discriminating ATS drugs from non-ATS (n-ATS) drugs. The remaining part of this study is structured as follows: In next section, an overview of ATS drug identification is given. Section 3 provides an overview of molecular structure representation. In Sect. 4, experimental setup describing the data source collection and experimental design is presented, and the results are discussed in Sect. 5, while conclusion and future works are drawn in Sect. 6.

2 ATS Drug Identification

Manual identification of ATS follows a set of standard methods outlined by United Nations Office of Drugs and Crime (UNODC). These standards, however, are not closely followed by chemists, causing the results obtained possibly ranging from one testing laboratory to another. Nevertheless, the most common method to identify a chemical substance is gas chromatography/mass spectrometry (GC/MS).

A recent study by forensic experts, however, shows that GC/MS is flawed while trying to identify several ATS drugs, most notably is methamphetamine [3]. Methamphetamine itself has two stereo-isomers, which is l-methamphetamine and d-methamphetamine. GC/MS is also increasingly incapable to determine that several chemical structures are actually ATS drugs. While l-methamphetamine has very little pharmacodynamics effect, *d*-methamphetamine on the other hand is a controlled substance that has high potential for abuse and addiction [4].

The traditional identification of ATS drugs defined by UNODC is fully dependent of the chemical composition of the drugs. However, due to the limitations of the current test kit to detect new brand or unfamiliar ATS drug, it presents a challenge to both national law enforcement authorities and scientific staff of forensic laboratories. In addition, existing drug test kits are sometimes prone to false positive detection. Therefore, the introduction of new chemical composition to ATS drugs will pose greater problems in the identification task. This problem can be resolved by relying on the shape of the chemical structure of the drug itself, because regardless of the chemical composition of ATS drugs, its core chemical structure will remain the same. Both 2D and 3D models are often used to depict the molecular structure. However, 2D model hides the properties of volume and surface which eliminates the ring substitutes in a molecule. Therefore, a 3D model is essential to show and differentiate the unique features at a ring substitute. Therefore, in-depth study is crucial to accurately identify the unique characteristics of molecular structure in ATS drugs.

Although chemical analysis is a matured field of science, very little research on the application of soft computing for chemical analysis was performed, especially on the drug similarity analysis and drug identification in chemoinformatics domain. However, chemoinformatics researches focus more on the development of chemical drugs that causes desired biological effect and less on the similarity search that can lead to identify unknown substances [5]. Another example of soft computing is the introduction of small molecule sub-graph detector (SMSD) [6].

However, most soft computing techniques introduced recently are applied in the drug design and drug discovery domain [7, 8]. Some researchers also explore the usage of principal component analysis (PCA) for screening drugs of abuse [9], neural network for structures to quantitative analysis [10, 11], neural network and *k*-nearest neighbor for quantitative analysis of Raman spectroscopy data [12], and the most recent, PCA and neural network to classify molecular data [13]. Literature has shown that this field is still not being extensively explored in the context of soft computing and computational intelligence, as evidenced by some of the most recent studies on drug identification and similarity searching [5, 6].

3 Molecular Structure Representation

To assess the similarity of these chemical substances, a numerical representation of chemical substance is required. A review of techniques that transform a chemical structure into numerical representations has been discussed by Nikolova and Jaworska [14]. The study concludes that the similarity computation may not correctly represent the similarity between two chemical structures, due to the properties of a chemical might not be implicit in its molecular structure, since it might not be fully measured and represented by a set of numbers.

Shape is one of the basic features used to describe image content [15], and thus, searching for an image by using the shape features giving challenges for many researches, since extracting the features that represent and describe the shape is a difficult task [16]. Invariants is the objects' descriptions by a set of measurable quantities that are insensitive regardless of any transformations, and thus remain possesses discriminative power to be distinguished from other objects which are instance of other classes [17]. Both 2D and 3D chemical structures are basically represented as shape. These representations are commonly referred as molecular descriptors, which are obtained when molecules, thought of as real objects, are transformed into a molecular representation enabling mathematical treatment [18].

There are simple molecular descriptors, usually called topological or 2D descriptors, and there are molecular descriptors derived from a geometrical representation that are called geometrical or 3D SD. Because a geometrical representation involves knowledge of the relative positions of the atoms in 3D space, geometrical descriptors usually provide more information and discrimination power than topological descriptors for similar molecular structures and molecule conformations. Searching for relationships between molecular structures and complex properties can often efficiently be performed by use of geometrical descriptors, by exploiting their large information content [18, 19].

Many molecular descriptors have been proposed; they are derived from different theories and approaches with the aim of predicting biological and physicochemical properties of molecules, such as 3D Zernike descriptors [20, 21], shape impact descriptor [22], local intersection volume [23], and path-space ratio [24].

3.1 2D United Moment Invariants

Moment invariants (MI), which are special functions of image moments, have been widely explored, and it is a very useful tool for pattern recognition research [1]. Moments are scalar quantities used to characterize a function and to capture its significant features. The first introduction of MI to pattern recognition and image processing was the employment of algebraic invariants theory by Hu [25], which derived his renowned seven invariants to the rotation of 2D objects, namely geometric moment invariants (GMI). Ding et al. [26] found that GMI lose its scale invariance in discrete condition. Yinan et al. [1] proposed UMI which are based on the moment invariants where the rotation, translation, and scaling can be discretely kept invariant to region, closed and unclosed boundary, and represents the eight formulae of UMI as the pattern to represent image shape as shown in (1) for 2D image shape:

$$\theta_1 = \frac{\sqrt{\phi_2}}{\phi_1} \qquad \theta_2 = \frac{\phi_6}{\phi_1 \phi_4} \qquad \theta_3 = \frac{\sqrt{\phi_4}}{\phi_4} \qquad \theta_4 = \frac{\phi_5}{\phi_3 \phi_4} \qquad \theta_5 = \frac{\phi_1 \phi_6}{\phi_2 \phi_3}$$
$$\theta_6 = \left(\phi_1 + \sqrt{\phi_2}\right) \frac{\phi_3}{\phi_6} \qquad \theta_7 = \frac{\phi_1 \phi_5}{\phi_3 \phi_6} \qquad \theta_8 = \frac{\phi_3 + \phi_4}{\sqrt{\phi_5}} \tag{1}$$

where ϕ_i are GMI. Two-dimensional UMI method has been explored in [16, 27–29]. Several studies have been conducted to verify the quality of UMI under various transformations, such as scaling, translation, and rotation, and it is proved that UMI is a robust and reliable image descriptor.

3.2 3D Zernike

Three-dimensional Zernike can be obtained by expanding 3D function $f(x)$ into a series in terms of Zernike-Canterakis basis [2] in (2).

$$Z_{nl}^m(r, \vartheta, \varphi) = R_{nl}(r)Y_l^m(\vartheta, \varphi) \tag{2}$$

with $-l < m < l$, $0 \le l \le n$, and $(n - l)$ even. Here, $Y_l^m(\vartheta, \varphi)$ are spherical harmonics [30]. Spherical harmonics is the angular portion of an orthogonal set of solutions to Laplace's equation [31], which is given in (3).

$$Y_l^m(\vartheta, \varphi) = N_l^m P_l^m(\cos \vartheta)e^{im\varphi} \tag{3}$$

where N_l^m is a normalization factor defined in (4).

$$N_l^m = \sqrt{\frac{2l + 1}{4\pi} \frac{(l - m)!}{(l + m)!}} \tag{4}$$

and P_l^m are the associated Legendre functions. $R_{nl}(r)$ is radial function [32], constructed so that $Z_{nl}^m(r, \vartheta, \varphi)$ are polynomials when written in Cartesian form [31]. The conversion between spherical coordinates and Cartesian x is defined in (5)

$$x = |x|\xi = r\xi = r(\sin \vartheta \sin \varphi, \sin \vartheta \cos \varphi, \cos \varphi)^T \tag{5}$$

and the harmonics polynomials e_l^m are defined in (6)

$$e_l^m(x) \equiv r^l Y_l^m(\vartheta, \varphi) = r^l c_l^m \left(\frac{ix - y}{2}\right)^2 z^{l-m} \times \sum_{\mu=2}^{\lfloor \frac{l-m}{2} \rfloor} \binom{l}{\mu}\binom{l - \mu}{m + \mu}\left(-\frac{x^2 + y^2}{4z^2}\right) \tag{6}$$

where c_l^m is normalization factor defined in (7)

$$c_l^m = c_l^{-m} = \frac{\sqrt{(2l + 1)(l + m)!(l - m)!}}{l!} \tag{7}$$

Using the harmonics polynomials e_l^m, 3D Zernike functions in (4) can be rewritten in Cartesian coordinates defined in (8).

$$Z_{nl}^m(r, \vartheta, \varphi) = R_{nl}(r)Y_l^m(\vartheta, \varphi) = \sum_{v=0}^{k} q_{kl}^v |x|^{2v} r^l Y_l^m(\vartheta, \varphi) = \sum_{v=0}^{k} q_{kl}^v |x|^{2v} e_l^m(x) \quad (8)$$

where $2k = n - l$ and the coefficient q_{kl}^v are determined in (9) to guarantee the orthonormality of the functions within the unit sphere [32].

$$q_{kl}^v = \frac{(-1)^k}{2^{2k}} \sqrt{\frac{2l + 4k + 3}{3}} \binom{2k}{k} (-1)^v \frac{\binom{k}{v}\binom{2(k+l+v)+1}{2k}}{\binom{k+l+v}{k}} \quad (9)$$

Now 3D Zernike moments of $f(x)$ are defined as the coefficients of the expansion in this orthonormal basis by the formula in (10).

$$\Omega_{nl}^m = \frac{3}{4\pi} \int\limits_{|x| \le 1} f(x)\overline{Z}_{nl}^m(x)dx \quad (10)$$

The moments are collected into $(2l + 1)$ dimensional vectors $\Omega_{nl} = (\Omega_{nl}^l, \Omega_{nl}^{l-1}, \Omega_{nl}^{l-2}, \Omega_{nl}^{l-3}, \ldots, \Omega_{nl}^{-l})$ to achieve rotation invariance and define the rotationally invariant F_{nl} as norms of vectors Ω_{nl} [31], defined in (11).

$$F_{nl} = \sqrt{\sum_{m=-1}^{m=l} \left(\Omega_{nl}^m\right)^2} \quad (11)$$

Index n is called the order of the descriptor. The rotational invariance of 3D Zernike descriptors means that calculating F_{nl} for an object and its rotated version would yield the same result [31].

4 Experimental Setup

In this section, a detailed description of the experimental method is provided in order to conduct an extensive and rigorous empirical comparative study.

4.1 Data Source Collection

This section describes the process of transforming molecular structure of ATS drug into 2D and 3D computational data representation. ATS dataset used in this research comes from [33], which contains 60 molecular structures which are commonly distributed for illegal use. On the other hand, 60 n-ATS drug chemical structures are also collected from various sources, which will be used as benchmarking dataset. The number of data is currently limited, due to the complexities in obtaining them. However, future works will include the dataset obtained from National Poison Centre, Malaysia.

These structures are drawn in 2D chemical structure format using MarvinSketch 6.3.0 [34]. After the 2D chemical structure is created, the structure will be saved as PNG image, which will be extracted as third-order moment using 2D UMI [1]. The sample of the 2D UMI features is shown in Table 1. After the 2D chemical structure is saved as PNG image, the structure will be cleaned and transformed to 3D chemical structure, also by using MarvinSketch [34]. The structure will be then saved as structure data format (SDF) file.

The SDF file must be then converted to protein data bank (PDB) format, because PDB format is the input type required for generating voxel data of 3D chemical structure. In order to convert SDF file to PDB file, Open Babel 2.3.2 [35] is required. PDB file will be then voxelized to grid resolution data with 256 voxel resolution using voxels class of 3D Zernike program [20]. After the voxel data has been generated, the eighth-order 3D Zernike will be calculated using Zernike class of 3D Zernike program [20]. The sample of the 3D Zernike is shown in Table 2.

As shown in Tables 1 and 2, the moments extracted from both 2D UMI and 3D Zernike are denoted in complex numbers. Therefore, these numbers must be transformed into real numbers, because most of the pattern recognition tasks only capable to handle real numbers. Every complex number can be expressed by

Table 1 Sample data of 2D UMI features for ATS and n-ATS chemical structures

Class	Structure name	F1	F2	...	F7	F8
ATS	2-(4-bromo-2, 5-dimethoxyphenyl) ethanamine	0.4592	−0.4114	...	2.3524	−2.0484i
n-ATS	Amikacin	0.1109	−0.1104	...	34.4864	−2.0910i

Table 2 Sample data of 3D Zernike features for ATS and n-ATS chemical structures

Class	Structure name	F1	...	F1457	F1458
ATS	2-(4-bromo-2, 5-dimethoxyphenyl) ethanamine	0.0	...	$-5.6139E{-}6$ $-1.1458E{-}6i$	$-7.2374E{-}6$ $-4.5976E{-}7i$
n-ATS	Amikacin	0.0	...	$2.9139E{-}7$ $+4.4098E{-}6i$	$-1.1301E{-}6$ $-1.3781E{-}6i$

specifying either the Cartesian coordinates or the polar coordinates. The complex number z can be represented in Cartesian coordinates as shown in (12)

$$z = x + yi \tag{12}$$

where i is the imaginary unit. The Cartesian coordinates x and y can be converted to polar coordinates r and φ with $r \geq 0$ and φ in the interval $[0, 2\pi)$ by using (13).

$$\begin{aligned} r &= \sqrt{x^2 + y^2} \\ \varphi &= \arctan 2(y, x) \end{aligned} \tag{13}$$

However, by representing the complex numbers as polar coordinates, the number of features will be doubled and the correlation between r and φ can be lost in the machine learning process. Therefore, these two values must be uniquely encoded into a single distinct value, and thus the correlation between the two values can be preserved. Pairing function (PF) can be used to perform this task. In this chapter, two well-known PFs are employed, which are Cantor [36] and Szudzik [37] PFs, defined in (14) and (15), respectively,

$$\langle k_1, k_2 \rangle = \frac{(k_1 + k_2)(k_1 + k_2 + 1)}{2} + k_2 \tag{14}$$

$$\langle k_1, k_2 \rangle = \begin{cases} k_2^2 + k_1 & k_1 \neq \max(k_1, k_2) \\ k_1^2 + k_1 + k_2 & k_1 = \max(k_1, k_2) \end{cases} \tag{15}$$

However, since PF can only be used to uniquely encode natural numbers [37, 38], both r and φ, which are commonly stored as 64-bit double-precision floating-number format, must be represented as natural numbers in order to be paired. Zuras et al. [39] defines IEEE 754 standard to represent the double-precision floating-number format as a binary string, which in turn can also be parsed as a long integer number. The value of a double-precision floating-number is given in (16).

$$(-1)^s \times \left(1 + \sum_{i=1}^{52} b_{52-i} \times 2^{-i} \right) \times 2^{e-1023} \tag{16}$$

where s is the sign of the floating-number, i is the index of bit in the binary string, and b_i is the bit in the specified index. The sample data for encoded values of 2D UMI and 3D Zernike are shown in Tables 3 and 4, respectively.

4.2 Experimental Design

The traditional framework of pattern recognition tasks, which are preprocessing, feature extraction, and classification, will be employed in this study. All extracted

Table 3 Sample data of encoded 2D UMI features for ATS and n-ATS chemical structures

PF	Class	Structure name	F1	F2	...	F7	F8
Cantor	ATS	2-(4-bromo-2, 5-dimethoxphenyl) ethanamine	1.0589E+37	4.2461E+37	...	1.0638E+37	4.2585E+37
	n-ATS	Amikacin	1.0546E+37	4.2383E+37	...	1.0719E+37	4.2586E+37
Szudzik	ATS	2-(4-bromo-2, 5-dimethoxphenyl) ethanamine	2.1178E+37	2.1291E+37	...	2.1275E+37	2.1317E+37
	n-ATS	Amikacin	2.1092E+37	2.1291E+37	...	2.1437E+37	2.1317E+37

Table 4 Sample data of encoded 3D Zernike features for ATS and n-ATS chemical structures

PF	Class	Structure name	F1	...	F1457	F1458
Cantor	ATS	2-(4-bromo-2,5-dimethoxyphenyl) ethanamine	0	...	4.1798E+37	4.1812E+37
	n-ATS	Amikacin	0	...	4.1736E+37	4.1745E+37
Szudzik	ATS	2-(4-bromo-2,5-dimethoxyphenyl) ethanamine	0	...	2.1296E+37	2.1293E+37
	n-ATS	Amikacin	0	...	2.1247E+37	2.1310E+37

instances are randomly divided into twelve datasets and the experiment has been performed using 12-fold cross-validation. In order to justify the quality of features produced by each SDs, the features are tested against various classifiers five times, which are Bayes Network (BayesNet) [40], Naïve Bayes/Decision-Tree Hybrid (NBTree) [41], Decision Table (DT) [42], Random Forest (RF) [43], and Multilayer Perceptron (MLP) [40].

5 Experimental Results and Discussion

The classification accuracy of both SDs is the primary consideration of this study. Table 5 shows the results of classification accuracy.

Based on the results shown in Table 5, the 3D Zernike produces the best average of classification accuracy in both Cantor and Szudzik PFs. However, to further validate the strength of 3D Zernike compared with 2D UMI, in-depth statistical validation using independent samples t-test must be conducted, by using SPSS 17 software. The result of independent samples t-test is shown in Table 6.

Based on the result shown on Table 6, there was a statistically significant difference in the accuracy using Cantor PF for 2D UMI ($\mu = 50.50$, $\sigma = 2.7386$) and 3D Zernike ($\mu = 69.00$, $\sigma = 4.8016$); $t(8) = -7.4836$, $p = 0.0001$ and Szudzik PF for 2D UMI ($\mu = 50.50$, $\sigma = 2.1731$) and 3D Zernike ($\mu = 24.30$, $\sigma = 4.3700$);

Table 5 Classification accuracy of moments functions

PF	SD	Bayes net (%)	NB tree (%)	DT (%)	RF (%)	MLP (%)	Mean (μ) (%)	SD (σ)
Cantor	2D UMI	50.00	47.50	50.00	50.00	50.00	50.50	2.7386
	3D Zernike	70.00	63.33	65.00	71.67	75.00	69.00	4.8016
Szudzik	2D UMI	50.00	48.33	50.00	54.17	50.00	50.50	2.1731
	3D Zernike	73.33	61.67	67.50	64.17	66.67	66.67	4.3700

Table 6 Independent sample t-test for 2D UMI versus 3D Zernike using Cantor and Szudzik PFs

PF	t	df	Sig. (2-tailed)
Cantor	−7.4836	8	0.0001
Szudzik	−7.4070	8	0.0001

$t(8) = -7.4070$, $p = 0.0001$. These results suggest that the 3D Zernike is capable to differentiate the unique features of ATS drugs and n-ATS drugs at a ring substitute.

This is because 2D UMI is designed as an image descriptor for 2D image, such as 2D chemical structure model. Taking on the example of methamphetamine isomers, UMI considers the 2D isomer structure as one similar chemical compounds, due to its translation invariant property. Therefore, this may lead to different classification result if it is used in pattern recognition domain. The distinguishing power comes when the chemical structure is represented in 3D model where the properties of the ring substitutes are completely presented. Few studies have been conducted to apply the concept of MI in 3D structures [44, 45]. Future work to improve the performance of 2D UMI by transforming it into 3D UMI is therefore required.

6 Conclusion and Future Works

An extensive comparative study on SD methods for representing drug molecular structure has been presented. This study compared the merits of 2D UMI and 3D Zernike. The experiments have shown that 3D Zernike is indeed performs better in representing the molecular structure compared to 2D UMI.

Hence, future work to develop a novel 3D UMI method to better represent the molecular structure based on this experimental study is required. The proposed feature extraction method is going to be compared again with 3D Zernike discussed in this chapter and more benchmarking 3D-based SDs, such as 3D Legendre and 3D Fourier transform, will also be compared using specifically tailored classifiers for shape representation, such as classifiers by Zhiyong et al. [46], Xu et al. [47]. Additional data from National Poison Centre, Malaysia, will also be used as dataset in the future works.

Acknowledgments This work was supported by Collaborative Research Programme (CRP)—ICGEB Research Grant (CRP/MYS13-03) from International Centre for Genetic Engineering and Biotechnology (ICGEB), Italy.

References

1. Yinan, S., Weijun, L., Yuechao, W.: United moment invariants for shape discrimination. In: International Conference on Robotics, Intelligent Systems and Signal Processing, Changsha, pp. 88–93. IEEE (2003)
2. Novotni, M., Klein, R.: 3D Zernike descriptors for content based shape retrieval. In: 8th ACM Symposium on Solid Modeling and Applications, Washington, USA, pp. 216–225. ACM (2003)
3. McShane, J.J.: GC-MS is not perfect: the case study of methamphetamine (2011)
4. Mendelson, J., Uemura, N., Harris, D., Nath, R.P., Fernandez, E., Jacob, P., Everhart, E.T., Jones, R.T.: Human pharmacology of the methamphetamine stereoisomers. Clin. Pharmacol. Ther. **80**(4), 403–420 (2006)
5. Monev, V.: Introduction to similarity searching in chemistry. Match-Commun. Math. Comput. Chem. **51**, 7–38 (2005)
6. Rahman, S., Bashton, M., Holliday, G., Schrader, R., Thornton, J.: Small molecule subgraph detector (SMSD) toolkit. J. Cheminform. **1**(1), 12 (2009)
7. Speck-Planche, A., V Kleandrova, V., Luan, F., Natalia, D.S., Cordeiro, M.: Chemoinformatics in multi-target drug discovery for anti-cancer therapy: in silico design of potent and versatile anti-brain tumor agents. Anti-Cancer Agents Med. Chem. **12**(6), 678 (2012). doi:10.2174/187152012800617722
8. Kothapalli, R., Khan, A.M., Basappa, Gopalsamy, A., Chong, Y.S., Annamalai, L.: Cheminformatics-based drug design approach for identification of inhibitors targeting the characteristic residues of MMP-13 hemopexin domain. PLoS ONE **5**(8), e12494 (2010). doi:10.1371/journal.pone.0012494
9. Praisler, M., Dirinck, I., Van Bocxlaer, J., De Leenheer, A., Massart, D.L.: Pattern recognition techniques screening for drugs of abuse with gas chromatography-fourier transform infrared spectroscopy. Int. J. Talanta **53**, 177–193 (2000)
10. Ting, H., Jingling, S., Meiyan, L.: Quantitative identification of illicit drugs by using SOM neural networks. Int. J. Measur. **44**(2), 391–398 (2011)
11. Bianucci, A.M., Micheli, A., Sperduti, A., Starita, A.: A novel approach to QSPR/QSAR based on neural networks for structures. Soft Comput. Approaches Chem. **120**, 265–296 (2003)
12. Madden, M.G., Ryder, A.G.: Machine learning methods for quantitative analysis of Raman spectroscopy data. In: Proceedings of SPIE, the International Society for Optical Engineering, pp. 1130–1139 (2002)
13. Gosav, S., Praisler, M., Birsa, M.L.: Principal component analysis coupled with artificial neural networks—a combined technique classifying small molecular structures using a concatenated spectral database. Int. J. Mol. Sci. **12**, 6668–6684 (2011)
14. Nikolova, N., Jaworska, J.: Approaches to measure chemical similarity—a review. QSAR Comb. Sci. **22**(9–10), 1006–1026 (2003)
15. Zhang, D., Lu, G.: Shape-based image retrieval using generic fourier descriptor. Sig. Process.: Image Commun. **17**(10), 825–848 (2002)
16. Muda, A.K.: Authorship Invarianceness for Writer Identification Using Invariant Discretization and Modified Immune Classifier. Universiti Teknologi Malaysia, Johor Bahru (2009)
17. Flusser, J., Suk, T., Zitová, B.: Moments and Moment Invariants in Pattern Recognition, vol. 1. Wiley, West Sussex (2009)

18. Todeschini, R., Consonni, V.: Descriptors from molecular geometry. In: Handbook of Chemoinformatics, pp. 1004–1033. Wiley-VCH Verlag GmbH, Weinheim (2008)
19. Kortagere, S., Krasowski, M.D., Ekins, S.: The importance of discerning shape in molecular pharmacology. Trends Pharmacol. Sci. **30**(3), 138–147 (2009)
20. Grandison, S., Roberts, C., Morris, R.J.: The application of 3D Zernike moments for the description of "model-free" molecular structure, functional motion, and structural reliability. J. Comput. Biol.: J. Comput. Mol. Cell Biol. **16**(3), 487–500 (2009). doi:10.1089/cmb.2008.0083
21. Kihara, D., Sael, L., Chikhi, R., Esquivel-Rodriguez, J.: Molecular surface representation using 3D Zernike descriptors for protein shape comparison and docking. Curr. Protein Pept. Sci. **12**, 520–530 (2011)
22. Axenopoulos, A., Daras, P., Papadopoulos, G., Houstis, E.N.: A shape descriptor for fast complementarity matching in molecular docking. IEEE/ACM Trans. Comput. Biol. Bioinform. **8**(6), 1441–1457 (2011)
23. Verli, H., Albuquerque, M.G., de Alencastro, R.B., Barreiro, E.J.: Local intersection volume: a new 3D descriptor applied to develop a 3D-QSAR pharmacophore model for benzodiazepine receptor ligands. Eur. J. Med. Chem. **37**(3), 219–229 (2002). doi:10.1016/S0223-5234(02)01334-X
24. Edvinsson, T., Arteca, G.A., Elvingson, C.: Path-space ratio as a molecular shape descriptor of polymer conformation. J. Chem. Inf. Comput. Sci. **43**(1), 126–133 (2002). doi:10.1021/ci020269x
25. Hu, M.K.: Visual pattern recognition by moment invariants. IRE Trans. Inf. Theor. **8**, 179–187 (1962)
26. Ding, M., Chang, J., Peng, J.: Research on moment invariants algorithm. J. Data Acquisition Process. **7**(2), 1–9 (1992)
27. Pratama, S.F.: Cheap computational cost class-specific swarm sequential selection for handwritten authorship. Universiti Teknikal Malaysia Melaka (2013)
28. Pratama, S.F., Muda, A.K., Abraham, A., Muda, N.A.: An alternative to SOCIFS writer identification framework for handwritten authorship. In: IEEE International Conference on Systems, Man, and Cybernetics, Manchester, UK. IEEE (2013)
29. Pratama, S.F., Muda, A.K., Choo, Y.-H., Muda, N.A.: SOCIFS feature selection framework for handwritten authorship. Int. J. Hybrid Intell. Syst. **10**(2), 83–91 (2013). doi:10.3233/HIS-130167
30. Dym, H., McKean, H.P.: Fourier Series and Integrals. Probability and Mathematical Statistics, vol. 14. Academic Press, New York (1972) (Accessed from http://nla.gov.au/nla.cat-vn1791862)
31. Sael, L., Li, B., La, D., Fang, Y., Ramani, K., Rustamov, R., Kihara, D.: Fast protein tertiary structure retrieval based on global surface shape similarity. Proteins: Struct. Funct. Bioinform. **72**(4), 1259–1273 (2008). doi:10.1002/prot.22030
32. Canterakis, N.: 3D Zernike moments and Zernike affine invariants for 3D image analysis and recognition. In: 11th Scandinavian Conference on Image Analysis, pp. 85–93 (1999)
33. United Nations Office of Drugs and Crime: Recommended Methods for the Identification and Analysis of Amphetamine, Methamphetamine and Their Ring-substituted Analogues in Seized Materials. United Nations, Vienna (2006)
34. ChemAxon: Marvin. http://www.chemaxon.com (2014)
35. O'Boyle, N., Banck, M., James, C., Morley, C., Vandermeersch, T., Hutchison, G.: Open Babel: an open chemical toolbox. J. Cheminform. **3**(1), 33 (2011)
36. Cantor, G.: Beiträge zur Begründung der transfiniten Mengenlehre. Math. Ann. **46**(4), 481–512 (1895). doi:10.1007/BF02124929
37. Szudzik, M.: An elegant pairing function. In: Wolfram Research (ed.) Special NKS 2006 Wolfram Science Conference. Complex Systems Publications, Washington DC (2007)
38. Lisi, M.: Some remarks on the Cantor pairing function. Le Matematiche **62**(1), 55–65 (2007)
39. Zuras, D. et al.: IEEE standard for floating-point arithmetic. IEEE Std 754-2008, 1–70 (2008). doi:10.1109/IEEESTD.2008.4610935

40. Hall, M., Frank, E., Holmes, G., Pfahringer, B., Reutemann, P., Witten, I.H.: The WEKA data mining software: an update. SIGKDD Explor. **11**, 10–18 (2009)
41. Kohavi, R.: Scaling Up the Accuracy of Naive-Bayes Classifiers: A Decision-tree Hybrid (1996)
42. Kohavi, R.: The power of decision tables. Paper presented at the Proceedings of the 8th European Conference on Machine Learning (1995)
43. Breiman, L.: Random forests. Mach. Learn. **45**, 5–32 (2001)
44. Suk, T., Flusser, J.: Tensor method for constructing 3D moment invariants. In: Berciano, A., Díaz-Pernil, D., Kropatsch, W.G., Molina-Abril, H., Real, P. (eds.) Computer Analysis of Images and Patterns, Sevilla, Spain 2011, pp. 213–219. Springer, Berlin
45. Mamistvalov, A.G.: n-dimensional moment invariants and conceptual mathematical theory of recognition n-dimensional solids. IEEE Trans. Pattern Anal. Mach. Intell. **20**(8), 819–831 (1998)
46. Zhiyong, W., Zheru, C., Feng, D.: Structural representation and BPTS learning for shape classification. In: Proceedings of the 9th International Conference on Neural Information Processing, 2002, ICONIP'02, vol. 131, pp. 134–138, 18–22 Nov 2002
47. Xu, J., Yang, G., Yin, Y., Man, H., He, H.: Sparse-representation-based classification with structure-preserving dimension reduction. Cogn. Comput. **6**(3), 608–621 (2014). doi:10.1007/s12559-014-9252-5

Risk Assessment for Grid Computing Using Meta-Learning Ensembles

Sara Abdelwahab and Ajith Abraham

Abstract Assessing risk associated with computational grid is an essential need for both the resource providers and the users who runs applications in grid environments. In this chapter, we modeled the prediction process of risk assessment (RA) in grid computing utilizing meta-learning approaches in order to improve the performance of the individual predictive models. In this chapter, four algorithms were selected as base classifiers, namely isotonic regression, instance base knowledge (IBK), randomizable filtered classified tree, and extra tree. Two meta-schemes, known as voting and multi schemes, were adopted to perform an ensemble risk prediction model in order to have better performance. The combination of prediction models was compared based on root mean-squared error (RMSE) to find out the best suitable algorithm. The performance of the prediction models is measured using percentage split. Experiments and assessments of these methods are performed using nine datasets for grid computing risk factors. Empirical results illustrate that the prediction performance is enhanced by predictive model using ensemble methods.

Keywords Risk assessment · Grid computing · Base prediction algorithms · Meta-schemes

S. Abdelwahab (✉)
Faculty of Computer Science and Information Technology,
Sudan University of Science and Technology, Khartoum, Sudan
e-mail: Saabdelghani@pnu.edu.sa

A. Abraham
Machine Intelligence Research Labs (MIR Labs), Scientific Network
for Innovation and Research Excellence, Auburn, WA, USA
e-mail: ajith.abraham@ieee.org

A. Abraham
IT4Innovations, VSB—Technical University of Ostrava, Ostrava, Czech Republic

S. Abdelwahab
College of Computer and Information Sciences, Princess Nora University, Riyadh, KSA

© Springer International Publishing Switzerland 2015   251
A. Abraham et al. (eds.), *Pattern Analysis, Intelligent Security
and the Internet of Things*, Advances in Intelligent Systems and Computing 355,
DOI 10.1007/978-3-319-17398-6_23

1 Introduction

Risk assessment (RA) is a wide concept that can be applied in many context of grid computing involving performance, resource failure, and security [1]. A grid is a collection of diverse computers and resources spread across several administrative domains with the purpose of resource sharing [2]. Currently, grid has been applied in many applications to solve large-scale scientific and e-commerce problems [3]. Therefore, risk reduction is needed to avoid security breaches. In order to offer reliable grid computing services, a mechanism is needed to assess the risks and make precaution measures to avoid them. RA is a set of methods that is applied in information system to investigate the probability of event that causes harm to assets [4]. RA has been studied extensively using different approaches to model it such as quantitative, qualitative, and hybrid approaches. Numerous RA models have been provided using different techniques to make RA more accurate and reliable. Our goal in this work was to enhance the performance in predicting risks in grid computing. Utilizing meta–schemes, we formulated an ensemble model. The idea of ensemble methodology is to build a predictive model by integrating multiple models each of which solves the same task and has comparable results. The aim of ensemble was to improve the prediction performance that can be obtained from any algorithm individually [5]. In this study, we used four prediction algorithms as base algorithm, namely isotonic regression, instance base knowledge (IBK), randomizable filter classifier, and extra tree. Then, we combined them by utilizing voting and multischeme as a type of meta-learning scheme. The rest of the chapter is structured as follows: Related work is presented in Sect. 2, followed by research methodology detailed in Sect. 3. Experimental results and discussions are described in Sect. 4. Section 5 offers concluding remarks.

2 Related Work

RA has been studied extensively in the literature, and there are many methodologies used in assessing risks such as factor analysis for information risk (FAIR) [6] and operationally critical threat, asset, and vulnerability evaluation (OCTAVE) [7]. The main drawback of the presented methodologies is that they do not include the human factor as a risk factor [8]. RA in grid computing has been addressed by many researchers [1, 9–11]. Although a significant number of researchers have proposed RA methods, the risk information in grid computing is limited, due to the dependability of RA efforts on the node or machine level [9]. The concept of RA in grid computing was introduced by Djemame et al. [4]. AssessGrid project supported RA and management for all three grid actors: end user, broker, and resource provider. However, AccessGrid did not provide any mechanism to determine the reason of the component failure and the influence of failure types on each other [1]. Assessing risk in grid computing has been done using stochastic processes; the RA problem is tackled at the node level as well as at the component level, all the suggested RA models were built on historical failure data.

Sangrasi and Djemame [9] provided a RA model at the component level on the basis of non-homogeneous poisson process (NHPP). In [9], they used grid failure data for the experimentation at the component level. Sangrasi and Djemame [1] proposed a probabilistic risk model at the component level; the suggested model involves series and parallel models.

On the other hand, Alsoghayer and Djemame [12] used a probabilistic RA method, where sufficient failure data are available. They analyzed the failure data by using a frequentist approach. And they estimate the parameters of the distribution by utilizing the maximum likelihood method. They take into consideration the failures that affect the whole system. Alsoghayer and Djemame [10] extended the model proposed by Alsoghayer and Djemame [12], and they introduced RA aggregation model build at the node level based on R-out-of-N model. The proposed model provides the risk estimates for any number of chosen nodes and estimates the risk for those failures. The provided model is built on assumption that when all the nodes fail, an SLA fails. However, the main drawback of the proposed model is that it is not applicable to all values of time in the given scenario.

The probability of resource failure plays a significant role in RA process. However the main drawback of the provided probability models that highlighted in the literature is that, all provided models are built on unrealistic assumption that the resource failure represents poisson process [13]. Alsoghayer and Djemame [13] proposed a mathematical model, by using historical and discrete time analytical model (Markov model), to predict the risk of resource failure in grid environment. However, most of proposed methods [9, 13] do not address the key issue of security risk that threatens the grid environment. A significant amount of the literature on grid computing addresses the problem of RA by providing hybrid model [14].

Carlsson and Fullér [11] developed a framework for resource management in grid computing by utilizing the predictive probabilistic approach. They introduced the upper limit of failure number and approximated the likelihood of successful of a specific computing task. They used a fuzzy nonparametric regression technique to estimate the possibility distribution of the future number of node failures. The proposed model is utilized by resource provider to get alternative RAs. Carlsson and Fullér [15] provided a model for assessing the risk of a SLA for a computing task in a grid environment based on node failures that have spare resources available. The provided hybrid model is constructed based on a probabilistic and possibilistic technique. The constructed hybrid model takes into account the possibility distribution for the maximal number of failures derived from a resource provider's observation. However, the proposed model focuses on node failure and ignores other factors that may cause a violation of the SLA. However, the proposed methods addressed RA in grid with the aspect of resource failure. In our work, we addressed the RA in grid computing in context of the security aspect.

3 Research Methodology

The methodology proposed in this chapter can be described as follows:

- Starting with a dataset of risk factors in grid computing, we utilized data mining techniques to pick the contributing attributes that improve the quality of the RA prediction process.
- Conduct an ensemble model by combing the prediction algorithms, using different meta-learning methods such as voting and multischeme.
- Assess the meta-schemes used to model the ensemble prediction model for assessing risk in grid computing.

3.1 Base Prediction Models

Four different algorithms were used as base predictors, each of which is the obtained outcome of learning algorithm applied to different dataset. We selected these methods based on the performance during the preliminary experiments. The base predictors were used for empirical testing of vote and multischeme, we present them as follows:

Isotonic Regression algorithm (Isoreg):
Isotonic regression is a regression method that uses the weighted least squares to evaluate linear regression models [16].

Instance-based Knowledge (IBK) Algorithm:
Instance-based knowledge uses the instances themselves from the training set to represent what are learned and be kept. When an unseen instance is provided, the memory is searched for the training instance [17].

Randomizable Filter Classifier (RFC) Algorithm:
This method used an arbitrary classifier on data that have been passed through an arbitrary filter. Like the classifier, the structure of the filter is based exclusively on the training data and test instances will be processed by the filter without changing their structure [18].

Extra Tree Algorithm (Etree):
This method is an extremely randomized decision tree that uses another randomization process. At each node of an extra tree, partitioned rules are depicted randomly, then on the basis of a computational score the rule that proceed well is selected to be linked with that node [19].

3.2 Meta-Schemes

Meta-learning has been developed in the field of data mining to aid experts in selecting the best algorithms to be used with certain datasets. Meta-level learning accumulates knowledge about the learning process itself and finds a relation between problem domains and learning strategies [20]. Utilizing meta-schemes available in WEKA, we found out possible combinations of ensemble, using vote and multischemes.

Voting: In vote combining schema, each algorithm has the same weight. A prediction of an unseen instance is performed according to the class that obtains the highest number of votes [5]. Based on the vote, the final predictor is conducted using a combination rule. In this work, we adopted the average of probability as a combination rule.

Multischeme: Using the performance on the training data, which is measured based on mean-squared error (regression), the classifier among several classifiers is selected.

4 Experimental Result and Discussion

4.1 Datasets and Percentage Split

We conducted an online survey with international experts to evaluate the risk factors associated with grid computing. We asked the experts to determine the influence of

Table 1 Risk factors (attributes)

Risk Factor	Abbreviation	Range
Service level agreement violation	SLAV	[0–1]
Cross-domain attacks	CDA	[1–3]
Job starvation	JS	[0–1]
Resource failure	RF	[0–1]
Resource attacks	RA	[0–1]
Privilege attack	PA	[0–1]
Confidentiality breaches	CB	[0–2]
Integrity violation	IV	[0–2]
DDoS attacks	DDoS	[1–3]
Data attack	DA	[0–2]
Data exposure	DE	[1–3]
Credential violation	CV	[0–1]
Man in the middle attack	MMA	[0–1]
Privacy violation	PV	[0–2]
Sybil attack	SA	[1–3]
Hosting illegal content	HIC	[0–1]
Stealing the input or output	SIO	[0–1]
Shared use threats	ShUTh	[1–3]
Stealing or altering the software	SS	[0–1]
Policy mapping	PM	[1–3]

these factors by categorizing it under three levels: severe, moderate, and marginal. At the next step, we assigned a numeric range to each included factor depending on its concept and chance of occurrence. Based on expert knowledge and some statistical approaches, we then simulated 1951 instances based on a generic grid environment. The original dataset consists of 20 input attributes (risk factors) and one output attribute (risk value). The attributes are summarized in Table 1.

After implementing the different attribute selection methods such as relief attribute evaluation and correlation-based feature selection subset evaluator (CFS Subset Eval), we obtained 8 different sub-datasets with different search methods. Table 2 illustrates the number of attributes in each dataset and summarizes the search method. We divided the datasets into training and testing data with different percentages to investigate the effectiveness of data splitting.

Table 2 Attributes selection methods

Dataset	Evaluator	Search method	Selected Attributes	Attributes
Original dataset	–	–	SLAV, CDA, JS, RF, RA, PA, CB, IV, DDoS, DA, DE, CV, MMA, PV, SA, HIC, SIO, ShUTh, SS, PM	20
1	Relief attribute evaluation	Ranker	DDoS, PM, DE, SA, ShUTh, HIC, CV, RA, SIO, CDA, RF, SLAV, JS, MMA, SS, PA, PV, IV, CB, DA	20
2	Relief attribute evaluation	Ranker	DDoS, PM, DE, SA, ShUTh, HIC, CV, RA, SIO, CDA, RF, SLAV, JS, MMA, SS, PA, PV, IV	18
3	Relief attribute evaluation	Ranker	DDoS, PM, DE, SA, ShUTh, HIC, CV, RA, SIO, CDA, RF, SLAV, JS, MMA, SS	15
4	Relief attribute evaluation	Ranker	DDoS, PM, DE, SA, ShUTh, HIC, CV, RA, SIO, CDA, RF, SLAV	12
5	Relief attribute evaluation	Ranker	DDoS, PM, DE, SA, ShUTh, HIC, CV, RA, SIO	9
6	CFS subset evaluation	Evolutionary search	SLAV, JS, RA, CV, HIC, SIO	6
7	CFS subset evaluation	Best first search backward	CV, HIC, SIO	3
8	CFS subset evaluation	Exhaustive search	RA, CV, HIC	3

- A: Split 60 % training, 40 % testing
- B: Split 70 % training, 30 % testing
- C: Split 80 % training, 20 % testing
- D: Split 90 % training, 10 % testing

4.2 Individual Prediction Model

Before starting to investigate the ensemble, the performance of individual algorithms was analyzed. In the preprocessing phase, the data are filtered to remove irrelevant and redundant features and to improve the quality. Table 3 reports the empirical results (for test data) illustrating the root mean squared error (RMSE) for the 4 selected datasets. As illustrated in Table 3, isotonic regression algorithm, IBK algorithm, randomizable filter classifier algorithm, and the extra tree algorithm performed well for all the training and testing combinations and for the 4 different datasets. All these algorithms exhibited the best performance for all the 4 selected datasets. It is noticed that the higher the percentage of training data (Dataset D), the better for achieving good results. The best result is accomplished with the correlation coefficient (CC) equal to 1 and the RMSE equal to 0.0015 for datasets 3 and 4.

Table 3 Evaluation of individual algorithms with different datasets

Algorithm	Data Split	9 Attributes	6 Attributes	3 Attributes	3 Attributes
		RMSE			
IsoReg	A	0.0023	0.0024	0.0023	0.0023
	B	0.002	0.002	0.002	0.002
	C	0.0018	0.0018	0.0018	0.0018
	D	**0.0017**	**0.0017**	**0.0017**	**0.0017**
IBk	A	0.0023	0.0022	0.0021	0.0021
	B	0.0019	0.0019	0.0018	0.0018
	C	0.0018	0.0018	0.0016	0.0016
	D	0.0016	0.0016	**0.0015**	**0.0015**
RFC	A	0.0023	0.0023	0.0022	0.0022
	B	0.002	0.002	0.0019	0.0019
	C	0.0018	0.0019	0.0017	0.0017
	D	0.0017	0.0018	**0.0016**	**0.0016**
Etree	A	0.0046	0.3677	0.0045	0.0045
	B	0.0038	0.0039	0.0038	0.0038
	C	0.0033	0.0034	0.0033	0.0033
	D	0.0032	0.0031	**0.0029**	**0.0029**

4.3 Ensemble with Two and Three Base Prediction Algorithm

In this stage, we combine the four base prediction algorithms (IsoReg, IBK, RFC, and ETree) to conduct an ensemble using vote and multischemes as combination methods. We use nine different datasets with four different splitting categories for training and testing. For simplicity reasons, only five datasets are shown in Table 4.

According to Table 4, comparing the possible combinations of two classifiers for risk prediction process, the best performance is achieved with dataset 2, dataset 4, and dataset 5 with RMSE equal to 0.0013 and 90 % for training and 10 % for testing. Tables 4 and 5 show that vote works better than multischeme in all possible combinations of the four base classifiers in all selected datasets. For combining the four base algorithms, the best result achieved is 0.0012 with dataset 1 and dataset 3, with vote meta-methods.

Table 4 RMSE of ensemble with two base classifiers

Method	Ensemble with 2 base predictors					
	Base predictors	Dataset 1 20 features	Dataset 2 9 features	Dataset 3 6 features	Dataset 4 3 features	Dataset 5 3 features
Voting	RMSE					
	IsoReg IBK	0.0014	**0.0013**	0.002	0.0014	0.0014
	IsoReg RFC	0.0013	0.0013	0.002	0.0014	0.0014
	IsoReg ETree	0.0017	0.0019	0.0028	0.0017	0.0017
	IBK RFC	0.0013	0.0015	0.0019	**0.0013**	**0.0013**
	IBK ETree	0.0018	0.0018	0.0027	0.0017	0.0017
	RFC Etree	0.0020	0.0014	0.0027	0.0017	0.0017
Multischeme	IsoReg IBK	0.0017	0.0017	0.0024	0.0017	0.0017
	IsoReg RFC	0.0017	0.0017	0.0024	0.0017	0.0017
	IsoReg ETree	0.0017	0.0017	0.0024	0.0017	0.0017
	IBK RFC	0.0017	0.0016	0.0022	0.0015	0.0015
	IBK ETree	0.0017	0.0016	0.0022	0.0015	0.0015
	RFC Etree	0.0017	0.0017	0.0023	0.0016	0.0016

Table 5 RMSE of ensemble with three base classifiers

Combination methods	Ensemble with 3 base classifiers					
	Base predictors	Dataset 1 20 features	Dataset 2 9 features	Dataset 3 6 features	Dataset 4 3 features	Dataset 5 3 features
Voting	IsoReg IBK RFC	**0.0012**	0.0013	**0.0012**	0.0013	0.0013
	IsoReg IBK ETree	0.0014	0.0014	0.0013	0.0014	0.0014
	IsoReg RFC Etree	0.0014	0.0015	0.0013	0.0014	0.0014
	IBK RFC ETree	0.0015	0.0015	0.0014	0.0014	0.0014
MultiScheme	IsoReg IBK RFC	0.0017	0.0017	0.0017	0.0017	0.0017
	IsoReg IBK ETree	0.0017	0.0017	0.0017	0.0017	0.0017
	IsoReg RFC Etree	0.0017	0.0017	0.0017	0.0017	0.0017
	IBK RFC ETree	0.0017	0.0016	0.0016	0.0015	0.0015

5 Conclusions

In this chapter, we investigated the performance of several machine-learning methods for the RA prediction process for the identified risk factors. We used different algorithms for feature selection, namely ranker, evolutionary search, best first search, and exhaustive search to obtain different datasets with different numbers of attributes. We considered four algorithms for grid risk factor prediction as base algorithms and then a comparison was made among different subset of features and different percentage split for training and testing. Using meta-methods of voting and multischemes, we formulated an ensemble for risk prediction process.

References

1. Sangrasi, A., Djemame, K.: Component level risk assessment in grids: a probabilistic risk model and experimentation. In: 2011 Proceedings of the 5th IEEE International Conference on Digital Ecosystems and Technologies Conference (DEST). IEEE (2011)
2. Foster, I., Kesselman, C., Tuecke, S.: The anatomy of the grid. Berman et al. [2] 171–197 (2003)
3. Foster, I., Kesselman, C., Nick, J.M., Tuecke, S.: The physiology of the grid. Grid Comput.: Making Glob. Infrastruct. Reality 217–249 (2003)
4. Djemame, K., Gourlay, I., Padgett, J., Birkenheuer, G., Hovestadt, M., Kao, O., Voss, K.: Introducing risk management into the grid. In: Second IEEE International Conference on e-Science and Grid Computing, 2006, e-Science'06. IEEE (2006)
5. Rokach, L.: Ensemble methods in supervised learning. In: Maimon, O., Rokach, L. (eds.) Data Mining and Knowledge Discovery Handbook, pp. 959–979. Springer, Berlin (2010)
6. Jones, J.: An introduction to factor analysis of information risk (fair). Norwich J. Inform. Assur. 2(1), 67 (2006)
7. Alberts, C.J., Dorofee, A.: Managing Information Security Risks: The OCTAVE Approach. Addison-Wesley Longman Publishing Co., Inc., Boston (2002)
8. Yadav, J.S., Jain, M.Y.A.: Risk assessment models and methodologies. Int. J. Sci. Res. Educ. 1(06) (2014)
9. Sangrasi, A., Djemame, K.: Risk assessment modeling in grids at component level: considering grid resources as repairable. In: Omatu, S. et al. (eds.) Distributed Computing and Artificial Intelligence, pp. 321–330. Springer, Berlin (2012)
10. Sangrasi, A., Djemame, K., Jokhio, I.A.: Aggregating node level risk assessment in grids using an R-out-of-N model. In: Chowdhry, B.S., Shaikh, F.K., Akbar Hussain, D.M., Aslam Uqaili, M. (eds.) Emerging trends and applications in information communication technologies, pp. 445–452. Springer, Berlin (2012)
11. Carlsson, C., Fullér, R.: Probabilistic versus possibilistic risk assessment models for optimal service level agreements in grid computing. Inf. Syst. e-Bus. Manage. 11(1), 13–28 (2013)
12. Alsoghayer, R., Djemame, K.: Probabilistic risk assessment for resource provision in grids. In: Proceedings of the 25th UK Performance Engineering Workshop, Leeds (2009)
13. Alsoghayer, R., Djemame, K.: Resource failures risk assessment modelling in distributed environments. J. Syst. Softw. 88, 42–53 (2014)
14. Carlsson, C., Fullér R.: Risk assessment in grid computing. In: Carlsson, C., Fullér R. (eds.) Possibility for Decision, pp. 145–165. Springer, Berlin (2011)
15. Carlsson, C., Fullér R.: Risk assessment of SLAs in grid computing with predictive probabilistic and possibilistic models. In: Greco, S., Pereira, R.A.M., Squillante, M., Yager, R.R., Kacprzyk, J. (eds.) Preferences and Decisions, pp. 11–29. Springer, Berlin (2010)
16. Wu, C.H., Su, W.H., Ho, Y.W.: A study on GPS GDOP approximation using support-vector machines. IEEE Trans. Instrum. Measur. 60(1), 137–145 (2011)
17. Chauhan, H., Kumar, V., Pundir, S., Pilli, E.S.: A comparative study of classification techniques for intrusion detection. In: 2013 International Symposium on Computational and Business Intelligence (ISCBI). IEEE (2013)
18. Hall, M., Frank, E., Holmes, G., Pfahringer, B., Reutemann, P., Witten, I.H.: The WEKA data mining software: an update. ACM SIGKDD Explor. Newslett. 11(1), 10–18 (2009)
19. Désir, C., Petitjean, C., Heutte, L., Salaun, M., Thiberville, L.: Classification of endomicroscopic images of the lung based on random subwindows and extra-trees. IEEE Trans. Biomed. Eng. 59(9), 2677–2683 (2012)
20. Vilalta, R., Giraud-Carrier, C., Brazdil, P.: Meta-learning-concepts and techniques. In: Maimon, O., Rokach, L. (eds.) Data Mining and Knowledge Discovery Handbook, pp. 717–731. Springer, Berlin (2010)

Modeling Cloud Computing Risk Assessment Using Ensemble Methods

Nada Ahmed and Ajith Abraham

Abstract Risk Assessment is a common practice in the information system security domain, besides that it is a useful tool to assess risk exposure and drive management decisions. Cloud computing has been an emerging computing model in the IT field. It provides computing resources as general utilities that can be leased and released by users in an on-demand fashion. It is about growing interest in many companies around the globe, but adopting cloud computing comes with greater risks, which need to be assessed. The main target of risk assessment is to define appropriate controls for reducing or eliminating those risks. The goal of this paper was to use an ensemble technique to increase the predictive performance. The main idea of using ensembles is that the combination of predictors can lead to an improvement of a risk assessment model in terms of better generalization and/or in terms of increased efficiency. We conducted a survey and formulated different associated risk factors to simulate the data from the experiments. We applied different feature selection algorithms such as best-first and random search algorithms and ranking methods to reduce the attributes to 4, 5, and 10 attributes, which enabled us to achieve better accuracy. Six function approximation algorithms, namely Isotonic Regression, Randomizable Filter Classifier, Kstar, Extra tree, IBK, and the multilayered perceptron, were selected after experimenting with more than thirty different algorithms. Further, the meta-schemes algorithm named voting is adopted to improve the generalization performance of best individual classifier and to build highly accurate risk assessment model.

N. Ahmed (✉)
Faculty of Computer Science and Information Technology,
Sudan University of Science, Technology, Khartoum, Sudan
e-mail: naessa@pnu.edu.sa

A. Abraham
Machine Intelligence Research Labs (MIR Labs), Scientific Network
for Innovation and Research Excellence, Auburn, WA, USA
e-mail: ajith.abraham@ieee.org

A. Abraham
IT4Innovations, VSB—Technical University of Ostrava, Ostrava, Czech Republic

© Springer International Publishing Switzerland 2015   261
A. Abraham et al. (eds.), *Pattern Analysis, Intelligent Security
and the Internet of Things*, Advances in Intelligent Systems and Computing 355,
DOI 10.1007/978-3-319-17398-6_24

Keywords Cloud computing · Risk factor · Ensemble method · Data mining ·
Meta-scheme

1 Introduction

The emergence of cloud computing represents a fundamental change in the way
information technology service is invented, deployed, developed, maintained,
scaled, updated, and paid [1]. Cloud computing provides an on-demand services
such as processing and storage to its consumers, and the consumers use these
services as they need and pay only for what is used [2–4].

On an operational level, cloud computing free up the resources and refocus them
on core business activities, thereby the potential for innovation is increased. A
recent Gartner research report predicts that the global cloud market is expected to
burst in the coming years [5].

It provides a level of abstraction between the physical infrastructure and the
owner of the information being stored and processed because the application
software and databases are stored in large data centers, where the management of
the data and services is not trustworthy [6]. In recent years, there is obvious
migration to cloud computing with end users, quietly handling a growing number of
personal data, such as photographs, music files, bookmarks, and much more, on
remote servers accessible via a network [7]. The use of cloud computing services
can cause great risks to consumers. Before consumers start using cloud computing
services, they must confirm whether the product satisfies their needs and under-
stands the risks involved in using this service [8].

This paper focuses on the use of ensemble methods to construct more accurate
risk assessment model. In this study, we used vote as a combination method, and
MLP, Isotonic Regression, Extra tree, RFC, Kstar, and IBK have been used as base
protectors. They combined by using a vote meta-learning algorithm to evaluate
effectiveness and extend the capabilities of these algorithms.

This paper is divided into 6 sections and organized as follows. The identified risk
factors of cloud computing are presented in Sect. 2. The meta-schema is described
in Sect. 3. The risk assessment algorithms that used in our experimental are presented in Sect. 4. Section 5 shows the experimental work using meta-schema
algorithm applied to datasets. Finally, in Sect. 6, the conclusions of this work are
presented.

2 Cloud Computing Risk Factors

We define the various risk factors associated with cloud computing as follows.

2.1 Authentication and Access Control (A&AC)

Organization's private and sensitive data must be secure, and only authenticated users can access it. When using the cloud, the data are processed and stored outside the premise of an enterprise, which brings a level of risk because outsourced services bypass the "physical, logical, and personnel controls," and any outsider or unwanted access is denied.

2.2 Data Loss (DL)

Data loss means that the valuable data disappear without a trace. Cloud customers need to make sure that this will never happen to their sensitive data.

2.3 Insecure Application Programming (IAP)

IAPs are an important and necessary part to the security and availability for whole cloud services. Building interfaces and injecting services will increase risk for some organization may in force relinquish their credentials to third parties in order to enable their agency.

2.4 Data Transfer (DT)

Sensitive data obtained from customers are processed and stored at the cloud provider end. All data flow over network needs to be secured in order to prevent seepage of customer's sensitive information. The application provided by the cloud provider to their customers has to be used and managed over the Web. The risk comes from the security holes in the Web applications.

2.5 Insufficient Due Diligence (IDD)

Before using the cloud services, the organization needs to fully understand the cloud environment and its associated risk.

2.6 Shared Environment (ShE)

Multi-tenancy is a key factor of cloud computing services. To achieve scalability, cloud provider provides shared infrastructure, platform, and application to deliver their services. This shared nature enables multiple users to share same computer resources, which may lead to leaking data to other tenants; also, if one tenant carried malicious activities, the reputation of other tenants may be affected.

2.7 Regulatory Compliance (RC)

If the provider is unable or unwilling to subject to external audits and security certification, they do not give their customers any information about the security controls that have been evaluated. It should only be considered for most trivial functions. Regardless of the location, the custodian is ultimately responsible for ensuring the security, protection, and integrity of the data, especially when they are passed to a third party.

2.8 Data Breaches (DB)

Breaching into a cloud environment will potentially attack all users' data. Those attackers can exploit a single flaw in one-client application to get to all other client's data as well, if the cloud service databases are not designed properly.

2.9 Business Continuity and Service Availability (BC&SA)

The nature of the business environment, competitive pressure, and the changes happening in it leads to some events that may affect the cloud service provider, such as a merger, bankruptcy, or its acquisition by another company. These things lead to loss or deterioration of service delivery performance and quality of service. Another important thing to the cloud computing provider is that their customers must be provided with service around the clock, but outages do occur and can be unexpected and costly to customers.

2.10 Data Location and Investigative Support (DL&IS)

Most cloud service providers have many data centers around the globe. Regarding privacy regulation in different jurisdictions, in different countries where the government restricts the access to data in their borders, or if the data stored in high-risk countries, all these things makes data location a big issue of concern. The investigation of an illegal activity may be impossible in a cloud computing environment, because multiple customer's data can be located in different data centers that are spread around the globe. If the enterprise relies on the cloud service for the processing of business records, then it must take into account the factor of the inability or unwillingness of the provider to support it.

2.11 Data Segregation (DS)

The risk arises here which comes from the failure of the mechanisms to separate data in storage, and memory, and from multiple tenants in the shared infrastructure.

2.12 Recovery (R)

Cloud users do not know where their data are hosted. Some events such as man-made or natural disaster may happen; in such events, customers need to know what happened to their data and how long the recovery process will take.

2.13 Virtualization Vulnerabilities (VV)

Virtualization is one of the fundamental components of the cloud service. However, it introduces major risks as every cloud provider uses it. Beside its own risks, it holds every risk posed by physical machines.

2.14 Third-Party Management (TPM)

There are many issues in cloud computing related to third party because the cloud service provider does not directly manage the client organizations. Some old concerns in information security appear with outsourcing such as integrity control and sustainability of supplier, and all risks that clients may take if it relies on a third party.

2.15 Interoperability and Portability (I&P)

Interoperability and portability become crucial because if the organization locks to a specific cloud provider, then the organization will be at the mercy of the service-level and pricing policies of that provider and it will not have the freedom to work with multiple cloud provider.

2.16 Resource Exhaustion (RE)

Cloud provider allocates resource according to the statistical projections. Inaccurate modeling of resource usage can lead to many issues such as service unavailability, access control compromised, economic and reputational losses, and infrastructure oversize.

2.17 Service-Level Agreement (SLA)

The organization needs to ensure that the terms of SLA are being met. Risk may appear with service-level application such as the data owner as some cloud provider includes explicitly some terms which state that the data stored is the provider's not the customer's.

2.18 Data Integrity (DI)

One of the most critical elements in all systems is DI. Cloud computing magnified the problem of DI and endangers the DI in transaction management, at the protocol level, which does not support transactions or guaranteed delivery. If DI is not guaranteed and there is a lack of integrity controls, this may result in deep problems.

3 Meta-schema

Machine learning aims to improve the generalization performance. Ensemble learning has gained considerable attention during the past two decades [9–11] owing to its good generalization capability. Ensemble method trains a set of base predictors, instead of a single one, and then combines their outputs with a fusion strategy. Many studies show that the combination of multiple predictors improves the generalization performance [12].

An important factor of an ensemble is in determining its generalization error and to be more accurate than any of its individual members that are also accurate and diverse [13, 14]. Diversity has been recognized as the most important feature of the combination of predictors, the main reason for this is that as many as predictors are

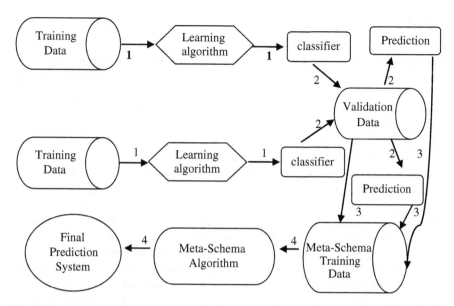

Fig. 1 Meta-schema scenario

different, and it is sensible to expect an increase in the overall performance [15]. Many methods have been developed for the construction of an ensemble. Some of them are meta-learners, and they can be applied to any learning algorithm. Others are specific to particular learners [16]. Meta-learning is defined as learning from learning knowledge. In meta-learning, a learning algorithm is used to learn how to integrate the learned predictor [14].

Figure 1 represents the different stages in a simplified meta-schema scenario:

1. The base predictors are trained from the training datasets.
2. Predictions are generated by the learned predictors algorithms on a validation or test dataset.
3. The validation dataset and the predictions generated by the predictors on the validation set are formed from the meta-level training set.
4. The final predictor (meta-predictor) is trained from the meta-level training set.

4 Risk Assessment Algorithms

Risk assessment involves building an accurate prediction model. Prediction comprises of two steps: in the first step, called the training phase, a predictor is built describing a predefined set of training data, and in the second step, the model is used for test data [15, 17]. Following are the predictors used.

4.1 Base Prediction Algorithms

Extremely Randomized Decision Trees: It is tree-based ensemble method for supervised classifier (commonly known Extra tree). Extra tree based on randomization process, the splitting rules are randomly drawn at each node of the Extra tree, and the base of chosen one of this rules to be associated with that node is the best performance according to a score computation. This allows increasing the speed of training, weakening the correlation between the induced decision trees, and reducing the complexity of the induction process [18].

Instance-Based Knowledge (IBK): IBK is an instance-based learning method and is an implementation of K-nearest neighbor classifier [19]. In its representation, it does not derive a rule set or decision tree and storing it; instead, it uses the instances themselves to represent what is learned [20]. IBK compares each new instance with existing ones using distance metric; most commonly, Euclidean distance and the closest existing instance are used to assign the class for the test sample [21].

Multilayered Perceptron (Artificial Neural Networks): A multilayer perceptron is a feed-forward artificial neural network model that maps sets of input data onto a set of appropriate outputs. The fundamental aspects that a neural network depends upon are input and activation function of the unit, network architecture, and the weight of each input connection [22].

K-Nearest Neighbors (K-NN or K*): K-NN is an instance-based learning algorithm that stores all training instances and does not build a model until a new instance needs to be classified [23], and they use some domain-specific distance function to retrieve single and most similar instance from the training set [24].

Isotonic Regression: It is a regression method, and it does its job by weighted least squares to evaluate linear regression models [25].

Randomizable Filter Classifier: It is used for running an arbitrary classifier on data that has been passed through an arbitrary filter. Like the classifier, the structure of the filter is based exclusively on the training data and test instances will be processed by the filter without changing their structure [26].

4.2 Meta-scheme Algorithms

Voting: It is a class used to combine multiple predictors, and different combinations of probability estimates for regression are available. In vote method, each predictor gets one vote and the majority wins [13].

5 Experimental Work

We conducted a survey to finalize the risk factors. In the survey, we asked the participants to categorize risk factors to three levels according to the likelihood of happening and their effect on cloud computing. These categories are as follows: important, neutral, and not important. Thirty-five international experts responded to the survey from different countries, and all of them agreed that the previously defined factors are important, which means that they have great effect over cloud computing. Next, we give each risk factor a numeric range of values, and finally, we formulated expert rules and use some statistical methods to generate the data based on the rules. The dataset contains 18 input attributes and comprises of 1940 instances. The 18 attributes were labeled as DI, IDD, RC, BC&SA, TPM, I&P, DL, IAP, DL&IS, R, RE, SLA, A&AC, ShE, DB, DS, VV, and DI. Risk factors are illustrated in Table 1 with their corresponding ranges for numeric values. We used percentage split to test and evaluate the algorithms. In percentage split, the dataset is randomly splitting the training and testing data as follows:

- 60–40 % (A)
- 70–30 % (B)
- 80–20 % (C)
- 90–10 % (D)

5.1 Ensemble Method

The experiments were implemented using WEKA software [18]. To implement ensembles to combine the base predictors, we use a voting combination method and the combination rule used is the average of probabilities. In our previous work, we did experiments and used more than thirty algorithms, six of them give us good results. In this study, we use those algorithms as the base predictors, and those

Table 1 Risk factors and their associated range values

Risk factor	Range value	Risk factor	Range value
DI	0–3	R	1–3
IDD	1–3	RE	0–2
RC	0–1	SLA	0–3
BC&SA	1–3	A&AC	0–3
TPM	0–2	ShE	1–3
I&P	0–1	DB	0–2
DL	0–3	DS	0–1
IAP	0–1	VV	1–3
DL&IS	0–3	DI	0–2

Table 2 RMSE from 2 predictors using the voting algorithm

Algorithms	Org	1st	2nd	3rd	4th	5th	6th	7th
IBK + Isreg	0.0014	0.0014	**0.0013**	0.0014	0.0014	0.0014	0.0014	0.0014
ET + K*	**0.0018**	0.009	0.0048	0.0024	0.0019	0.0019	0.0021	**0.0018**
MLP + RFC	0.0011	**0.0009**	0.0012	0.0059	0.0011	0.001	0.0016	0.0011
ET + RFC	0.0019	**0.0018**	**0.0018**	**0.0018**	**0.0018**	0.0019	0.002	0.0019
IBK + K*	**0.0011**	0.0089	0.0048	0.0019	0.0012	0.0013	0.0015	**0.0011**
IBK + ET	0.0019	0.0019	0.0019	0.0019	**0.0018**	0.0019	0.0019	0.0019
IBK + MLP	0.0011	**0.001**	0.0012	0.006	**0.001**	**0.001**	0.0016	0.0011
IBK + RFC	**0.0014**	**0.0014**	**0.0014**	**0.0014**	**0.0014**	**0.0014**	0.0015	**0.0014**
Isreg + ET	0.0019	0.0019	**0.0018**	0.0019	0.0019	0.0019	0.0019	0.002
Isreg + K*	0.0012	0.0088	0.0047	0.0018	**0.0011**	0.0012	0.0015	**0.0011**
Isreg + MLP	0.001	**0.0009**	0.0011	0.0058	0.001	0.001	0.0016	0.001
Isreg + RFC	0.0014	**0.0013**	**0.0013**	0.0014	0.0014	0.0014	0.0014	0.0014
ET + MLP	0.0019	**0.0016**	0.002	0.006	0.0017	0.0017	0.0022	0.0018
K* + MLP	0.0011	0.009	0.005	0.0064	0.0011	0.0012	0.0021	**0.001**
K* + RFC	**0.0011**	0.0089	0.0048	0.0019	**0.0011**	0.0013	0.0014	**0.0011**

Table 3 RMSE from 3 predictors using the vote algorithm

Algorithm	Org	1st	2nd	3rd	4th	5th	6th	7th
BK + K* + RFC	**0.001**	0.006	0.0033	0.0014	**0.001**	0.0011	0.0012	**0.001**
IBK + K* + MLP	**0.0009**	0.006	0.0034	0.0043	**0.0009**	0.001	0.0015	**0.0009**
IBK + K* + ET	**0.0013**	0.0061	0.0033	0.0017	0.0014	0.0014	0.0015	**0.0013**
IBK + K* + Isreg	0.0011	0.0014	0.0032	0.0014	**0.001**	0.0011	0.0012	**0.001**
IBK + RFC + MLP	**0.001**	**0.001**	0.0011	0.004	**0.001**	**0.001**	0.0013	**0.001**
IBK + RFC + ET	**0.0014**	0.0015	**0.0014**	**0.0014**	**0.0014**	0.0015	0.0015	0.0015
IBK + RFC + Isreg	**0.0012**	**0.0012**	**0.0012**	**0.0012**	**0.0012**	**0.0012**	0.0013	**0.0012**
IBK + MLP + ET	0.0014	0.0015	0.0015	0.0041	**0.0013**	**0.0013**	0.0016	0.0014
IBK + MLP + Isreg	0.001	**0.0009**	0.001	0.0039	0.001	0.001	0.0014	0.001
IBK + ET + Isreg	0.0015	0.0015	**0.0014**	0.0015	**0.0014**	0.0015	0.0015	0.0015
K* + RFC + MLP	**0.0009**	0.006	0.0034	0.0044	**0.0009**	0.001	0.0015	**0.0009**
K* + RFC + ET	**0.0013**	0.006	0.0033	0.0017	0.0014	0.0014	0.0015	**0.0013**
K* + RFC + Isreg	0.0011	0.0059	0.0032	0.0014	**0.001**	0.0011	0.0012	**0.001**
K* + MLP + ET	**0.0013**	0.0061	0.0035	0.0044	**0.0013**	0.0014	0.0018	**0.0013**
K* + MLP + Isreg	0.0009	0.0059	0.0033	0.0044	0.0009	0.0009	0.0016	**0.0007**
K* + ET + Isreg	**0.0014**	0.006	0.0032	0.0017	**0.0014**	**0.0014**	0.0015	**0.0014**
RFC + MLP + ET	0.0013	**0.0012**	0.0014	0.004	0.0013	0.0013	0.0016	0.0014
RFC + MLP + Isreg	0.001	**0.0009**	0.001	0.0019	0.001	0.001	0.0014	0.001
RFC + ET + Isreg	0.0015	**0.0014**	**0.0014**	**0.0014**	0.0015	0.0015	0.0015	0.0015
MLP + ET + Isreg	0.0014	**0.0013**	0.0014	0.004	0.0014	**0.0013**	0.0017	0.0014

Table 4 RMSE from 4 predictors using the vote algorithm

Algorithm	Org	1st	2nd	3rd	4th	5th	6th	7th
IBK + K* + RFC + MLP	**0.0008**	0.0046	0.0026	0.0033	**0.0008**	0.0009	0.0013	0.0009
IBK + K* + RFC + ET	**0.0011**	0.0033	0.0025	0.0014	**0.0011**	0.0012	0.0013	**0.0011**
IBK + K* + RFC + Isreg	**0.001**	0.0045	0.0025	0.0012	**0.001**	**0.001**	0.0011	**0.001**
IBK + K* + MLP + ET	0.0011	0.0046	0.0026	0.0034	**0.001**	0.0011	0.0014	0.0011
IBK + K* + MLP + Isreg	**0.0009**	0.0045	0.0025	0.0033	**0.0009**	**0.0009**	0.0014	**0.0009**
IBK + K* + ET + Isreg	**0.0012**	0.0046	0.0025	0.0014	**0.0012**	**0.0012**	0.0013	**0.0012**
IBK + RFC + MLP + ET	**0.0011**	0.0012	0.0012	0.0031	**0.0011**	0.0012	0.0013	0.0012
IBK + RFC + MLP + Isreg	0.001	**0.0009**	0.001	0.003	0.001	0.001	0.0013	0.001
IBK + RFC + ET + Isreg	**0.0012**	0.0013	**0.0012**	0.0013	**0.0012**	0.0013	0.0013	0.0013
IBK + RFC + MLP + Isreg	**0.0012**	0.0012	0.0012	0.003	0.0012	**0.0011**	0.0014	0.0012
K* + RFC + MLP + ET	**0.0011**	0.0046	0.0026	0.0034	**0.0011**	**0.0011**	0.0014	**0.0011**
K* + RFC + MLP + Isreg	0.0009	0.0045	0.0025	0.0033	**0.0008**	0.0009	0.0013	0.0009
K* + RFC + ET + Isreg	**0.0011**	0.0045	0.0025	0.0013	0.0012	0.0012	0.0012	0.0012
K* + MLP + ET + Isreg	**0.001**	0.0045	0.0026	0.0033	**0.001**	0.0011	0.0015	0.0011
RFC + MLP + ET + Isreg	0.0012	**0.0011**	0.0012	0.003	0.0012	**0.0011**	0.0014	0.0012

Table 5 RMSE from 5 predictors using the vote algorithm

Algorithm	Org	1st	2nd	3rd	4th	5th	6th	7th
IBK + K* + RFC + MLP + ET	**0.001**	0.0037	0.0022	0.0027	**0.001**	**0.001**	0.0012	**0.001**
IBK + K* + RFC + MLP + Isreg	0.0009	0.0036	0.0021	0.0027	**0.0008**	0.0009	0.0012	0.0009
IBK + K* + RFC + ET + Isreg	**0.001**	0.0037	0.002	0.0012	0.0011	0.0011	0.0011	0.0011
IBK + Kstar + MLP + ET + Isreg	**0.001**	0.0037	0.0021	0.0027	**0.001**	**0.001**	0.0013	**0.001**
IBK + RFC + MLP + ET + Isreg	0.0011	0.0011	0.0011	0.0024	**0.001**	0.0011	0.0012	0.0011
K* + RFC + MLP + ET + Isreg	**0.001**	0.0037	0.0021	0.0027	**0.001**	**0.001**	0.0013	**0.001**

algorithms are as follows: Isotonic Regression, Randomizable Filter Classifier, Kstar, Extra tree, IBK, and the multilayered perceptron.

Performance statistics are calculated across all datasets using root-mean-square error (RMSE) and correlation coefficient (CC), but since CC is almost (0.9999 or 1), we did not include it in the tables. Table 2 represents the root-mean-square error of the ensemble with two base predictors applied to all 8 datasets. In this table, we tried all possible combinations of the six base predictors and then applied them to all percentages of each dataset. Table 3 represents the RMSE of ensembles with three base predictors. Similarly, Tables 4 and 5 represent the RMSE of ensembles with 4 and 5 base predictors, respectively. The bold numbers in these tables represent the best RMSE for each dataset obtained.

6 Discussions

The main aim of this study was to improve the performance by combining the outputs of multiple predictors. In order to do that, several numbers of combinations of basic predictors (2, 3, 4, and 5) were applied by using the vote combination method and then implemented it on all datasets. Table 6 represents the best results for each dataset obtained from all probability of a combination of base predictors. The bold numbers show the best result obtained from each combination in all datasets. It shows that the best result is attained when we combine 3 algorithms.

Table 6 The best RMSE from different combinations

Algorithm	Org	1st	2nd	3rd	4th	5th	6th	7th
2 Algorithms	0.001	**0.0009**	0.0011	0.0014	0.001	0.001	0.0014	0.001
3 Algorithms	0.0009	0.0009	0.001	0.0012	0.0009	0.0009	0.0012	**0.0007**
4 Algorithms	**0.0008**	0.0009	0.001	0.0012	0.0008	0.0009	0.0011	0.0009
5 Algorithms	0.0009	0.0011	0.0011	0.0012	**0.0008**	0.0009	0.0011	0.0009

7 Conclusions

In this paper, an investigation of using meta-scheme algorithms was performed. The main goal of this experimental work is the use of ensemble methods to improve the performance and increase the efficiency. We examined the performance of the combination of (2, 3, 4, and 5) predictors by using vote combination method.

The effect of subsets of training and testing data is also illustrated here by splitting the sub-dataset randomly into four different groups and taking the best RMSE from all percentages. The empirical results show that the combination of 3 algorithms gives the best result with the 7th dataset.

References

1. Avram, M.: Advantages and challenges of adopting cloud computing from an enterprise perspective. Proc. Technol. **12**, 529–534 (2014)
2. Paquette, S., Jaeger, P.T., Wilson, S.C.: Identifying the security risks associated with governmental use of cloud computing. Gov. Inf. Q. **27**, 245–253 (2010)
3. Carroll, M., Van Der Merwe, A., Kotze, P.: Secure cloud computing: benefits, risks and controls. In: Information Security South Africa (ISSA), vol. 2011, pp. 1–9 (2011)
4. Sun, D., Chang, G., Sun, L., Wang, X.: Surveying and analyzing security, privacy and trust issues in cloud computing environments. Proc. Eng. **15**, 2852–2856 (2011)
5. Brender, N., Markov, I.: Risk perception and risk management in cloud computing: results from a case study of Swiss companies. Int. J. Inf. Manag. **33**, 726–733 (2013)
6. Subashini, S., Kavitha, V.: A survey on security issues in service delivery models of cloud computing. J. Netw. Comput. Appl. **34**, 1–11 (2011)
7. Zissis, D., Lekkas, D.: Addressing cloud computing security issues. Future Gener. Comput. Syst. **28**, 583–592 (2012)
8. Chandran, S., Angepat, M.: Cloud computing: analyzing the risks involved in cloud computing environments. Proc. Nat. Sci. Eng., 2–4 (2010)
9. Chen, H., Tiho, P., Yao, X.: Predictive ensemble pruning by expectation propagation. IEEE Trans. Knowl. Data Eng. **21**, 999–1013 (2009)
10. Freund, Y., Schapire, R.E.: A desicion-theoretic generalization of on-line learning and an application to boosting. In: Computational Learning Theory, pp. 23–37 (1995)
11. Partalas, I., Tsoumakas, G., Vlahavas, I.: An ensemble uncertainty aware measure for directed hill climbing ensemble pruning. Mach. Learn. **81**, 257–282 (2010)
12. Li, L., Hu, Q., Wu, X., Yu, D.: Exploration of classification confidence in ensemble learning. Pattern Recogn. **47**, 3120–3131 (2014)
13. Prodromidis, A., Chan, P., Stolfo, S.: Meta-learning in distributed data mining systems: Issues and approaches. In: Kargupta, H., Chan, P. (ed.) Advances in Distributed and Parallel Knowledge Discovery, AAAI/MIT Press (2000)
14. Dietterich, T.G.: Ensemble methods in machine learning. In: Multiple Classifier Systems, pp. 1–15. Springer, Berlin (2000)
15. Canuto, A.M., Abreu, M.C., de Melo Oliveira, L., Xavier, J.C., Santos, A.D.M.: Investigating the influence of the choice of the ensemble members in accuracy and diversity of selection-based and fusion-based methods for ensembles. Pattern Recogn. Lett. **28**, 472–486 (2007)
16. Melville, P.: Creating Diverse Ensemble Classifiers. Computer Science Department, University of Texas at Austin (2003)

17. Han, J., Kamber, M.: Data Mining, Southeast Asia Edition: Concepts and Techniques. Morgan Kaufmann, Massachusetts (2006)
18. Désir, C., Petitjean, C., Heutte, L., Salaun, M., Thiberville, L.: Classification of endomicroscopic images of the lung based on random subwindows and extra-trees. IEEE Trans. Biomed. Eng. **59**, 2677–2683 (2012)
19. Witten, I.H., Frank, E., Trigg, L.E., Hall, M.A., Holmes, G., Cunningham, S.J.: Weka: practical machine learning tools and techniques with Java implementations (1999). http://www.cs.waikato.ac.nz/ml/publications/1999/99IHW-EF-LT-MH-GH-SJC-Tools-Java.ps
20. Chauhan, H., Kumar, V., Pundir, S., Pilli, E.S.: A comparative study of classification techniques for intrusion detection. In: 2013 International Symposium on Computational and Business Intelligence (ISCBI), pp. 40–43 (2013)
21. Ali, S.S., Kate, A.: On learning algorithm selection for classification. Appl. Soft Comput. **6**, 119–138 (2006)
22. Kotsiantis, S.B., Zaharakis, I.D., Pintelas, P.E.: Machine learning: a review of classification and combining techniques. Artif. Intell. Rev. **26**(3), 159–190 (2007)
23. Phyu, T.N.: Survey of classification techniques in data mining. In: Proceedings of the International MultiConference of Engineers and Computer Scientists, pp. 18–20 (2009)
24. Cleary, J.G., Trigg, LE.: K*: an instance-based learner using an entropic distance measure. In: ICML, pp. 108–114 (1995)
25. Wu, C.H., Su, W.H., Ho, Y.W.: A study on GPS GDOP approximation using support-vector machines. IEEE Trans. Instrum. Measur. **60**, 137–145 (2011)
26. http://weka.sourceforge.net/doc.dev/weka/classifiers/meta/package-summary.html

Design Consideration for Improved Term Weighting Scheme for Pornographic Web sites

Hafsah Salam, Mohd Aizaini Maarof and Anazida Zainal

Abstract Illicit Web content filtering is a content-based analysis technique, applied to censor inappropriate contents on the Internet. Web content filtering can recognize undesirable contents through the application of AI techniques, linguistic analysis, or machine learning to classify Web pages into a set of predefined categories. However, the capacity to distinguish between useful and harmful Web content remains a major research challenge, which usually leads to the problem of under-blocking and over-blocking. Further, the extraction of best term representation for classifier presents a major limitation due to curse of dimensionality, where a feature can have the same term frequency (TF) in two or more categories but has different semantic meanings such as illicit pornography and sex education context also known as ambiguous issues. Besides, the high dimensionality of features on a Web page, even for moderate size, it has made the term representation value for classifier more complex, which affects the performance of classification. Thus, this research proposes a modified term weighting scheme (TWS) for narrative and discrete Web in order to increase the classification performance. Characteristics of pornography Web site were extracted and significant characteristics were identified and mapped against term weighting factors. Initial result revealed that other criteria such as rare feature have potential to be regarded as significant criteria in TWS technique to distinguish high-similarity Web content.

Keywords Text categorization · Web pages classification · Textual content analysis · Term weighting scheme · Feature selection

H. Salam (✉) · M.A. Maarof · A. Zainal
Information Assurance and Security Research Group (IASRG), Faculty of Computing,
Universiti Teknologi Malaysia, 81310 Skudai, Johor, Malaysia
e-mail: nurhidayahkasih@gmail.com

M.A. Maarof
e-mail: aizaini@utm.my

A. Zainal
e-mail: anazida@utm.my

© Springer International Publishing Switzerland 2015
A. Abraham et al. (eds.), *Pattern Analysis, Intelligent Security
and the Internet of Things*, Advances in Intelligent Systems and Computing 355,
DOI 10.1007/978-3-319-17398-6_25

1 Introduction

With the rapid growth of the Internet, Web pages have become a major source for people to look for information. However, information on the Internet is not always useful, some information is classified as healthy and some are unhealthy. Health information such as news, knowledge, and sports, while unhealthy information such as pornography, deviant teaching, bullying and violence, comprises wealth of information on the Internet. Datuk Seai Kie (Malaysia Deputy Minister of Women, Family and Community Development) (2011) asserts that the impacts of unhealthy information toward Malaysian society are strong as the number of cases such as raping, child molestation, pedophilia, prostitution, domestic violence and sexual harassment is on the rise [1]. Factors that lead to such negative scenario can be attributed to the no-restriction policies in Web services [2], large amount of pornography resources [3, 4], and vulnerability of human to reveal unethical online behavior, such as seeking online pornography [4]. Therefore, the security of information also needs to be considered when accessing information on the Web pages.

Security of the information on the Web pages does not only consider confidentiality, integrity, and availability of the information, but also the usefulness of the information. Usefulness in information security can be defined as valuable, helpful, and positive [5, 6]. If usefulness is not considered, it can increase the cases of violence against women [7], situations where kids have possibility to act out sex against other kids [4] and kids' attitudes and development changes badly [8]. Thus, this fact shows that a proper system is needed which can help to block Malaysian citizens from accessing harmful Web pages. One of the available methods that can be used to censor the inappropriate Web content is by using Web filtering system. Usually, researchers use four different techniques: uniform resource locator (URL) blocking, platform for Internet content selection (PICS) blocking, keyword matching and Web content analysis to filter the Web content.

However, these techniques have their own limitations in classifying Web pages. The drawbacks of URL blocking are difficult to maintain up-to-date list, high maintenance, and the need to store offline data because new Web sites are continuously and rapidly emerging [9]. Meanwhile, PICS blocking requires metadata element to describe the content of Web pages, which relies mainly on *"trust"* because publishers are allowed to label metadata information and only a small fraction of Web page follows the PIC standard [10]. Keyword matching technique cannot deal with the Web sites that contain misspelled words, similar terminologies, and with the Web sites written in different languages [3]. Finally, the fourth common technique used is the Web content analysis. It requires substantial time-consumption process in order to process the data in-depth analysis of Web sites [9, 11].

Among these techniques, the Web content filtering can deal with frequently changing Web sites and can understand the semantic meaning of the Web sites [9, 11]. This technique is reliable in classifying pornographic Web sites due to its ability to analyze various types of element in a Web page for classification including text, images, metadata, links, and scripts. However, this technique has a limitation to solve ambiguous issues. An ambiguous issues refer to Web pages that have high similar content but different terminologies, e.g., pornography and sex education Web pages.

For instance, Lee et al. [12] addressed this challenge using modified entropy (M. Entropy) with 90 % accuracy rate based on the assumption that the document length for pornography Web sites is short, while comprising majorly, visual–multimedia information. However, it may be insufficient if such harmful Web sites are designed in a narrative form. Besides, M. Entropy opines to only consider terms that are frequent in most documents which obeys the concept of entropy method [13]. Conversely, Wang and Zhang [14] opine to consider the distribution of term categories, rather than among documents.

Further, Hammami et al. [15] have proposed a WebAngels to address these limitations with 89 % classification accuracy rate by considering the number of violent words in the URL and combining textual and structural analysis. However, it faces difficulty when tested with unknown Web sites because WebAngels require profile list to launch. Additionally, it requires high maintenance due to high processing time, periodic update of URL checklist as well as high dimensionality of feature space of textual analysis.

However, Santos et al. [16] proposed WR, which utilizes text content analysis to filter pornography Web sites through TFIDF and Dynamic Markov compression (DMC). WR possess high advantage in term of vulnerability to attack from illicit Web sites. As known, TFIDF is a common approach with a disadvantage of cannot differentiate between pornography and sex education Web sites [12].

Thus, the improvements of textual content analysis technique are desired in order to increase the classification accuracy of pornography Web sites. This can be achieved by considering the criteria TWS factor. Without properly selecting a significant term, this problem can introduce high dimensionality of feature space and cause heavy overhead to build the document classifier, which automatically influences the categorization accuracy. Therefore, this ambiguous issue can be reduced by enhancing the existing term weighting scheme (TWS) factor by using the idea of considering the characteristics of pornography Web sites based on the criteria of TWS factor in order to select most relevant features to represent the original content.

This paper is organized as follows. Section 2 discusses the criteria of TWS factor and the related work. Section 3 explains the initial result of the characteristics of pornography Web sites based on the criteria of TWS factor. The analysis and conclusion of the results are discussed in Sect. 4.

2 Term Weighting Scheme (TWS)

As previously discussed in Sect. 1, the text representation requires to reduce dimensionality of feature space in text content analysis by selecting only a significant subset of features to represent the Web pages. This process occurs in dimension reduction phase for text classification, which is also known as feature selection method. The main challenge in representing Web pages as faced by the existing TWS methods is to solve ambiguity in Web pages with high similar content but different terminologies. TWS is a statistical measure used to rank the importance of the term and weight by quantifying the contribution of each term to the document [12]. Table 1 shows the review summary of the existing techniques of TWS based on the criteria of TWS factors. Based on Table 1, twelve existing TWS had been considered and mapped against term weighting factor. These methods includes term frequency (TF), document frequency (DF), TFIDF, entropy, M. Entropy, information gain (IG), chi-square (χ^2), mutual information (MI), glasgow, term strength (TS), gain ratio, and odd ratio. In order to associate a weight to each term, three factors need to be considered which includes TF factor, collection frequency factor and normalization frequency factor [17]. Further explanation of criteria for each TWS factor for Table 1 is discussed.

2.1 Term Frequency Factor

Previous research works defined TF factor (*tf*) as the total number of term repeated in a document to closely represent the content of the document [14]. This factor can be categorized as local weight by considering the frequency of term occurrence in the document. TF factor has three common criteria, which are *favoring common, absence,* and *rare features.* In favoring common features criteria, the importance of the term in the document depends on the frequency of the term used. This feature is included in existing TWS methods such as TF, DF, entropy, modified entropy, IG, χ^2, MI, and odds ratio. Erenel and Altincay [18] used TF method to define their weight (w) by considering these common criteria.

Based on Table 1, TF, DF, Entropy, M. Entropy, IG, χ^2, MI, and odds ratio are methods that favor common features in their weight. Most of these methods consider the percentage of frequency of common features in determining their weight. For instance, TF, entropy, M. Entropy, χ^2, MI, and odds ratio is the method with concept of "the higher occurrence of a term, the more important it is in that document" [14, 17, 19, 20]. This is in contrast to DF, which considers common features, but ignores the real contribution of a word within a document. This prevents it from examining the relative importance of a document. Thus, the technique used in selecting the features can be different even when the same criteria are considered. However, due to data ambiguity, this criterion may be insufficient to represent the content of the document.

Meanwhile, favoring absent feature is second criterion of the TF factor and it is used to measure the document based on the presence or absence of a term in a document [21]. As shown in Table 1, IG and χ^2 are methods that consider this criterion by knowing the presence and absence of a term in a document. For instance, IG measures the amount of information obtained for category prediction by knowing the presence or absence of a term in a document [19, 21].

Table 1 Criteria of term weighting scheme (TWS)

TWS factors	Criteria	Existing term weighting scheme (TWS)											
		Term frequency (TF)	Document frequency (DF)	TFIDF	Entropy	M. Entropy	Information gain (IG)	CHI square (χ^2)	Mutual information (MI)	Glasgow	Term strength (TS)	Gain ratio	Odds ratio
Term frequency	Favoring common features	✓	✓	X	✓	✓	✓	✓	✓	X	X	X	✓
	Using feature absence	X	X	X	X	X	✓	✓	X	X	X	X	X
	Favoring rare features	X	X	✓	X	X	X	X	✓	✓	✓	✓	✓
Collection frequency factor	Using term frequency information	✓	X	✓	✓	✓	✓	X	✓	X	✓	✓	✓
	Favoring features in document	X	✓	✓	✓	✓	✓	✓	✓	✓	✓	✓	✓
	Favoring frequent term in a collection, it is estimated to be very relevant for the document	X	X	X	✓	✓	X	X	X	X	✓	X	X
	Favoring infrequent term in a collection, it is estimated to be very relevant for the document	X	X	✓	X	X	X	X	X	✓	X	X	X
	Favoring features in a category	X	X	X	X	X	✓	✓	✓	X	X	✓	✓
Normalization frequency	Normalize term length	✓	✓	✓	✓	X	X	✓	✓	✓	✓	✓	✓
	Favor short documents	X	X	X	X	✓	X	X	X	X	X	X	X
	Penalizes too long document	X	X	X	X	✓	X	X	X	✓	X	X	X
Total		3	3	5	5	6	5	5	6	5	5	5	6

On the other hand, favoring rare feature focuses on low frequency of term that appears in the document, which is the opposite of the first criterion [22]. According to Table 1, methods that determine their weight by using favoring rare TFIDF define the words in the document to be very relevant if a word is infrequent in text collection by penalizing a term if it frequently appears in most of document using variables log (N/df_i) [19, 21, 22].

2.2 Collection Frequency Factor

By considering the TF alone, TWS may not be able to select all the relevant documents. Therefore, another important aspect is the collection frequency factor also known as idf factor or global weight by considering the distribution of the documents containing the term [20]. It is important to increase the discrimination power to only select relevant terms to get a better result by the multiplication of tf and idf factors. Based on Table 1, the criteria for collection frequency factor include TF information, favoring features in document, favoring frequent term in a collection, favoring infrequent term in a collection and favoring features in a category.

Lan et al. [23] have examined TWS by comparing idf with the TF approach such as tf:idf, log (1 + tf):idf, tf, only idf *prob* and idf *alone* that is related to collection frequency factor. The result has no significant difference among them in some instance, it even decreases the discriminating power of selecting relevant terms when combining idf with the common approach of TF factor. Other research findings such as Deng et al. [24] and Debole and Sebastiani [25] replace this collection factor with metrics such as IG, gain ratio, χ^2 and odds ratio but need to know the information in the category relationship to make a better result as shown in Table 1. Thus, this factor will be further discussed in future work.

2.3 Normalization Frequency Factor

Another important factor that needs to be considered is normalization factor. This factor takes into account the length of the documents [12]. Three common approaches in normalization frequency factor are normalizing the documents length, favoring short document and penalizing longer document [26]. As shown in Table 1, M. Entropy and glasgow considered this factor to penalize the length whether short or too long document. Based on Table 1, MI and odds ratio are methods that are used in the six criteria of TWS factor to represent the original content. Thus, those methods will be further researched in order to examine the accuracy when classifying pornography Web content.

3 Characteristic of Pornography Web sites

Pornography Web sites have distinguishable characteristics which can be used to explicitly represent pornographic content. The purpose of this analysis is to distinguish high-similarity Web content within pornography and sex education Web sites by considering the TWS factors. Pornographic Web sites can be in the form of narrative or discrete or a mixture of both. A manual inspection is first considered before applying the technicality of content analysis method, by observing the characteristics of pornography Web sites in order to provide learned opinion to address data ambiguity. This analysis is done by categorizing the Web pages into narrative type, or discrete type. This initial study used 300 Web sites. The corpus consists of three categories of Web sites; narrative pornographic Web sites, discrete pornographic Web sites, and healthy Web sites, 100 Web sites for each category, respectively. The data were passed through preprocessing phase in order to clean the data before manually analyzing the data. The three steps of preprocessing include HTML parsing, stemming and stopping. In this study, Porter2 Stemmer is used for stemming process.

This section discusses the characteristics of pornography Web sites based on the criteria used in TWS factors as discussed in previous section. The objective of this paper is to determine which criterion has the potential to solve data ambiguity challenge.

The result for the first criterion of TF factor is shown in Table 2. D1 until D3 represents 100 documents of narrative pornography Web sites which are subdivided into three types; D1 represents 35 Web pages that have no image, D2 represents 35 Web pages that have not more than three images and D3 represents 30 Web pages that have not more than 5 images. D4 through D6 represents 100 documents of discrete pornography Web sites that are subdivided into three types; D4 represents 35 Web sites that have less than 30 words, D5 represents 35 Web sites that have between 31 and 60 words, and D6 represents 30 Web sites that have 300 words. D7 until D9 represents 100 documents of healthy Web sites which are divided into three types includes; D7 represents 35 narrative healthy Web sites, D8 represents 35 discrete healthy Web sites and D9 represents 30 discrete healthy Web sites with not more than 5 images.

Table 2 Common features criteria

Term (%)	Pornography Web sites						Healthy Web sites		
	Narrative form			Discrete form					
	D1	D2	D3	D4	D5	D6	D7	D8	D9
Illicit term	32	36	38	38	40	44	8	6	4
Healthy term	68	64	62	62	60	56	92	94	96

(a) D1–D3 represent 100 document of narrative pornography Web sites
(b) D4–D6 represent 100 document of discrete pornography Web sites
(c) D7–D9 represent 100 document of healthy Web sites

Table 3 Absence features criteria

Term	Frequency (%)
Cumshot	3
Shaft	2
Tits	2
Asshole	1
Clit	3

The result shows about 40 % of *common features*, which consists of illicit terms, whether in narrative or discrete pornography Web sites. Meanwhile, less than 8 % of common features consist of illicit terms in healthy Web sites. Thus, illicit term is observes more frequent features in pornography document than healthy document. Thus, *common feature* criterion can significantly differentiate pornography from non-pornography Web sites due to illicit terms appearance in common features since pornography Web sites possesses higher features than healthy Web sites.

The second criterion for the TF factor is *favoring absence* features. Table 3 shows the terms occurrence for narrative and discrete form of pornography Web sites but absent in healthy Web sites. This feature is able to distinguish between pornography and healthy Web sites if it contains the words shown in Table 3. Thus, this criterion can also be considered in order to represent pornographic document.

Finally, the last criterion is *rare features* which is defined as the feature that appears infrequently in a document. D1 through D9 represents document of pornography and healthy Web sites similar to notation in Table 2. From the result in Table 4, about 3 % of *rare features* content is illicit terms in pornography Web sites which are D1 through D6. However, the results of healthy Web sites (D7, D8 and D9) are significantly different from pornography Web sites, which contains non-illicit term as observed in the Web sites. As shown in the results, some of the healthy Web sites use illicit term, but the term occurrence is very high which is the primary source of data ambiguity challenge. However, a closer observation of the result in Table 4 reveals that there are no terms for illicit features. Therefore, this criterion is potentially useful for data ambiguity mitigation since rare features can differentiate between pornography and healthy Web sites by using the appearance of illicit terms in rare features.

Table 4 Rare features criteria

Term (%)	Pornography Web sites						Healthy Web sites (100)		
	Narrative (100)			Discrete (100)					
	D1	D2	D3	D4	D5	D6	D7	D8	D9
Illicit	3	1	2	2	2	3	0	0	0
Healthy	97	99	98	98	98	97	100	100	100

(a) D1–D3 represent 100 document of narrative pornography Web sites
(b) D4–D6 represent 100 document of discrete pornography Web sites
(c) D7–D9 represent 100 document of healthy Web sites

Table 5 Characteristics of Pornography Web site

No.	TWS factors	Criteria	Web sites		
			Narrative form of pornography	Discrete form of pornography	Healthy
A.	Term frequency factor	Favoring common features	✓	✓	✗
		Using feature absence	✓	✓	✓
		Favoring rare features	✓	✓	✓
B.	Collection frequency factor	Using term frequency information	✗	✗	✗
		Favoring features in document	✓	✓	✗
		Favoring frequent term in a collection, it is estimated to be very relevant for the document	✓	✓	✓
		Favoring infrequent term in a collection, it is estimated to be very relevant for the document	✗	✗	✗
		Favoring features in a category	✓	✓	✓
C.	Normalization frequency factor	Normalize term length	✓	✗	✓
		Favor short documents	✗	✓	✗
		Penalizes too long document	✗	✗	✓
Total			7	7	6

As in Tables 2, 3, and 4, the characteristics of pornography Web site were extracted and significant characteristics were identified and then mapped against term weighting factors with the result come out as shown in Table 5. Based on Table 5, there are six out of eleven criteria of TWS factor that represent the narrative form of pornography Web sites, seven criteria for discrete form of pornography Web sites, and six criteria for healthy Web sites. These criteria will be considered in order to produce TWS, which are capable of representing term in ambiguous Web content and discrete Web pages with low-dimensional feature space.

4 Conclusion and Future Work

The problems of new Web sites that continuously and rapidly emerge are factors that influence the performance of existed Web filtering techniques, which are not significantly efficient. Thus, accuracy rate of illicit Web content filtering can still be improved by considering all criteria in TWS factors in order to select significant features to represent term for ambiguous Web contents. Based on the finding obtained from Tables 1 and 5, we considered M. Entropy, MI and odds ratio to be useful feature for dimension reduction, feature selection, and data ambiguity challenge. This is because, the total number of criteria of TWS factor for these methods (6 in this case) in Table 1 is be very close to total number of the characteristics of narrative and discrete pornography Web sites as shown in Table 5. Thus, criterion favoring absent features and rare features will be highlighted in future research in order to solve data ambiguity issues.

Accuracy rate of TWS can still be improved by considering all criteria in TWS factors in order to solve ambiguous issues. In addition, we plan to examine existing TWS especially M. Entropy, MI, and odds ratio by expanding the class of data set to huge corpus in order to study which methods have higher accuracy in order to classify narrative and discrete pornography Web sites. Other factors in TWS such as *DF* factor and *normalization frequency* factor are not discussed in this paper. These topics are strongly recommended for future works. Further, a better feature selection method can be adapted to solve data ambiguity issues.

Acknowledgments The authors would like to acknowledge the Ministry of Higher Education (MOHE) and Universiti Teknologi Malaysia (UTM) for supporting this research. We would also be grateful for the comments from the PARS'10 and reviewers.

References

1. Utusan: 152,182 Orang Anak Luar Nikah 2008–2010. Arkib 16/11/2011 (in Malay) (2011)
2. Maulana, F.A., Abdulmana, S., Alfariti, F.: Collaborative internet content filtering on the internet infrastructure in Malaysia. In: 2011 International Conference on Uncertainty Reasoning and Knowledge Engineering (URKE) (2011)
3. Akbulut, A., Patlar, F., Bayrak, C., Mendi, E., Hanna, J.: Agent based pornography filtering system. In: 2012 International Symposium on Innovations in Intelligent Systems and Applications (INISTA) (2012)
4. Markey, P.M.: Online Pornography Seeking Behaviors (2011)
5. Kahn, B.K., Strong, D.M., Wang, R.Y.: Information quality benchmarks: product and service performance. Commun. ACM **45**(4), 184–192 (2002)
6. Eyono Obono, S.D.: An information and system quality evaluation framework for Tribal portals: the case of selected Tribal portals from cameroon. In: 2010 2nd International Conference on Computer Technology and Development (ICCTD) (2010)
7. Denram, U.R.: WAO annual statistics 2012. Retrieved on 21 May 2013 (2012)
8. Maulana, F.A., Abdulmana, S., Alfariti, F., Nong, R.A.: Collaborative internet threat detection on internet infrastructure in Malaysia (2011)
9. Chen, T.M., Wang, V.: Web filtering and censoring. Computer **43**(3), 94–97 (2010)
10. Banday, M.T., Shah N.A.: A concise study of web filtering (2010)
11. Niharika, S., Latha V.S., Lavanya, D.R.: A survey on text categorization (2012)
12. Lee, Z.S., Maarof, M.A., Selamat, A., Shamsuddin, S.M.: Enhance term weighting algorithm as feature selection technique for illicit web content classification (2010)
13. Selamat, A., Omatu, S.: Web page feature selection and classification using neural networks. Inf. Sci. **158**, 69–88 (2004)

14. Wang, D., Zhang, H.: Inverse-category-frequency based supervised term weighting schemes for text categorization (2013)
15. Hammami, M., Guermazi, R., et al.: Automatic violent content web filtering approach based on the KDD process. Int. J. Web Inf. Syst. 4(4), 441–464 (2008)
16. Santos, I., Galán-García, P., Santamaría-Ibirika, A., Alonso-Isla, B., Alabau-Sarasola, I., Bringas, P.: Adult content filtering through compression-based text classification. In: Herrero Á., Snášel V., Abraham A. et al. (eds.) International Joint Conference CISIS'12-ICEUTE'12-SOCO'12 Special Sessions, vol. 189, pp. 281–288. Springer, Berlin (2012)
17. Liu, Y., Loh, H.T., Sun, A.: Imbalanced text classification: a term weighting approach. Expert Syst. Appl. 36(1), 690–701 (2009)
18. Erenel, Z., Altincay, H.: Nonlinear transformation of term frequencies for term weighting in text categorization. Eng. Appl. Artif. Intell. 25(7), 1505–1514 (2012)
19. Yang, Y., Pederson, J.O.: Comparative study on feature selection in text categorization (1997)
20. Largeron, C., Moulin, C., Gery, M.: Entropy based feature selection for text categorization. In: Proceedings of the 2011 ACM Symposium on Applied Computing, pp. 924–928 (2011)
21. Yan, X., Lin, C.: Term-frequency based feature selection methods for text categorization. In: 2010 Fourth International Conference on Genetic and Evolutionary Computing (ICGEC) (2010)
22. Benjamin, C.M.X., et al.: Customized term weighting scheme for document classification (2008)
23. Lan, M., Sung, S.Y., Low, H.B., Tan, C.L.: A comparative study on term weighting schemes for text categorization (2005)
24. Deng, Z.H., Tang, S.W., Yang, D.Q., Zhang, M., Li, L.Y., Xie, K.Q.: A comparative study on feature weight in text categorization. In: Yu J.X., Lin X.M., Lu H.J., Zhang Y.C. (eds.) Advanced Web Technologies and Applications, vol. 3007, pp. 588–597. Springer, Berlin (2004)
25. Debole, F., Sebastiani, F.: Supervised term weighting for automated text categorization. In: Proceedings of the 2003 ACM Symposium on Computing, vol. 2003, pp. 784–788. ACM Press, New York (2003)
26. Lan, M., Tan, C.L., Su, J.: Supervised and traditional term weighting methods for automatic text categorization (2007)

A Novel Secure Two-Party Identity-Based Authenticated Key Agreement Protocol Without Bilinear Pairings

Seyed-Mohsen Ghoreishi, Ismail Fauzi Isnin, Shukor Abd Razak and Hassan Chizari

Abstract Many Identity-Based two-party Key Agreement protocols have been proposed in recent years. Some of them are built on pairing maps, whereas some others could eliminate the pairings in order to decrease the complexity of computation. In this paper, we proposed a secure pairing-free Identity-Based two-party Key Agreement protocol which besides supporting security requirements uses less computational cost in comparison with existing related works.

Keywords Identity-based · Pairing-free · Two-party · Key Agreement · Efficiency

1 Introduction

A Key Agreement protocol enables two or more entities to establish a shared secret through an unsecure channel. In an Identity-Based Key Agreement protocol, the public key of involving entities is driven from their public identity. Since providing a secure session key in an unsecure channel is one of the most significant challenging issues, Key Agreement protocols received widespread attention in cryptography research community. It is worth to note that the focus of this paper is on two-party Identity-Based Key Agreement protocols.

S.-M. Ghoreishi (✉) · I.F. Isnin · S.A. Razak · H. Chizari
Faculty of Computing, Univeristi Teknologi Malaysia (UTM), 81310 Johor, Malaysia
e-mail: mohsen.gh100@gmail.com

I.F. Isnin
e-mail: ismailfauzi@utm.my

S.A. Razak
e-mail: shukorar@utm.my

H. Chizari
e-mail: chizari@utm.my

© Springer International Publishing Switzerland 2015
A. Abraham et al. (eds.), *Pattern Analysis, Intelligent Security
and the Internet of Things*, Advances in Intelligent Systems and Computing 355,
DOI 10.1007/978-3-319-17398-6_26

In order to avoid complex certificate management in traditional public key cryptosystems (PKC), Shamir in [1] introduced a novel idea named identity-based cryptography. In this category of PKC, users' public key is their identity (e.g., telephone number, image, and email address). Therefore, both communicating entities should have knowledge about each other's identifier before starting the communication.

However, making this theory functional remained an open problem until 2001 that Boneh and Franklin in [2] could propose a fully functional identity-based encryption scheme.

Following the work of Boneh and Franklin, various identity-based cryptosystems including Key Agreement protocols have been published based on bilinear pairings [3–6]. Bilinear pairing is a cryptographic function that maps a pair of elements of two elliptic curve-based algebraic groups to an element of a determined finite field [7]. However, pairing operations have been considered as an expensive cryptographic function by consuming about twenty times more expensive computational cost than scalar multiplication over an elliptic curve group [7]. Hence, to avoid high computational cost of pairings, several Identity-Based pairing-free Key Agreement protocols have been proposed recently (refer to Sect. 2).

To improve the efficiency, we proposed a pairing-free Identity-Based Key Agreement protocol, named PF_{ID} KA.

The rest of this paper is organized as follows. Some related works are reviewed in the Sect. 2. In Sect. 3, preliminaries including utilized notations and description of main phases of Identity-Based Key Agreement protocols are described. Section 4 assigns to our proposed pairing-free Key Agreement protocol in detail. In Sect. 5, analysis over security and efficiency of the proposed protocol is provided. At last, we draw the conclusion.

2 Related Works

There exist many pairing-free two-party Key Agreement protocols over elliptic curve-based algebraic groups. In 2010, Cao et al. in [8] proposed a pairing-free Identity-Based authenticated Key Agreement protocol with two message exchanges. They could reduce the required message exchange in comparison with previous related works presented in [9, 10]. However, as shown in [11], the proposed protocol by Cao et al. in [10] was not secure against known session-specific temporary information attack and key offset attack. Islam and Biswas in [11] could propose an improved version that does not suffer from mentioned security flaws. Their proposed scheme requires less computational cost by the use of three scalar multiplication and one point addition.

Besides the proposed protocols above, Farash and Attari in [12] have tried to modify the proposed protocol of Cao et al. [10] by considering different private key generators.

3 Preliminaries

In this section, we are going to present the required preliminaries for this article.

3.1 Notations

The suggested notations and assumptions, which are needed to realize following sections, are listed as follows:

q	A large prime number
$\mathbb{F}_q$	A finite field over q
$E/\mathbb{F}_q$	An elliptic curve over $\mathbb{F}_q$
G	A subgroup of $E/\mathbb{F}_q$
P	A generator of the group G
s	A randomly chosen element of $\mathbb{Z}_q^*$
P_{pub}	sP
H_1, H_2	Two collision-free one-way hash functions
ID_i	Identity of user i
k_s	Session key

Next section explains the main phases of Key Agreement protocols in the context of identity-based cryptosystems in detail.

3.2 Main Phases of Identity-Based Key Agreement Protocols

A possible way to define an Identity-Based two-party Key Agreement protocol is to partition four sub-protocols as main phases. Based on this categorization, these phases are named SETUP, EXTRACTION, EXCHANGE, and COMPUTATION.

SETUP
In this phase, the corresponding algorithm takes the security parameter to generate Params and master key. A trusted third party named private key generator (PKG) keeps master key confidential, whereas Params must be publicly known to all entities.

EXTRACTION
In this phase, each entity can obtain his private key by interacting with the PKG.

EXCHANGE
In this phase, communicating parties compute a trapdoor one-way function of a randomly chosen value and exchange it.

COMPUTATION

In this phase, communicating parties can compute the considered session key as a function of Params and other possessing public and secret parameters.

4 Our Proposed Identity-Based Key Agreement Protocol

In this section, we propose our efficient pairing-free Identity-Based Key Agreement protocol (named PF_{ID} KA) which can satisfy all security requirements. The outline of current section is to investigate this protocol in detail.

SETUP

This algorithm generates the master key $s \in_r \mathbb{Z}_q^*$ randomly and then outputs Params $\langle q, \mathbb{F}_q, E/\mathbb{F}_q, G, P, P_{\text{Pub}}, H_1, H_2 \rangle$ by the use of taken security parameter. In Params, $H_1: \{0,1\}^* \times G \to \mathbb{Z}_q^*$ and $H_2: \{0,1\}^* \times \{0,1\}^* \times G \times G \times G \to \mathbb{Z}_q^*$. The rest elements are introduced in Sect. 3.

EXTRACTION

In this phase, an entity such as the one who possesses ID_i identifier refers to PKG to take corresponding private key. The PKG first randomly chooses $r_i \in_r \mathbb{Z}_q^*$, then computes $R_i = r_i P$ and $h_i = H_1(ID_i, R_i)$. Finally, the entity's private key would be $\langle R_i, s_i \rangle$ where $s_i = r_i + h_i s \pmod q$.

Now assume that two entities, A and B, are going to agree on a session key. The EXCHANGE and COMPUTATION phases are as follows:

EXCHANGE

To explain the EXCHANGE phase, mentioned entities do the following:

1. A chooses a random $a \in_r \mathbb{Z}_q^*$, computes the key token $T_A = a(s_A P) = a((r_A + h_A s \pmod q))P)$ and sends T_A, R_A to the entity B.
2. B chooses a random $b \in_r \mathbb{Z}_q^*$, computes the key token $T_B = b(s_B P) = b((r_B + h_B s \pmod q))P)$ and sends T_B, R_B to the entity A.

COMPUTATION

In this phase, mentioned entities are able to compute the shared secret as follows:

A computes $K_{AB} = [a(r_A + h_A s \pmod q))]T_B$
B computes $K_{BA} = [b(r_B + h_B s \pmod q))]T_A$

Following equation proves that the two computed values for this shared secrets would be the same.

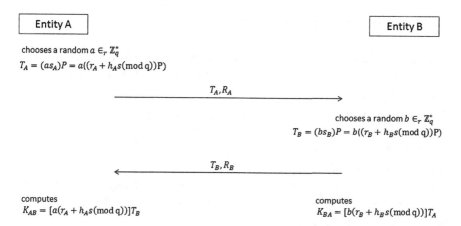

Entity A

chooses a random $a \in_r \mathbb{Z}_q^*$
$T_A = (as_A)P = a((r_A + h_A s(\mathrm{mod}\, q))P)$

Entity B

chooses a random $b \in_r \mathbb{Z}_q^*$
$T_B = (bs_B)P = b((r_B + h_B s(\mathrm{mod}\, q))P)$

T_A, R_A

T_B, R_B

computes
$K_{AB} = [a(r_A + h_A s(\mathrm{mod}\, q))]T_B$

computes
$K_{BA} = [b(r_B + h_B s(\mathrm{mod}\, q))]T_A$

Fig. 1 Our proposed protocol

$$
\begin{aligned}
K_{AB} &= [a(r_A + h_A s(\mathrm{mod}\, q))]T_B \\
&= (as_A)[b((r_B + h_B s(\mathrm{mod}\, q))P)] \\
&= (as_A)(bs_B)P \\
&= [b(r_B + h_B s(\mathrm{mod}\, q))]T_A \\
&= K_{BA}
\end{aligned}
$$

Finally, the agreed session key, k_s, is a key derivation function of K_{AB}:

$$
\begin{aligned}
k_s &= H_2(ID_A, ID_B, T_A, T_B, K_{AB}) \\
&= H_2(ID_A, ID_B, T_A, T_B, K_{BA})
\end{aligned}
$$

Figure 1 illustrates $PF_{ID} KA$ protocol in a general form.

5 Security and Efficiency Analysis

In this section, we will explain the required security considerations for a Key Agreement protocol. Moreover, we represent the computational cost of existing related works to compare them with our proposed protocol from computational efficiency viewpoint.

5.1 Security Considerations

In order to evaluate the security of Key Agreement protocols, one common approach is the use of following security features explained in [13, 14].

Known-Key Security (KKS)
The KKS indicates that any knowledge about past secret session keys do not lead to finding future ones. The main reason is that the secret session key is unique and independent from past established ones.

Forward Secrecy (FS)
A protocol can support this property if in the condition of leakage of entities' long-term private keys, the previously established session keys remain secret.

Perfect Forward Secrecy (PFS)
A protocol has this property if in the condition of leakage of entities' long-term private keys including PKG, the previously established session keys remain secret.

Key-Compromise Impersonation (KCI)
In the condition of compromising the long-term key of one of the entities, adversary can impersonate the victim to others but not vice versa.

Unknown Key-Share Resilience (UKSR)
Unknown key-share happens if an adversary could convince an entity to share a secret session key with him instead of a legitimate entity. A Key Agreement protocol should be resilient against this type of attack.

Key Control (KC)
This security property indicates that the secret session key would be generated by both communicating entities together. It means the session key should not be predetermined by one of them alone.

Known Session-Specific Temporary Information (KSSTI)
If the session key can be computable by the adversary in the condition of the leakage of a and b (refer to the EXCHANGE phase in Sect. 4), the protocol would be vulnerable to this attack.

   It is worth to note that our proposed protocol supports all mentioned security attributes. In addition, it can provide key confirmation and prevent key offset attack if the entities A and B exchange message authentication code (MAC) of a significant message which is generated based on the session key (for more information refer to [11]).

5.2 Efficiency Considerations

As mentioned in the second section, related to our proposed protocol, several two-party Identity-Based Key Agreement protocols without bilinear pairings have been

Table 1 Efficiency comparisons of different protocols

Authors	Exchange and computation from A entity viewpoint	Computed exponentiation (scalar multiplication)	Computed point addition	Efficiency consideration
Cao et al. [8]	$T_A = aP, T_B = bP$ $K_{AB}^1 = s_A T_B + a S_B$ $K_{AB}^2 = a T_B$	$aP, s_A T_B, a S_B, a T_B,$	$(s_B T_A) + (b S_A)$	4 exponentiation (scalar multiplication) 1 point addition
Islam and Biswas [11]	$T_A = a S_A, T_B = b S_B$ $K_{AB} = s_A [T_B + a S_B]$	$a S_A, a S_B, s_A [T_B + a S_B]$	$T_B + (a S_B)$	3 exponentiation (scalar multiplication) 1 point addition
PF_{ID} KA	$T_A = a(s_A P)$ $K_{AB} = a(s_A [T_B])$	$a(s_A P), s_A T_B, a(s_A [T_B])$	–	3 exponentiation (scalar multiplication)

proposed. Cao et al. in [8] proposed a pairing-free Key Agreement protocol that has four scalar multiplications and one point addition. The proposed protocol by Islam and Biswas in [11] has only three scalar multiplications and one point addition. Moreover, in 2014, another pairing-free two-party Identity-Based Key Agreement scheme has been proposed by Farash et al. in [12] that has four scalar multiplications.

To clear our claim, Table 1 depicts details of proposed protocols in [8, 11] and the assigned computational costs.

As illustrated in Table 1, our proposed pairing-free Identity-Based Key Agreement protocol is quite efficient because it just requires three scalar multiplications without any point addition performed by each communicating participant.

6 Conclusion

In recent years, various pairing-free cryptosystems have been designed in order to reduce high cost of computation resulted by utilizing pairing maps. In this area, several pairing-free Key Agreement protocols in the context of Identity-Based cryptosystems have been proposed. In this paper, we could propose an authenticated Identity-Based two-party Key Agreement protocol without using pairing maps. The proposed protocol is efficient in comparison with existing related works.

Acknowledgments Authors would like to thank Universiti Teknologi Malaysia and Ministry of Higher Education, Malaysia, for sponsoring this research under vote number 01G98.

References

1. Shamir, A.: Identity-based cryptosystems and signature schemes. In: Advances in Cryptology—Crypto 1984. Lecture Notes in Computer Science 196, Springer, Berlin (1984)
2. Boneh, D., Franklin, M.: Identity based encryption from the weil pairing. In: Advances in Cryptology—Crypto (2001)
3. Smart, N.P.: An identity based authenticated key agreement protocol based on the Weil pairing. Electron. Lett. **38**, 630–632 (2002)
4. Chen, L., Kudla, C.: Identity based authenticated key agreement from pairings. In: IEEE Computer Security Foundations Workshop, pp. 219–233 (2003)
5. Yuan, Q., Li, S.A.: A new efficient ID-based authenticated key agreement protocol. Cryptology ePrint Archive, Report 2005/309 (2005)
6. Wang, Y.: Efficient Identity-based and authenticated key agreement protocols. Transactions on Computational Science Xvii (2013)
7. Chen, L., Cheng, Z., Smart, N.P.: Identity-based key agreement protocols from pairings. Int. J. Inf. Secur. **6**, 213–241 (2007)
8. Cao, X., Kou, W., Du, X.: A pairing-free identity-based authenticated key agreement protocol with minimal message exchanges. Inf. Sci. **180**, 2895–2903 (2010)
9. Zhu, R.W., Yang, G., Wong, D.S.: An efficient identity-based key exchange protocol with KGS forward secrecy for low-power devices. Theor. Comput. Sci. **9**, 198–207 (2007)
10. Cao, X., Kou, W., Yu, Y., Sun, R.: Identity-based authentication key agreement protocols without bilinear pairings. IEICE Trans. Fundam. **91**(12), 3833–3836 (2008)
11. Islam, S.K.H., Biswas, G.P.: An improved pairing-free identity-based authenticated key agreement protocol based on ECC. Procedia Eng. **30**, 499–507. ISSN 1877-7058 (2012)
12. Farash, M.S., Attari, M.A.: A pairing-free ID-based key agreement protocol with different PKGs. Int. J. Netw. Secur. **16**(2), 143–148 (2014)
13. Cheng, Z., Nistazakis, M., Comley, R., Vasiu, L.: On the indistinguishability-based security model of key agreement protocols-simple cases. Cryptology ePrint Archieve, Report 2005/129 (2005)
14. Blake-Wilson, S., Johnson, D., Menezes, A.: Key agreement protocols and their security analysis. In: Proceedings of the 6th IMA International Conference on Cryptography and Coding, vol. 1335, pp. 30–45, LNCS, Springer, Berlin (1997)

An Efficient Pairing-Free Certificateless Authenticated Two-Party Key Agreement Protocol Over Elliptic Curves

Seyed-Mohsen Ghoreishi, Ismail Fauzi Isnin, Shukor Abd Razak and Hassan Chizari

Abstract Due to the high computation cost of bilinear pairings, pairing-free cryptosystems have received widespread attention recently. Various pairing-free two-party key agreement protocols in the context of public key cryptography (PKC) have been studied. To avoid complex certificate management in traditional PKC and key escrow problem in identity-based ones, several certificateless cryptosystems have been proposed in this research area. In this paper, we proposed a secure and efficient certificateless pairing-free two-party key agreement protocol. In comparison with related works, our protocol requires less computational cost.

Keywords Pairing-free · Certificateless · Two-party key agreement · Efficiency

1 Introduction

One of the fundamental cryptographic primitives in the context of public key cryptography (PKC) which enables two or more entities to create a shared session key is called key agreement protocol. Due to exchanging key tokens over a public channel before deriving the shared session key, key agreement protocols became one of critical challenging issues in PKC research area. If roughly speaking, public key cryptosystems can be categorized in three classes named traditional, identity-based, and certificateless ones. In a traditional PKC, digital certificates issued by

S.-M. Ghoreishi (✉) · I.F. Isnin · S.A. Razak · H. Chizari
Faculty of Computing, Universiti Teknologi Malaysia, Johor Bahru, Johor, Malaysia
e-mail: mohsen.gh100@gmail.com

I.F. Isnin
e-mail: ismailfauzi@utm.my

S.A. Razak
e-mail: shukorar@utm.my

H. Chizari
e-mail: chizari@utm.my

© Springer International Publishing Switzerland 2015
A. Abraham et al. (eds.), *Pattern Analysis, Intelligent Security
and the Internet of Things*, Advances in Intelligent Systems and Computing 355,
DOI 10.1007/978-3-319-17398-6_27

Certificate Authority are required in order to provide trustable relationship among users' public key. Complex management of certificates is the main drawback of this group of cryptosystems (for more details, refer to [1]). Shamir in [2] introduced the theory of identity-based cryptography (IBC) that could resolve mentioned problem by considering users' identity (e.g., telephone number, image, email address) as their public key. However, this theory remained nonfunctional for several years, until 2001 that a fully functional identity-based encryption scheme has been proposed by Boneh and Franklin in [3]. Although this category of cryptosystems could solve the problem of traditional PKC, it suffers from key escrow problem resulted from private key generation process. More precisely, in IBC, the users' private key is generated by a trusted third party named private key generator (PKG). Hence, PKG is able to impersonate users or eavesdrop the transmitted messages between them. To overcome mentioned problems in traditional PKC and IBC, Al-Riyami and Paterson introduced a new concept named certificateless public key cryptography (CL-PKC) [4]. To generate private key, each user interacts with a trusted third party called key generation center (KGC) and takes partial keys which is depended on master key (only known to KGC) and the users' identity. Afterward, considered entity chooses a secret value. The final private key would be driven from the possessing public and private items and mentioned chosen secret.

In continuous to what mentioned above, it is necessary to point out that various certificateless key agreement protocols have been proposed based on the use of bilinear pairings [5–8]. The functionality of bilinear pairings is to map two points of two algebraic groups over an elliptic curve to an element of a determined finite field [9]. To avoid high computational cost of pairing operations, several certificateless pairing-free key agreement protocols have been proposed recently (refer to Sect. 2).

Improving the efficiency persuaded us to propose a certificateless key agreement protocol without bilinear pairings, named $E_{CL}KA$.

The rest of this paper is organized as follows. Section 2 reviews some related works. In the third section, preliminaries including utilized notations and description of main phases of certificateless key agreement protocols are described. In Sect. 4, we present our pairing-free key agreement protocol in detail. In the fifth section, analysis over security and efficiency of the proposed protocol is provided. Finally, we draw the conclusion.

2 Related Works

In the context of certificateless two-party key agreement schemes, Hou and Xu in [10] proposed a pairing-free certificateless authenticated key agreement protocol inspired by the certificateless encryption scheme proposed by Baek et al. [11]. They claimed that mentioned work supports all security requirements. At the same year, Geng and Zhang in [12] proposed another CL-PKC key agreement protocol. These authors could prove the security of their work under Gap Diffie–Hellman assumption. However, Yang and Tan in [13] who showed that both mentioned works above are insecure proposed a CL-PKC key agreement protocol with strong security proof. Although their protocol was secure, it failed to be efficient by requiring nine ECC scalar multiplications. The proposed scheme by He et al. in [14] seems to be efficient in comparison with previous protocols, but it is insecure as pointed out by He et al. in [15]. Beside of what mentioned before, the authors of the proposed protocol in [16] have tried to present an efficient and secure key agreement protocol in the context of pairing-free CL-PKC. However, they could not reach this goal based on the discussion provided in two different documents [16, 17]. More precisely, the proposed protocol by He et al. in [16] is vulnerable to key replacement attack and ephemeral key compromise attack.

3 Preliminaries

In this section, we are going to present the required preliminaries for this paper.

3.1 Required Notations

For the sake of simplicity, we standardized utilized notations and assumptions for the proposed and investigated schemes in the rest of this paper. Table 1 depicts these items in more detail.

Next subsection represents detailed explanations of the main phases of key agreement protocols, in the context of certificateless cryptosystems.

3.2 Main Phases of Certificateless Key Agreement Protocols

A certificateless key agreement protocol can be defined in five phases, which are setup, partial-private extract, set-private-public keys, exchange, and computation.

The setup algorithm is responsible to generate params and master key, after taking the security parameter. It is worth to note that params is publicly known to all entities, whereas the master key is known only for KGC.

Table 1 Suggested notations and assumptions

Notation	Description
q	A large prime number
$\mathbb{F}_q$	A finite field over q
$E/\mathbb{F}_q$	An elliptic curve over $\mathbb{F}_q$
G	A cyclic subgroup of $E/\mathbb{F}_q$
P	A generator of the group G
s	A randomly chosen element of $\mathbb{Z}_q^*$

The KGC sends a partial-private key to a corresponding user where he made a request in the partial-private extract phase. Thereupon, in set-private-public keys phase, an entity chooses a random value in order to generate his public and private keys. Eventually, an entity can communicate with other ones to share the corresponding session key in the two last phases.

4 Our Proposed Certificateless Key Agreement Protocol

In this section, we introduce our proposed efficient key agreement protocol, named $E_{CL}KA$, which can satisfy all security requirements. The outline of current subsections is to investigate the protocol in detail.

Setup: This algorithm takes the security parameter and returns a master key $s \in \mathbb{Z}_q^*$ and params

$\langle q, \mathbb{F}_q, E/\mathbb{F}_q, G, P, P_{\text{Pub}}, H_1, H_2 \rangle$ that $H_1 : \{0, 1\}^* \times G \to \mathbb{Z}_q^*$ and $H_2 : \{0, 1\}^* \times \{0, 1\}^* \times G \times G \times G \to \mathbb{Z}_q^*$ are two one-way collision-free hash functions.

Partial-private extract: This algorithm randomly chooses $r_i \in_r \mathbb{Z}_q^*$ and then computes $R_i = r_iP$ and $h_i = H_1(ID_i, R_i)$. The partial-private key of the user i will be $s_i = r_i + h_is \pmod{q}$.

Set-public-private keys: This algorithm randomly chooses $x_i \in_r \mathbb{Z}_q^*$ and computes $X_i = x_iP$, $z_i = x_i + h_i's_i \pmod{q}$, and $Z_i = z_iP$. Here, $h_i' = H_1(ID_i, X_i)$. The private and public key of the user i will be $SK_i = (s_i, x_i)$ and $PK_i = (R_i, S_i, X_i)$, respectively. Here, the value of $S_i = (R_i + h_iP_{\text{pub}}) = s_iP$ will be publicly computable by all entities.

Exchange: To explain the exchange phase, mentioned entities do the following:

1. A chooses a random $a \in_r \mathbb{Z}_q^*$, computes the key token $T_A = (az_A)P = a((x_A + h_A's_A \pmod{q}))P$, and sends T_A to the B entity.
2. B chooses a random $b \in_r \mathbb{Z}_q^*$, computes the key token $T_B = (bz_B)P = b((x_B + h_B's_B \pmod{q}))P$, and sends T_B to the A entity.

Computation: In this phase, the entities A and B are able to compute the shared secret as follows:

$$A \text{ computes } K_{AB} = [a(x_A + h_A's_A \pmod{q}))]T_B$$
$$B \text{ computes } K_{BA} = [b(x_B + h_B's_B \pmod{q}))]T_A$$

Following equation proves that the two computed values for this shared secrets would be the same.

$$K_{AB} = [a(x_A + h'_A s_A(\text{mod q}))]T_B$$
$$= (az_A)[b((x_B + h'_B s_B(\text{mod q}))P)]$$
$$= (az_A)(bz_B)P$$
$$= [b(x_B + h'_B s_B(\text{mod q}))]T_A$$
$$= K_{BA}$$

Finally, the agreed session key, k_s, is a key derivation function of K_{AB}:

$$k_s = H_2(ID_A, ID_B, T_A, T_B, K_{AB})$$
$$= H_2(ID_A, ID_B, T_A, T_B, K_{BA})$$

5 Security and Efficiency Analysis

In this section, the proposed protocol is investigated from two different dimensions: security and efficiency. Our proposed protocol could achieve all security attributes. Moreover, we claim that our scheme is nicely efficient in comparison with current related works.

5.1 Security Considerations

To evaluate a key agreement protocol, one well-known method is the use of following security properties as defined by Cheng et al. and Blake-Wilson et al. in [18, 19].

5.1.1 Known-Key Security (KKS)

To support this security property, peer entities should create a unique secret session key, which is independent from previous secret session keys. Hence, deduction of future secret session keys cannot be done by any knowledge about past secret session keys.

5.1.2 Forward Secrecy (FS)

The past session keys must be secret under a condition that long-term private keys of the entity(ies) are compromised.

5.1.3 Perfect Forward Secrecy (PFS)

This attribute is similar to FS. The difference is that the long-term private key of the KGC is also considered to be compromised.

5.1.4 Key Compromise Impersonation

The condition that leads to resisting a protocol against key compromise impersonation attack is that although compromising the long-term key of an entity helps the adversary to impersonate the victim to others, he must be unable to impersonate others to the victim entity.

5.1.5 Unknown Key-Share Resilience

In unknown key-share attack, the adversary convinces an entity to share a secret session key with him, while the entity believes that a secret has been shared with a legitimate entity.

5.1.6 Key Control

To satisfy this security attribute, it must be impossible to predetermine the shared session key by one of the communicating parties alone.

5.1.7 Known Session-Specific Temporary Information

Vulnerability against this attack means that any leakage of randomly chosen values in exchange phase leads to compromising the session key by the adversary.

We claim that our protocol is secure against above-mentioned issues as the generated session key satisfies all of them. Moreover, to support key conformation in our protocol and make it secure against key offset attack, it is possible to just consider that communicating entities exchange a message authentication code (MAC) of a meaningful message by the use of the agreed key.

5.2 Efficiency Considerations

Related to our proposed protocol, in the context of certificateless key agreement protocols without pairings, Hou et al. proposed a protocol with four scalar multiplications [10]. The proposed protocol by Geng et al. in [12] computes five scalar

Table 2 Efficiency comparisons of different protocols

Authors	Exchange and computation from A entity viewpoint	Computed exponentiation (scalar multiplication)	Computed point addition	Efficiency Consideration
He et al. [15]	$T_A = aP, T_B = bP$ $K_{AB}^1 = (x_A + s_A)T_B + a(X_B + S_B)$ $K_{AB}^2 = aT_B$	$aP, (x_A + s_A)T_B,$ $a(X_B + S_B), aT_B$	$[(x_A + s_A)T_B] + [a(X_B + S_B)]$	4 Exponentiation (scalar multiplication) 1 point addition
He et al. [16]	$T_A = aP, T_B = bP$ $K_{AB}^1 = (a + s_A)[T_B + S_B]$ $K_{AB}^2 = (a + x_A)[T_B + X_B]$ $K_{AB}^3 = aT_B$	$aP, (a + s_A)[T_B + S_B],$ $(a + x_A)[T_B + X_B], aT_B$	$(T_B + S_B), (T_B + X_B)$	4 Exponentiation (scalar multiplication) 2 point addition
$E_{CL}KA$	$T_A = (az_A)P$ $K_{AB} = [az_A]T_B$	$aZ_A, a(z_A(T_B))$	–	3 Exponentiation (scalar multiplication)

multiplications. In 2011, He et al. in [15] proposed a protocol that computes four scalar multiplications and one point addition for key computation. Moreover, another scheme is proposed by He et al. in [16] based on computing four scalar multiplications and two point additions. Yang and Tan in [13] proposed a CL-PKC key agreement protocol that requires nine ECC scalar multiplications.

To clear our claim, Table 2 depicts details of proposed protocols by He et al. in [15, 16] and He et al. in [15, 16] and the assigned computational costs.

As illustrated in Table 2, our proposed pairing-free certificateless key agreement protocol ($E_{CL}KA$) is quite efficient because it just requires three scalar multiplications without point addition computation for each communicating entities.

6 Conclusion

Since pairing-based cryptosystems suffer from high computational cost of bilinear pairings, the pairing-free cryptography became an active research area in recent years. In this area, a subset of pairing-free key agreement protocols in the context of certificateless cryptosystems has been proposed. In this paper, we could propose a secure and authenticated certificateless two-party key agreement protocol that does not require bilinear pairings. The proposed protocol is computationally more efficient in comparison with related works.

Acknowledgments Authors would like to thank Universiti Teknologi Malaysia and Ministry of Higher Education, Malaysia, for sponsoring this research under vote number 01G98.

References

1. Adams, C., Lloyd, S.: Understanding Public-Key Infrastructure: Concepts, Standards, and Deployment Considerations, 2nd edn. Pearson education, Boston (2003)
2. Shamir, A.: Identity-based cryptosystems and signature schemes. Advances in Cryptology—Crypto. Lecture Notes In Computer Science. Springer, Berlin (1984)
3. Boneh, D., Franklin, M.: Identity based encryption from the weil pairing. Advances in Cryptology—Crypto. Springer, Berlin (2001)
4. Al-Riyami, S.S., Paterson, K.G.: Certificateless public key cryptography. In: Laih, C.S. (ed.) Advances in Cryptology C Asiacrypt 2003. Lecture Notes in Computer Science, pp. 452–473. Springer, Berlin (2003)
5. Wang, S., Cao, Z., Dong, X.: Certificateless authenticated key agreement based on the MTI/CO protocol. J. Inf. Comput. Sci **3**, 575–581 (2006)
6. Mandt, T., Tan, C.: Certificateless authenticated two-party key agreement protocols. In: Proceedings of the ASIAN 2006. LNCS, vol. 4435, pp. 37–44. Springer, Berlin (2006)
7. Shi, Y., Li, J.: Two-party authenticated key agreement in certificateless public key cryptography. Wuhan Univ. J. Nat. Sci. **12**(1), 71–74 (2007)
8. Lippold, G., Boyd, C., Nieto, J.: Strongly secure certificateless key agreement. Pairing 2009, pp. 206–230. Springer, Berlin (2009)
9. Chen, L., Cheng, Z., Smart, N.P.: Identity-based key agreement protocols from pairings. Int. J. Inf. Secur. **6**, 213–241 (2007)
10. Hou, M., Xu, Q.: A two-party certificateless authenticated key agreement protocol without pairing. In: 2nd IEEE international conference on computer science and information technology, pp. 412–416, 2009
11. Baek, J., Safavi-Naini, R., Susilo, W.: Certificateless public key encryption without pairing. In: Proceedings of the 8th international conference on information security. Volume 3650 of LNCS, pp. 134–148. Springer, Berlin doi: 10.1007/11556992, 2005
12. Geng, M., Zhang, F.: Provably secure certificateless two-party authenticated key agreement protocol without pairing. In: International conference on computational intelligence and security, pp. 208–212, 2009
13. Yang, G., Tan, C. Strongly secure certificateless key exchange without pairing. In: 6th ACM symposium on information, computer and communications security, pp. 71–79, 2011
14. He, D., Chen, J., Hu, J.: A pairing-free certificateless authenticated key agreement protocol. Int. J. Commun. Syst. **25**, 221–230 (2011). doi:10.1002/dac.1265
15. He, D., Chen, Y., Chen, J., Zhang, R., Han, W.: A new two-round certificateless authenticated key agreement protocol without bilinear pairings. Math. Comput. Model. **54**(11–12), 3143–3152 (2011)
16. He, D., Padhye, S., Chen, J.: An efficient certificateless two-party authenticated key agreement protocol. Comput. Math. Appl. **64**(6), 1914–1926 (2012)
17. Sun, H., Wen, Q., Zhang, H., Jin, Z.: A novel pairing-free certificateless authenticated key agreement protocol with provable security. Front. Comput. Sci. **7**, 544–557 (2013)
18. Cheng, Z., Nistazakis, M., Comley, R., Vasiu, L.: On the indistinguishability-based security model of key agreement protocols-simple cases. Cryptology ePrint Arch. **2005**, 129 (2005)
19. Blake-Wilson, S., Johnson, D., Menezes, A. Key agreement protocols and their security analysis. In: Proceedings of the 6th IMA international conference on cryptography and coding, LNCS, Springer, vol. 1335, pp. 30–45 (1997)

Selection of Soil Features for Detection of Ganoderma Using Rough Set Theory

Nurfazrina Mohd Zamry, Anazida Zainal, Murad A. Rassam, Majid Bakhtiari and Mohd Aizaini Maarof

Abstract Ganoderma boninense (G. boninense) is one of the critical palm oil diseases that have caused major loss in palm oil production, especially in Malaysia. Current detection methods are based on molecular and non-molecular approaches. Unfortunately, both are expensive and time consuming. Meanwhile, wireless sensor networks (WSNs) have been successfully used in precision agriculture and have a potential to be deployed in palm oil plantation. The success of using WSN to detect anomalous events in other domain reaffirms that WSN could be used to detect the presence of G. boninense, since WSN has some resource constraints such as energy and memory. This paper focuses on feature selection to ensure only significant and relevant data that will be collected and transmitted by the sensor nodes. Sixteen soil features have been collected from the palm oil plantation. This research used rough set technique to do feature selection. Few algorithms were compared in terms of their classification accuracy, and we found that genetic algorithm gave the best combination of feature subset to signify the presence of Ganoderma in soil.

Keywords Precision agriculture · Palm oil · Ganoderma boninense · Rough set · Wireless sensor networks · Anomaly detection and feature selection

N.M. Zamry (✉) · A. Zainal · M. Bakhtiari · M.A. Maarof
Faculty of Computing, Universiti Teknologi Malaysia, 81310 Johor, Malaysia
e-mail: nurfazrina.mohdzamry@gmail.com

A. Zainal
e-mail: anazida@utm.my

M. Bakhtiari
e-mail: majid@utm.my

M.A. Maarof
e-mail: aizaini@utm.my

M.A. Rassam
Faculty of Engineering and Information Technology, Taiz University,
Taiz, The Republic of Yemen
e-mail: murad.utm@gmail.com

© Springer International Publishing Switzerland 2015
A. Abraham et al. (eds.), *Pattern Analysis, Intelligent Security and the Internet of Things*, Advances in Intelligent Systems and Computing 355,
DOI 10.1007/978-3-319-17398-6_28

1 Introduction

The major palm oil loss in southeast Asia, especially in Malaysia, is caused by Basal Stem Rot (BSR) disease. BSR attacks the root system of oil palm, thereby causing damage to the internal palm tissues through mushroom-like fungus called Ganoderma. As the fungus destroys the internal tissue of the palm tree, it may immobilize the distribution of water and nutrient to the upper level of the tree and cause the tree to die. Ganoderma disease is more likely to spread in soil through root or through airborne by spores with wind, rain, and insect as the agent. There are various species of Ganoderma, namely Ganoderma boninense, Ganoderma zonatum, Ganoderma miniatotinctum, and Ganoderma tornatum, which can be differentiated based on the locality, characteristic, and the aggressiveness of the infection. Currently, a number of research works have been reported on the detection of Ganoderma boninense fungus and the BSR [1].

A healthy palm will get infected when they are contacted with the infected palm root from the soil or with the dead root. When the trunk of the infected palm is cut out, they will show lesion with brown to gray color depending on the severity of the infection. The infected palm then causes the destruction of the inner stem which leads to the malfunction of water and nutrient distribution to the upper part of the tree. With the nutrient deficiencies, the leaves show several symptoms based on the severity of malnutrition. These symptoms on the other hand lead to modification of the canopy properties of palm tree. Unfortunately, the experiment on foliar of canopy does not get much attention than the sample from stem tissues. Other than that, the symptom can be observed when mushroom-like fungus appears on the bottom of the tree or on the roots.

Detecting Ganoderma is challenging as the infected tree is hard to be detected. It can be detected when the tree shows severe infection like appearance of fruiting body and several leaves symptom or foliage symptom like "skirt-like" shape of the crown due to leaves declination, unopened spears, and yellowing and drying of the leaves [2]. According to Lelong et al. [2], Ganoderma visual symptom can be categorized into three levels of severity as shown in Table 1. The symptom can be found by evaluating the stem and also the leaf or canopy of the palm tree.

Table 1 Visual symptoms of G. boninense according to the infection level of palm's stem and canopy [2]

Infection level	Evolution of stem conditions	Evolution of canopy structure
1	Presence of mycelium in the stem bark or crumbly wood	Yellowing or drying of some leaves. One or two new leaves remain as unopened spears
2	Presence of fruiting bodies (mushrooms) at the bottom of the stem	Apparition of leaf necrosis. Three to five new leaves remain as unopened spears. Declination of older leaves
3	Rotten stem	Largely spread leaf necrosis. No new leaf. No new bunch. Skirt-like shape of the crown due to the total leaf declination

Survival of palm tree can depend on the severity of the infection and the plantation age. The young infected plant may die 1–2 years later, after getting infected, and matured infected tree survives only another 3–4 years. As the disease is slowly developed, infection may take 15–25 years and the disease level is up to 50 and 85 %, respectively, according to Shakaff et al. [3]. Several researches have been done to detect G. boninense. Generally, palm oil detection can be classified into molecular and non-molecular methods. Molecular method involves biological or chemical techniques, while non-molecular detection is mostly done with the aid of electronic tools or computer system.

Most of the G. boninense detections are done using molecular-based method where fresh sample such as stem tissue or soil is from the plantation to the laboratory. Data collection poses difficult challenges due to hilly terrain of palm plantation. Nevertheless, data accuracy and detection efficiency still need to be enhanced for better detection. In this research, soil data were manually extracted and laboratory experiment was done to get the nutrient reading by the palm oil organization. A total of 16 features were collected.

Even though there are some devices used to detect G. boninense like e-nose which captures the gas odor that surround the tree or hyperspectral remote sensing, cost factor and the need for physical work to capture the data present additional challenges. For e-nose, rough terrain and close contact with palm oil fungus may jeopardize the safety of e-nose operator. Meanwhile, hyperspectral remote sensing collects data with humongous size. As mentioned earlier, wireless sensor networks (WSNs) have been used in the agricultural domain to monitor and also detect emergency event and therefore provide an alternative to the existing method of G. boninense detection.

Meanwhile, WSNs have been used in agricultural domain to help the industry to produce better agricultural products whereby WSN help farmer or agriculturist to deal with the challenges. Markom et al. [4] stated among of the palm tree challenges are the competition with crop pests to get nutrient and water from soil, the palm tree diseases like G. boninense.

Anomaly detection has been used in various domains such as fraud detection, intrusion detection, performance analysis, forecasting, and many more. Hence, the use of anomaly detection for monitoring soil composition can help to mitigate changes that occur in palm soil in early stage and treat the disease earlier as well as to enhance crop production.

The rest of the paper is arranged as follows: related works on detecting Ganoderma as well as the use of WSNs in agricultural domain are highlighted in Sect. 2. Section 3 provides an overview of challenges and requirement of using WSNs for detection. In Sect. 4, the initial result of implementing data filtering and feature selection is provided. The experimental result of this research is discussed in Sect. 5. Finally, conclusion is presented in Sect. 6.

2 Related Works

A lot of effort has been done in order to detect the G. boninense disease. Hushiarian et al. [5] classified the G. boninense detection methods into molecular and non-molecular method. Ganoderma and Disease Research for Oil Palm (GanoDROP) Unit is the special unit in the Malaysian Palm Oil Board that focuses on Ganoderma disease research and development. They have achieved lot of improvement on G.boninense detectionespecially on molecular method from GanoDROP units including [6–9]. Meanwhile, some researches have been done on the agricultural domain based on WSNs.

2.1 Molecular Method

Molecular methods are based on biological or chemical laboratory experiment. In early G. boninense detection, diagnosis is observed either on leaf wilting and falling or on the appearance of pathogen in the stem. In early molecular detection, most research attempts immunoassay or the DNA of the palm tree. Some of the molecular methods are presented in [5, 6, 8, 10].

2.2 Non-molecular Method

Meanwhile, as long as molecular methods remain complex and time consuming, other methods such as e-nose, tomography imagery, and hyperspectral reflectance data continue to be explored. G. boninense diseases can be detected in a lesser time-consuming manner while enhancing the detection accuracy. Table 2 shows research on non-molecular detection method for detecting G. boninense in palm oil field.

Non-molecular method detection is done with the aid of electronic device or computer application. Lelong et al. [1] and Hushiarian et al. [5] implemented electronic nose (e-nose) to detect plant disease by mimicking the human olfaction system to replace the human expert in classification process or differentiate the types of product. Shafri and Hamdan [11] used airborne hyperspectral imagery data to differentiate the healthy and infected oil palm tree. Further, Shafri [12] used hyperspectral remote sensing to discriminate the spectral reflectance of oil palm tree. Mohd Su'ud [9] proposed the usage of tomography image combined with expert's knowledge to detect the Ganoderma infection. Tomography image can be constructed based on the sound propagation in the stem. This approach is called real-time approach, since detection can be done on site. Similarly, Rees et al. [10] introduced GanoSken software which constructs tomography image from the sound wave captured using sound sensor generated. The combination of expert knowledge with mathematical models provides visualization of sound-line intersection points.

Table 2 Research on non-molecular detection methods

Author(s)	Detection area	Parameter(s)
Abdullah et al. [1]	Odor	Gasoline and diesel engine; air contaminants; ammonia; carbon dioxide; solvent vapor; hydrogen sulfide; methane; LP gas/propane; water vapor; temperature; humidity
Markom et al. [4]	Odor	Bored trunk; surrounding trunk; soil
Shafri and Hamdan [11]	Hyperspectral airborne imagery data	Six vegetation indices (normalized difference vegetation indices (NDVI), Renormalized Vegetation Index (RDVI), simple ratio index (SRI), modified simple ratio (MSR), soil-adjusted vegetation index (SAVI), optimized soil-adjusted vegetation index (OSAVI) and four red edge techniques (Lagrangian interpolation technique, Vogelmann red edge (VOG1), linear four point interpolation, and maximum first derivative)
Shafri and Anuar [12]	Palm oil canopy; leaves	Spectral signature; chlorophyll content and moisture content
Mohd Su'ud et al. [9]	Tomography stem image	Expert knowledge to set rules: infection pattern's eccentricity, infection pattern's orientation; infection pattern's solidity; infection pattern's roundness; infection pattern's area and size; infection pattern's position in the cross section; color of the patter
Idris et al. [13]	Palm trunk	Sound wave

Most research findings are carried out on palm stem, trunk, or leaf. However, other research on palm infection is carried out on the soil of the plantation. Hushiarian et al. [5] stated that infected tissues in the soil are more likely to spread the disease to healthy roots than airborne spores. These non-molecular methods are still carried out manually and need human intervention. But it is expensive, and the collected data are huge especially for hyperspectral techniques. The idea of this research is to analyze the soil in the palm oil plantation to give an early detection of G. boninense. With the promising ability of WSNs used for collecting data from the field-related studies, anomaly detection in WSN presents a promising technique, to automate the data collection process for research on detecting G. boninense.

2.3 Wireless Sensor Network in Agricultural Domain

Some research has been conducted in different agricultural domains in WSN such as in [9, 12, 14, 15]. Those researches concentrated on monitoring the crop. For instance, to monitor the irrigation, the pathology symptoms and the microclimate in the crop field. Recently, Liao et al. [16] explored WSN for early earning purpose to

Table 3 Soil features collected from The Palm Oil Institution

No.	Features
1	Position
2	Depth
3	Available phosphorus
4	Total nitrogen
5	Organic carbon
6	C/N ratio
7	Total phosphorus
8	Field
9	Available phosphorus
10	Exchangeable cation K
11	Exchangeable cation Mg
12	Exchangeable cation Ca
13	Cation exchangeable capacity
14	Exchangeable bases Al
15	Exchangeable bases Hg
16	Block

inform the farmer to take precautionary actions in time before major pest outbreaks cause an extensive crop loss, as well as to schedule maintenance tasks to repair faulty devices. The outbreak detection is done by using machine learning techniques (self-organizing maps and support vector machines).

This research aims to conduct an early detection for oil palm tree using information captured from sensor nodes. In this regard, the sample of soil measurement will be taken to test the abnormal or anomalous data which has the potential to reveal infected tree. This research will apply rough set theory to filter the data and select the significant features from 16 features stated in Table 3. Afterward, selected features for anomaly detection can be carried out on the sensor nodes, with considers that sensor nodes are limited in resources and it will be our future work.

3 Challenges and Requirement Using WSNs for Detection

The constraints of sensor nodes are limited energy, low computing power, and storage. Hence, the major challenges to be addressed in WSNs are to balance the security needs to protect the node during data transferring. Some of the data packets may contain erroneous or malicious data which can affect the accuracy of the data, and isolated node failures can bring down the entire network, which is harmful to the reliability of WSN [17]. This corrupted data or malicious data are called anomalous data which can be defined as the patterns that deviate from the rest of the data. These anomalous patterns are usually called outliers, noise, anomalies, exceptions, faults, defects, errors, damage, surprise, novelty, or peculiarities in different application domains [18]. These raw data packets can detect by sensor nodes which are often inaccurate and incomplete because of the resource constraints and environmental factors. Moreover, data accuracy sensed by the node is very important in order to make important decisions in wireless applications or systems. As for solution, anomaly detection that is a branch of the intrusion detection can be used to detect those anomalous data sense by the nodes. On the other hand, to tackle the energy depletion of sensor nodes, data aggregation is suggested during the data transmission to energy saving.

4 Data Filtering and Feature Selection

Data filtering was performed to find the significant features that will be used to detect anomalies. Feature selection is the process of creating a subset of the original features that makes machine learning easier and less time consuming [19]. Elimination of redundant data and irrelevant features can improve the detection while maximizing time and energy consumption. The performance of the feature selection can be measured based on the accuracy, precision, and recall of the classification. Experiment of data filtering and feature selection and classifications of G. boninense were done using rough set theory.

4.1 Rough Set

Rough set was first introduced by Pawlak in 1982 which is a mathematical approach with the capability to distinguish between objects. The rough set philosophy is founded on the assumption that we associate some information (data, knowledge) with every object of the universe of discourse [20]. The basic definition of rough set is given below.

Definition 1 Rough set theory can be defined as decision system or information system which is represented as $S = (U, Q, V, f)$ where U is the non-empty finite called universe of N object $\{x_1, x_2, ..., x_N\}$, while Q is non-empty set of n attribute $\{q_1, q_2, ..., q_n\}$. $V = U_q \in_Q V_q$, where V_q is the value of the attribute q and f is a function of information, $f:(U, A) \rightarrow V$ where $f(x, q) \in V_q$ for every $q \in Q, x \in U$. The decision system or decision table is presented as $A = C \cup D$ and $C \cap D = \Phi$ if the features in A can be separated into condition set C and decision feature set D.

Definition 2 The indiscernibility can be presented as $IND_A (B) (U \times U)$, given by: $IND_A(B) = \{(x, x'): a B a(x) = (x')\}$ a reduct of A is the least $B \subset A$ that is equivalent to A up to indiscernibility, i.e., $IND_A (B) = IND_A (A)$ [21].

Reduction attribute called reduct is the one of the important properties in rough set. By calculating reducts, we identify this part of data (features) which carries most of the essential informations [22]. Four types of algorithms including genetic algorithm, Johnson's algorithm, exhaustive algorithm, and dynamic reduct are used during the data reduction process. Generally, Genetic algorithm is used to select the minimum hitting sets, while Johnson's algorithm only computes single reduct which uses a simple greedy algorithm. Genetic algorithm-based concepts are used to explore the solution candidates by constructing the generation [23]. Meanwhile, the concept of exhaustive algorithm is incrementally to examine the subsets of attributes and the return the required reduct. Lastly, dynamic reduct tries to find the reduct from several subtables from original decision table.

The main steps in the rough set analysis are preprocessing, data splitting, data reduction, and classification. In preprocessing step, the problem of missing data collection is taken into account. Bazan and Szczuka [22] addressed the problem of missing value using RSES software with four approaches, including (1) removal of objects with missing values, (2) filling the missing part of data, (3) analysis without taking account the missing data, and (4) treating missing data with regular value. The missing part of data will be fixed using the second approach in this research. Meanwhile, data splitting is used to split data into training and testing set. Data reduction is used to filter redundant and select significant attributes for classification step. Meanwhile, rule induction is the step for obtaining decision rules such as if-then statement from reduce set. Lastly, classification steps are used to classify the objects based on the rules generated in the previous step.

4.2 Dataset

A total of 88 data have been collected from palm oil organization from two plantations. In Malaysia, oil palm cultivation is under the sunny and rainy climate throughout the year in Malaysia. The climate changes on the other hand can affect the soil reading other than the replanting of the field. Palm oil will be replanted around 25 years after the cultivation. Table 3 shows the 16 soil features collected from the palm oil plantation.

The features are based on the reading of nutrients (phosphorus, Mg, K, Ca) and other physical factor (position of soil from palm tree, depth of the soil sample taken and which plantation field and block). On the other hand, Table 4 shows the splitting of dataset for training and testing. Data are split into Dataset 1, Dataset 2, and Dataset 3. Each of the dataset contains 50 % and 50, 60, and 40 % as well as 70 and 30 % training and testing.

Table 4 Dataset used on rough set

Dataset	Training data	Testing data
1	44 data	44 data
2	53 data	35 data
3	62 data	26 data

5 Results

In this section, 16 features or attributes are taken from palm oil organization for analysis. Meanwhile, four algorithms were used to obtain reducts and they are genetic algorithm (GA), Johnson's algorithm (JA), exhaustive calculation (EC), and dynamic reducts (DR) as mentioned before.

5.1 Reduction

Table 5 shows the reduction result of the four algorithms used in the experiments. It can be observed that reduct value changes when datasets are changes in three of the algorithm, but Johnson's algorithm gave the same reduct value for different dataset. Reducts are rules that use a subset of all available attributes such that the bulk of accurate prediction [20] is gained. There is also repetition of the features in some of the algorithm. Therefore, this reduced feature can be considered to be used in developing anomaly detection model for WSN.

After reduction process, each of the algorithms has produced minimum reduct as possible. Out of the four algorithms, genetic algorithm, exhaustive calculation, and dynamic reduct give maximum of 3 reduct's length. Genetic algorithm has produced nine significant reducts. This is due to the nature of both algorithms for thoroughly finding the reduct. Even it can find the more stable reduct, but it may lead to high computational cost. Meanwhile, exhaustive calculation and dynamic reduct presented only eight features with more than ten combinations. However, Johnson's algorithm only produces one maximum length with five proposed features. From the result, six unions' reduct of these four algorithms are selected. This

Table 5 Reduction process result

Algorithm	Dataset	Reduct	Max. reduct's length
GA	1, 2,3	Exchangeable cation K, exchangeable bases Hg, total nitrogen exchangeable bases Al, depth, position, CN ratio, exchangeable bases Mg, cation exchangeable capacity	3
JA	1, 2, 3	pH, CN ratio, total nitrogen, total phosphorus, organic carbon	1
EC	1, 2, 3	Exchangeable cation K, exchangeable bases Hg, Total nitrogen, exchangeable bases Al, depth, position, cation exchangeable capacity, exchangeable bases Mg	3
DR	1, 2, 3	Exchangeable cation K, exchangeable bases Hg, total nitrogen, exchangeable bases Al, depth, position, cation exchangeable capacity, exchangeable bases Mg	3

Table 6 Result of accuracy, precision, and recall before and after reduction process

Dataset	Algorithm	GA (%)		JA (%)		EC (%)		DR (%)	
		Before	After	Before	After	Before	After	Before	After
1	Accuracy	63.6	88.5	20.5	81.8	63.6	88.6	68.1	93.2
	Precision	71.4	94.4	68.6	96.4	71.4	96.8	73	94.1
	Recall	80.6	94.4	77.4	93.1	80.6	91	87.1	97
2	Accuracy	65.7	88.6	25.7	80	65.7	88.6	62.9	91.4
	Precision	67.7	95.8	64.5	95.2	67.7	95.8	64.7	92.3
	Recall	91.3	92	87	91	91.3	92	95.7	96
3	Accuracy	69.2	88.5	38.5	80.8	73.1	88.5	69.2	93.2
	Precision	70.8	94.4	68.2	93.8	73.9	94.4	69.2	94.1
	Recall	94.4	94.4	83.3	88.2	94.4	94.4	100	97

reducts are known as most significant features which are used in classification process. The six features are position, depth, total nitrogen, exchangeable cation K, exchangeable bases Al, and exchangeable bases Hg.

5.2 Classification

Classification was performed before and after feature subsets were obtained. The classification accuracy result of the four algorithms is shown in Table 6. The result shows that genetic algorithm gives better accuracy result than the other algorithm.

From the classification result, we can see that the accuracy, precision, and recall result increase after significant features were selected. Again Exhaustive Calculation and Dynamic Reduct give higher result. Unfortunately, due to the basic nature of these algorithms, it may give high computational cost and consumed more time to calculate the reduct. Hence, genetic algorithm can consider to be used for feature selection in this research. Moreover, the selected features in genetic algorithm more match with exhaustive and dynamic algorithms than Johnson's algorithm.

6 Conclusions

Feature selection was used to select only significant features. By implementing rough set theory for data filtering and feature selection, the research achieved reasonable accuracy for feature selection classification. This result shows that Genetic algorithm is acceptable which is used for feature selection and for other algorithms. In conclusion, adopting WSNs to monitor G. boninense in palm oil plantation is one of the promising alternatives for early detection of the diseases. In our future work, data aggregation approach will be used to detect anomalies (G. boninense) found in the soil properties.

Acknowledgments The authors would like to acknowledge Ministry of Higher Education (MOHE) and Universiti Teknologi Malaysia (UTM) for supporting this research under Research University Grant (RUG) vote 06H03. We would also like to thank Kulim (M) and MPOB for their knowledge sharing.

References

1. Abdullah, A.H., Adom, A.H., Shakaff, A.Y.M., Ahmad, M.N., Zakaria, A., Saad, F.S.A., Kamarudin, L.M.: Hand-held electronic nose sensor selection system for basal stamp rot (BSR) disease detection. In: Third International Conference on Intelligent Systems Modelling and Simulation, pp. 737–742 (2012)
2. Lelong, C.C.D., Roger, J.-M., Brégand, S., Dubertret, F., Lanore, M., Sitorus, N.A., Caliman, J.-P.: Evaluation of oil-palm fungal disease infestation with canopy hyperspectral reflectance data. Sensors **10**(1), 734–747 (2010)
3. Shakaff, A.Y.M., Ahmad, M.N., Markom, M.A., Adom, A.H., Hidayat, W., Abdullah, A.H., Fikri, N.A.: E-nose for basal stem rot detection. UniMAP, pp. 4–6 (2009)
4. Markom, M.A., Shakaff, A.Y.M., Adom, A.H., Ahmad, M.N., Hidayat, W., Abdullah, A.H., Fikri, N.A.: Intelligent electronic nose system for basal stem rot disease detection. Comput. Electron. Agric. **66**(2), 140–146 (2009)
5. Hushiarian, R., Yusof, N.A., Dutse, S.W.: Detection and control of Ganoderma boninense: strategies and perspectives. SpringerPlus **2**(1), 555 (2013)
6. Idris, A.S, Yamaoka, M., Hayakawa, S., Basri, M.W., Noorhasimah, I., Ariffin, D.: PCR technique for detection of Ganoderma. MPOB Information Series, pp. 1–4 (2003)
7. Idris, A.S., Kushairi, A., Ismail, S., Ariffin, D.: Selection for partial resistance in oil palm progenies to Ganoderma basal stem rot. J. Oil Palm Res. **16**(2), 12–18 (2004)
8. Zain, N., Seman, I.A.B.U., Kushairi, A., Ramli, U.M.I.S.: Metabolite profiling of oil palm towards understanding basal stem rot (BSR) disease. J. Oil Palm Res. **25**(1), 58–71 (2013)
9. Mohd Su'ud, M., Loonis, P., Idris, A.S.: Towards automatic recognition and grading of Ganoderma infection pattern using fuzzy systems, pp. 30–35 (2007)
10. Rees, R.W., Flood, J., Hasan, Y., Potter, U., Cooper, R.M.: Basal stem rot of oil palm (Elaeis guineensis); mode of root infection and lower stem invasion by Ganoderma boninense. Plant. Pathol. **58**(5), 982–989 (2009)
11. Shafri, H.Z.M., Hamdan, N.: Hyperspectral imagery for mapping disease infection in oil palm plantation using vegetation indices and red edge techniques H. Am. J. Appl. Sci. **6**(6), 1031–1035 (2009)
12. Shafri, H.Z.M., Anuar, M.I.: Hyperspectral signal analysis for detecting disease infection in oil palms. In: International Conference on Computer and Electrical Engineering, pp. 312–316 (2008)
13. Idris, A.S., Mazliham, M.S., Madihah, A.Z., Kushairi, A.: Field recognition and diagnosis tools of Ganoderma disease in oil palm. In: Second International Seminar Oil Palm Disease Advance in Ganoderma Research and Management, pp. 1–13 (2010)
14. Morais, R., Valente, A., Serôdio, C.: A Wireless sensor network for smart irrigation and environmental monitoring: a position article, July, pp. 845–850 (2005)
15. Baggio, A.: Wireless sensor networks in precision agriculture. In: ACM Workshop on Real-World Wireless Sensor Networks (REALWSN 2005) (2005)
16. Liao, M.-S., Chuang, C.-L., Lin, T.-S., Chen, C.-P., Zheng, X.-Y., Chen, P.-T., Jiang, J.-A.: Development of an autonomous early warning system for Bactrocera dorsalis (Hendel) outbreaks in remote fruit orchards. Comput. Electron. Agric. **88**, 1–12 (2012)
17. Xie, M., Han, S., Tian, B., Parvin, S.: Anomaly detection in wireless sensor networks : a survey. J. Netw. Comput. Appl. **34**(4), 1302–1325 (2011)

18. Gogoi, P., Bhattacharyya, D.K., Borah, B., Kalita, J.K.: A survey of outlier detection methods in network anomaly identification. Comput. J. **54**(4), 570–588 (2005)
19. Vesterlund, J.: Feature selection and classification of cDNA microarray samples in ROSETTA, April (2008)
20. Olson, D.L., Delen, D.: Rough sets. In: Advanced Data Mining Techniques, pp. 87–109. Springer, Berlin (2008)
21. Shrivastava, S.K.: Effective anomaly based intrusion detection using rough set theory and support vector machine. Int. J. Comput. Appl. **18**(3), 35–41 (2011)
22. Bazan, J.G., Szczuka, M.: The rough set exploration system, pp. 37–56. Springer, Berlin (2005)
23. Zuhtuogullari, K., Allahverdi, N., Arikan, N.: Strategy for feature reduction of medical systems (2013)

Category-Based Graphical User Authentication (CGUA) Scheme for Web Application

Mohd Zamri Osman and Norafida Ithnin

Abstract Graphical user authentication (GUA) is an alternative replacement for traditional password that used text-based form. Even though GUA has high usability and security, it is also facing security attacks that legitimate from the traditional password such as brute force, shoulder surfing, dictionary attack, social engineering, and guessing attacks. The proposed category-based graphical user authentication (CGUA) scheme is developed for web application and based on image category. This category image is inspired from the Hanafuda Japanese card game. The scheme also involved several security features such as decoys, randomly assigned, hashing, limited login attempts, and random characters to strengthen the CGUA scheme. Overall, the proposed CGUA scheme was able to mitigate known attacks based on the security features analysis.

Keywords Graphical user authentication · Graphical password · Knowledge-based authentication

1 Introduction

Security in computer and network area is crucial and important to be protected from the outsider such as attacker, hackers, and spammers at all work. Information system may contain sensitive information that needs to be secure. For this reason, password authentication is an important mechanism in information security. Information system must allow only authentic users to gain access into the system.

M.Z. Osman (✉) · N. Ithnin
Information Assurance and Security Research Group (IASRG), Faculty of Computing, Universiti Teknologi Malaysia, 81310 Skudai, Johor, Malaysia
e-mail: mzamri27@gmail.com

N. Ithnin
e-mail: afida@utm.my

© Springer International Publishing Switzerland 2015
A. Abraham et al. (eds.), *Pattern Analysis, Intelligent Security and the Internet of Things*, Advances in Intelligent Systems and Computing 355,
DOI 10.1007/978-3-319-17398-6_29

This can be done by identifying legitimate users, and once users are identified, they can access resources according to user level. In current practice, alphanumeric or text-based password is a widely used mechanism to authenticate the users. This traditional authentication mechanism requires users to enter a username and password. According to Adam and Sasse [1], as the number of passwords per user increases, the rate of forgetting them also increases. In order to have multiple passwords in different systems, users tend to apply unsafe strategies such writing them down, using the same password, and share with others. De Angeli [2] stated that there are three fundamentals of authentication information as shown in the Table 1. These three fundamentals are to provide security protection.

A secure password should be random, hard to guess, frequently changed, and different from the other systems. Typically, to have a strong password, combination of digits, symbols, and letters are required. However, strong password would be hard to memorize. If the password is hard to guess, then it is hard to remember [3]. Instead of having the security guidelines provided to have a good password, traditional password are still faced several attacks such as brute force, dictionary, guessing, spyware, and social engineering attack. In addition, the password can be easy to forget or shared. These challenges lead the researchers to come with a new authentication mechanism where using image representation as the password. This authentication is normally called as image-based authentication or graphical user authentication (GUA). GUA is an alternative for replacing traditional password due to certain criteria such as easy to remember. Previous researchers studied that pictures, especially photographs, are easier to remember than text-based passwords [3]. However, Chowdhury et al. [4] reviewed that memorability still remain unsolved even when the GUA is employed.

GUA can be categorized into two—recognition-based and recall-based. In recognition-based, user needs to recognize and identify a set of images and the user passes selected during the registration. In recall-based, user needs to create the pass image that the user created during the registration. Recall-based scheme also can be divided into two parts—pure recall-based and cued recall-based [4, 5].

This study is organized as follows. Several related works are compared in the next section. In Sect. 3, the design and implementation of the proposed GUA scheme are presented. Section 4 presents the experiment with security result and analysis. Discussion and conclusion are provided in the Sect. 5.

Table 1 Three fundamentals of authentication information

Fundamental	Correspondents to	Example
What you have	Token-based authentication	Smart card, ATM
What you know	Knowledge-based authentication	Password, user ID
What you are	Biometric authentication	Iris, fingerprint, palm

2 Background Work

A great need in security mechanism for web-based application has led the researchers to come out with the possible solution by proposing GUA as an alternative. GUA offers great security based on the study done by the previous researchers. Since the proposed solution for the web-based application, recognition-based scheme is selected as the scheme for developing GUA as a web-based authentication system. For this reason, 90 % of the users can memorize their graphical pass after certain period. In addition, most of the researchers found that users are not familiar in drawing the graphical password in recall-based technique and need to use the mouse and other input device such stylus pen [6]. Several recognition-based GUA schemes have been studied and can be found in Table 2.

PassFaces is a commercial product developed and patented by PassFaces Corporation [7]. PassFaces reported to have high usability by providing easy to remember password than the traditional password. However, a study by [6, 8] shows that PassFaces is prone to guessing attack and also to shoulder surfing attack

Table 2 Recognition-based GUA scheme

Schemes	Technique/method	F1	F2	F3	Layout		
					F4	F5	F6
Picture password [17]	For PDA			X		X	6 × 5
Takada and Koike [18]	For mobile			X		X	11 × 11
Story [19]	Sequence of portfolio		X			X	3 × 3
Passfaces™ [7]	Face portfolio		X			X	n rounds of 3 × 3
Weinshall [20]	The use of the keyboard (up, right, left, down)	X				X	8 × 10
ColorLogin [21]	Color representation for Windows XP		X			X	5 × 8
GUABRR [12]	Rotation, resizing, and random character	X				X	5 × 5
Jetafida [10]	–		X			X	4 × 5
Lashkari [22]	Multi-line grid		X		X	X	5 × 5
ImagePass [13]	One-time password	X				X	3 × 4
Lashkari [23]	Secure the image gallery using watermarking	X				X	5 × 5
Goa and Liu [24]	CAPTCHA	X				X	5 × 10
TwoStep [25]	Hybrid (text password and recognition-based)	X	X			X	n rounds of 6 × 8
Ayannuga [26]	Hybrid	X	X			X	–

F1 keyboard input, *F2* mouse click, *F3* devise (stylus pen), *F4* grid, *F5* set of images, *F6* matrix
F1–F6 are the features in the GUA scheme

and has slower time to complete the authentication [9]. Meanwhile, Jetafida scheme was developed to gather all the usability features such as easy to use, easy to create, easy to memorize, and easy to learn [10]. This scheme used mouse click event, and the clicked picture will display a red border, to make it easier for the user to recognize. Farmand and Omar [11] found that Jetafida scheme have limited images. In addition, red border of clicked pass images is visible to other people and make it easier for the shoulder surfer. Hence, no security analysis is being conducted.

On the other hand, Lashkari [12] proposed GUA based on rotation and resizing (GUABRR). This GUA scheme is developed for web application. The proposed GUABRR security features are gathered based on the ISO standard definition of usability and the security tested by proposing three different techniques—login by rotation, resizing, and both rotation and resizing.

Mihajlov and Jerman-Blažič [13] proposed ImagePass scheme. The scheme is proposed based on the human ability to memorize the pass image by conducting a survey using three sets of sample abstracts, faces, and single object. From the survey, they found that the majority of respondent prefers single object instead of abstract or false images due to easy to remember and recognize. During the authentication, 12 images were displayed instead of 30 images during the enrollment phase. Decoys security feature was displayed during the authentication process. If the pass image is 5 images, then 7 random images will be displayed as decoys.

To have a good system, security and usability cannot be separated, and it is hard to have both of them at high level. Farmand and Omar [11] suggested several security requirements to build high-level security such as

   (i)   Randomly assigned
   (ii)   Unique to the application
  (iii)   Robust again known attacks
  (iv)   Simple
   (v)   Reliable—no fallback needed
  (vi)   Not sharable casually or easily
 (vii)   Lacks social vulnerabilities
(viii)   Useable anywhere
  (ix)   Two-way authentication.

To build GUA scheme, these security requirements must take into consideration in order against known attacks such as shoulder surfing, phishing, brute force, dictionary, and spyware attacks. Due to image-based representation, GUA is more vulnerable to shoulder surfing than traditional password [14]. Only a few recognition-based scheme is proposed to resist shoulder surfing. However, none of the recall-based GUA scheme reported resistant to shoulder surfing attack. This is due to a mouse click or stylus pen that can be easily observed by the attacker. For this reason, Jebriel and Poet [15] found that keyboard-based GUA scheme is more secure than using mouse clicks or a stylus pen.

3 Design and Implementation

The main aim of the study was to propose GUA scheme with several security features based on the image category. Adding security features to the application means adding complexity in the proposed GUA scheme. We believe that a combination of security features would provide a degree of security level. Table 3 shows the security features that are added to the proposed GUA scheme for web application. Large password space and repeated verification are not selected to be used in the proposed category-based graphical user authentication (CGUA). Most of recognition-based scheme has smaller password space compared to recall-based scheme [16].

There are five security features selected to be used in the proposed scheme which are decoys, randomly assigned, hashing, limited login attempts, and random character. In the proposed GUA scheme, two phases are involved—registration and authentication phase. Registration phase is for new users, while authentication phase is for the registered user.

3.1 Registration Phase

During the registration phase, new user is required to create an account by entering valid information into the application. New users will go through two parts—Enrolment and Memorize. Figure 1 shows the steps to register in CGUA scheme. The user will be redirected to the username and secret image selection stage. In this step, user needs to choose only one secret image from the image sets. The secret image is randomly displayed 40 pictures from the database application. Only 8 pictures will be displayed, and these pictures will be randomized each time a new user registration. The chosen secret image will be displayed every time registered

Table 3 Security features in the proposed GUA scheme

Security feature	Justification of selection
Decoys	Fake image. Confuses the observer and for user identification
Randomly assigned	Confuse the observer
Large password space	*Not selected*
Hashing	The image category will be hashed to keep the text secret from the attacker. The proposed GUA scheme will be used SHA1 as the hashing algorithm with salt function
Limited login attempts	Brute force issue. Prevent the user after several unsuccessful login attempts. Its block attacker launch dictionary, guessing, and brute force attack
Random character	Shoulder surfing issue. Keyboard-based scheme is more vulnerable than mouse click-based
Repeated verification	*Not selected*

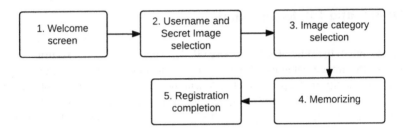

Fig. 1 Steps to register with CGUA scheme

user login. In the example of Fig. 2, picture of 'Koala' is selected as the secret image. Secret image helps the user to identify whether the user is accessing a valid application. In addition, it also assists the user to verify their valid username. Each time the user enters valid username, picture of 'Koala' will be appear.

After successfully inserting the username and choosing the secret image, user needs to select their image category. In the proposed CGUA scheme, we provide twenty categories of images which are as follows: Flower, House, Drinks, Apple, Abstract, Horse, Bird, Coins, Sweets, Cat, Fish, Car, Insect, Colours, Keys, Bicycle, Numbers, Face, Alphabets, and Ball. Figure 3 shows the phase of image category selection. Here, the concept of Hanafuda Japanese Card game is taken. We borrow the concept of category-based GUA scheme to be applied in the proposed GUA scheme. Selection of image categories is made by entering the random code of the image categories. For instance, the user had chosen Sweets, Cat, and Face as image category. Hence, the random code is 'e2' (Sweets), 'c1' (Cat), and 'qz' (Face). The code 'e2', 'c1', and 'qz' must be entered into a random code field in sequence order. These random codes are generated by the CGUA algorithm, and it is changed every time the user tries to login. The proposed scheme allows the user to memorize the image category only instead of a random code. The motivation of using random code is to avoid the scheme from user click such as a mouse click or touch. This is to prevent the shoulder surfer to take advantage of the user event.

Furthermore, to increase the password space of CGUA scheme, repeated image categories is accepted, for instance, like the Apple, Apple, and Apple image category. However, users must select at least three image categories. Finally, once valid combination code is accepted, the user will be redirected to memorize phase. This

Fig. 2 Registration—
username and secret image
selection

Fig. 3 Registration—image category selection

phase requires the user to memorize the secret image and also the image category. It is important to remember the image categories because the users will use these image categories. Hence, email notification will be forwarded to the email user.

3.2 Authentication Phase

The authentication process consists of two steps—identifying the secret image and the authentication step. Figure 4 shows the flowchart of the proposed CGUA

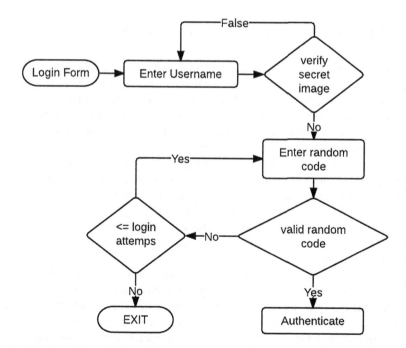

Fig. 4 Flowchart of the CGUA scheme

scheme. During identification step, if the user enters a valid username, CGUA scheme will display an authentic secret image of the user. However, if the user enters an invalid username, CGUA will display a decoy (fake) image. This is where a decoys security feature is implemented. For this reason, the user knows that he or she has entered the wrong username. In this application, the decoy is displayed from the ten sets of pictures. Then, the user needs to verify the secret image by entering a valid combination of random codes.

User only has three limited try to enter the combination of random codes. If the user enters invalid combination code, the image category set is randomized again and the images are different. Then, the user will be directed to the exit page. This is to prevent automated search by brute force attack. Once the random code is accepted, authentic user is accessed.

4 Experiment and Analysis

This study is implemented in a web-based form, and the experiments were conducted online. The experiment was conducted on 60 students and graduate students in our university. The respondents are from the Faculty of Computing, Universiti Teknologi Malaysia.

4.1 Validation Parameter of Security Features

In this study, we use questionnaire and laboratory experiments to evaluate the proposed CGUA. Questionnaires are available online inside the CGUA scheme after they are successfully registered. The questionnaire is structured into four main sections according to the security features of the propose CGUA scheme. The sections are limited login attempts, random characters, randomly assigned secret image (decoy), and a few usability studies.

From the analysis of decoy (a secret image) security feature, 98 % of respondents agreed that decoy helps the user to identify as a legitimate user. In addition, all the respondents agreed that decoy can help to secure the application. These results show that decoy help to secure the application by identifying the real user based on identification of secret image. This result shows that high acceptance that decoys can assist the user either he or she access a genuine application by secret image identification (Fig. 5).

For randomly assigned security feature analysis, results are represented in Fig. 6. We analyze the random set of image and random positions. From the figure, 'Yes' indicates that the respondents get confused for the first time, while 'No' indicates that they do not get confused. For a random set of image result, majority with 71 % of the respondents get confused for the first time when using the application. 29 % do not get confused with the random set image. Meanwhile, for random position

Fig. 5 Result of decoy security feature (Yes, agree; No, not agree)

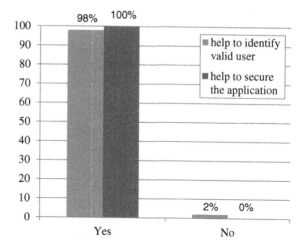

Fig. 6 Result of random sets of image and random position (No, not confuse; Yes, confuse)

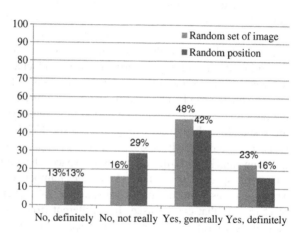

feature, 58 % of respondents also get confused. Nearly half with 42 % said that they get confused with the random position.

The motivation of this security feature is to make the people feel confused with the random position and images displayed. In other word, it confuses the observer by preventing them to remember the exact position or images chosen by the user. In terms of usability based on the random set of image and random position, we manage to achieve the target to make the user confuse for the first time. However, from the observation, users can adapt with the application after using it several times. As a result, randomly assigned feature would strengthen the CGUA application by mitigating the shoulder surfing attack.

Limited login attempts is another security features added in the proposed CGUA. A majority of 94 % said that this security feature must be added to the application. On the other hand, 87 % agreed that the limited login attempts can reduce the

Fig. 7 Usability study on random code (no, not confuse; yes, confuse)

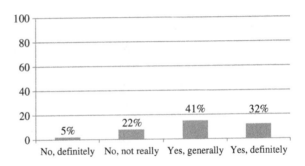

guessing attack. This is because the attacker can gain access into the application by guessing the image categories of the user.

Figure 7 represents the usability study on the easiness of the random code feature. From the result, a majority of the respondents, with 73 %, said they feel confused and uncomfortable with the random code for the first time. Meanwhile, only 27 % said they are not confused. It shows that the random code will make the users get confused for the first time during their first attempt using this CGUA scheme, which reduced the usability level of the scheme.

During the laboratory experiment, we conducted guessing attack by inviting registered users to guess the pass image category of a valid username. Since the CGUA scheme can be accessed online, we also invited anonymous person to break the pass image category within a period of seven days. From the analysis, none of the respondents had successfully entered the CGUA application using the given registered username. It is shown that the proposed CGUA is resistant from the guessing attack. In terms of password space, the proposed CGUA have smaller password space, 20^N, where N is the number of chosen image categories. Traditional password has password space of 94^N, where 94 is the possible number of characters that can be used. It is different from recognition-based GUA where they always have a smaller password space than traditional password. However, password space can be increased by choosing more image categories as shown. For instance, choosing 5 image categories will result in 20^N (3,200,000) password space. For GUA scheme, password space can be enlarged by the increasing the number of image categories. Unfortunately, this is not practical because it would make the web page more crowded, and slow internet is a major concern for delivery during network transmission.

5 Conclusion and Future Work

In this work, we proposed a GUA called as CGUA. This scheme allows the users to go through a two-step process—identifying and authentication. From the analysis, the proposed CGUA scheme has a degree of security level. We believe that CGUA scheme can mitigate the known attacks such as shoulder surfing, guessing attack, phishing attack, and brute force attack. We combine several security features in order to have high security level. However, we found that certain security feature in the CGUA lowers the usability. Random code and randomly assigned image make the user confused for the first time. Users are more convenient using mouse click rather than keyboard-based scheme. In future works, we would like to study the effectiveness of using different types of image that are suitable for the user and also using watermarking techniques to secure the image library in the future research.

Acknowledgments This research work is supported and funded by Universiti Teknologi Malaysia (UTM) under Research University Grant (RUG) vote No. Q.J130000.2528.05H76.

References

1. Adams, A., Sasse, M.: Users are not the enemy. Commun. ACM **42**, 40–46 (1999)
2. De Angeli, A., Coventry, L., Johnson, G., Renaud, K.: Is a picture really worth a thousand words? Exploring the feasibility of graphical authentication systems. Int. J. Hum. Comput. Stud. **2**(63), 128–152 (2005)
3. Suo, X., Zhu, Y., Owen, G.S.: Graphical passwords: a survey. In: Proceedings of the 21st Annual Computer Security Applications Conference, pp. 463–472 (2005)
4. Chowdhury, S., Poet, R., Mackenzie, L.: Do graphical authentication systems solve the password memorability problem? In: Tryfonas T., Askoxylakis I. (eds.) Human Aspects of Information Security, Privacy, and Trust SE—13, vol. 8533, pp. 138–148. Springer, Berlin (2014)
5. Zakaria, O., Zangooei, T., Mohd-Shukran, M.-A.: Graphical password authentication: review and analysis. AISS **4**(15), 25–32 (2012)
6. Komanduri, S., Hutchings, D.R.: Order and entropy in picture passwords. In: Proceedings of Graphics Interface 2008, pp. 115–122 (2008)
7. Passfaces, Graphical password technology [Online] (2011). Available: http://www.realuser.com
8. Ekeke, E., Ugochukwu, K., Jusoh, Y.Y.: A review on the graphical user authentication algorithm: 2. Categories of graphical user authentication algorithm. Int. J. Inf. Process. Manag. **4**, 238–252 (2013)
9. Furkan, T., Ozok, A.A., Holden, S.H.: A comparison of perceived and real shoulder-surfing risks between alphanumeric and graphical passwords. In: Proceedings of the Second Symposium on Usable Privacy and Security, pp. 56–66 (2006)
10. Eljetlawi, A.M.: Study and Develop A New Graphical Password System. Universiti Teknologi Malaysia, Malaysia (2008)
11. Farmand, S., Omar, B.Z.: Improving graphical password resistant to shoulder-surfing using 4-way recognition-based sequence reproduction (RBSR4). In The 2nd IEEE International Conference on Information Management and Engineering (ICIME), pp. 644–650 (2010)
12. Lashkari, A.H.: Graphical user authentication algorithm based on rotation and resizing (GUABRR) [Online] (2011). Available: http://graphicalpassword.net/
13. Mihajlov, M., Jerman-Blažič, B.: On designing usable and secure recognition-based graphical authentication mechanisms. Interact. Comput. **23**, 582–593 (2011)
14. Tari, F., Ozok, A., Holden, S.: A comparison of perceived and real shoulder-surfing risks between alphanumeric and graphical passwords. In: Proceedings of the Second Symposium on Usable Privacy and Security (SOUPS '06), pp. 56–66 (2006)

15. Jebriel, S.M., Poet, R.: Preventing shoulder-surfing when selecting pass-images in challenge set. In: International Conference on Innovations in Information Technology (IIT), pp. 437–442 (2011)
16. Hu, W., Wu, X., Wei, G.: The security analysis of graphical password. In: International Conference on Communications and Intelligence Information Security (ICCIIS) (2010)
17. Jansen, W., Gavrilla, S., Korolev, V., Ayers, R., Swanstrom, R.: Picture password: a visual login technique for mobile devices (2003)
18. Takada, T., Koike, H.: Awase-E: image-based authentication for mobile phones using user's favorite images. In: Human-Computer Interaction With Mobile Devices, pp. 347–351 (2003)
19. Davis, D., Monrose, F., Reiter, M.: On user choice in graphical password schemes. In Proceedings of the 13th Conference On USENIX Security Symposium (SSYM'04), p. 11 (2004)
20. Weinshall, D., Kirkpatrick, S.: Password you'll never forget, but can't recall. In: Proceeding of the Conference on Human Factors in Computing System (CHI), pp. 1399–1402 (2004)
21. Gao, H., Liu, X., Dai, R., Wang, S., Chang, X.: Analysis and evaluation of the colorlogin graphical password scheme. In: Proceedings of the 2009 Fifth International Conference on Image and Graphics, pp. 722–727 (2009)
22. Lashkari, A.H., Gani, A., Ghasemi Sabet, L., Farmand, S.: A new algorithm on graphical user authentication (GUA) based on multi-line grids. Acad. J. **5**, 3865–3875 (2010)
23. Lashkari, A.H., Manaf, A.A., Masrom, M.: A secure recognition based graphical password by watermarking. In: 2011 IEEE 11th International Conference on Computer and Information Technology (CIT), pp. 164–170 (2011)
24. Gao, H., Liu, X.: A new graphical password scheme against spyware by using CAPTCHA. In: Proceedings of the 5th Symposium on Usable Privacy and Security (2010)
25. van Oorschot, P.C., Wan, T.: TwoStep: an authentication method combining text and graphical passwords. In E-Technologies: Innovation in an Open World, pp. 233–239. Springer, Berlin (2009)
26. Ayannuga Olanrewaju, O., Olusegun, F.: Evalution of a usable hybrid authentication system. Int. J. Comput. Appl. **17**(8), 27–31 (2011)

An Improved Certificateless Public Key Authentication Scheme for Mobile Ad Hoc Networks Over Elliptic Curves

Shabnam Kasra-Kermanshahi and Mazleena Salleh

Abstract Due to the resource constrained property of mobile ad hoc networks (MANETs), making their application more lightweight is one of the challenging issues. In this area, there exists a large variety of cryptographic protocols especially in the context of public key cryptosystems. The difficulty of managing complex public key infrastructures in traditional cryptosystems and the key escrow problem in identity-based ones persuaded many researchers to propose appropriate certificateless cryptosystems for such an environment. In this area, a protocol, named ID-RSA, has been proposed to authenticate the public key in RSA based schemes in the basis of certificateless cryptosystems. Although the security of this protocol is proved, the use of bilinear pairings made it computationally expensive. In this paper, we improved the performance of ID-RSA by the use of elliptic curve based algebraic groups instead of multiplicative ones over finite fields as the output of bilinear pairings. Our results show that our secure protocol is significantly more lightweight than ID-RSA.

Keywords Certificateless PKC · Elliptic curves · Lightweight · MANETs

1 Introduction

The popularity of mobile devices and especial characteristics of mobile ad hoc networks (MANETs) led to various applications from education to military over this type of network. Generally, MANET can be defined as a kind of wireless network, which does not rely on any fixed-infrastructure. In addition, nodes are allowed to join/leave the network at any time; hence, the topology of these

S. Kasra-Kermanshahi (✉) · M. Salleh
Faculty of Computing, Universiti Teknologi Malaysia, Johor Bahru, Johor, Malaysia
e-mail: shabnam.kasra@gmail.com

M. Salleh
e-mail: mazleena@fsksm.utm.my

© Springer International Publishing Switzerland 2015
A. Abraham et al. (eds.), *Pattern Analysis, Intelligent Security*
and the Internet of Things, Advances in Intelligent Systems and Computing 355,
DOI 10.1007/978-3-319-17398-6_30

327

networks is unstable. It should be noted that mentioned features beside the nodes cooperation in running the network functions make MANETs prone to many attacks (e.g. wormhole, blackhole, impersonation, and modification). Since providing security for such environment is a challenging issue, this type of networks became an interesting research area in the recent years [1–3].

In order to provide security for MANETs, two approaches could be considered; attack-oriented and cryptography. Earlier, most of the proposed works were based on the first approach [4–8]. More precisely, the author(s) tried to propose new solutions for possible threats or improve existing works. However, attack-oriented schemes could not resist all types of attacks or combination of them at the same time [1]. The second approach has been used extensively to support security for MANETs in the form of general design framework [1]. In this way, the developers struggled to propose lightweight cryptosystem that could satisfy limitation of resources in MANETs. While the use of symmetric cryptography (SC) is usually acceptable for resource constrained environments [9], the difficulty of key management in SC, persuade developers to apply public key cryptography (PKC) instead [10, 11].

However, traditional PKC could not be the best choice for MANETs because of the need to certificates and public key infrastructure. In 1986, Adi Shamir introduced identity-based cryptography that considered the identity of users as their public key [12]. This idea became practical in 2001 by the illustrious work of Boneh and Franklin [13]. To avoid the inherent problem of identity-based PKC named key escrow, Al-Riyami and Paterson in [14] introduced the concept of certificateless PKC. Afterward, some researchers have been tried to propose schemes based on certificateless PKC in the context of MANETs [15, 16]. Since both of them utilized bilinear pairings which is considered as an expensive cryptographic map, it seems they are not lightweight enough to be used in MANETs. In this paper, we propose a new certificateless scheme that is an enhancement of the proposed scheme in Eissa et al. [15]. More accurately, we could reduce the complexity of computation by the use of elliptic curve based algebraic groups instead of multiplicative ones over finite fields as the output of bilinear pairings.

The rest of this paper is organized as follows. Section 2 explains the required preliminaries. In Sect. 3, ID-RSA protocol has been reviewed. Our proposed scheme is described in Sect. 4 while in Sect. 5, the comparison is presented in the terms of efficiency. Finally, the conclusions are provided in the Sect. 6.

2 Preliminaries

The basis of this section is explaining elliptic curves and their significant role as strong algebraic groups in the cryptography research area. The followed subsection briefly introduces elliptic curves in more detail. Then, a subsection represents the significance of this category of algebraic groups.

2.1 Elliptic Curve Cryptography

Elliptic curves are one of the significant scientific topics in the cryptographic literatures. The significant property of this kind of curves is that a subset of the points over them, beside of a binary operation can generate a beneficial algebraic group.

In the sake of simplicity, it is possible to claim that the elements of mentioned algebraic group are a subset of points of the equation $y^2 z = x^3 + axz^2 + bz^3$ with nonzero discriminant $(\Delta = -16[4a^3 + 27b^2])$. Here, a and b are two elements of the determined finite field. Without loss of generality, assume that the coefficients and variables of mentioned equation above are elements of the finite field F_{p^n}. Here, the elements of considered algebraic group are members of the set which is constructed of the points such as (x_0, y_0, z_0). To introduce the elements of this group, it is possible to partition the solutions of mentioned equation into two classes. By assuming the points $(0, y_0, 0)$ as the class of identity element, other ones would be the points of the equation $y^2 = x^3 + ax + b$.

Beside of what explained above, the operation of mentioned algebraic group, named "+", is introduced as following. Without loss of generality, assume that we need to add two points (x_1, y_1) and (x_2, y_2). The final result will be calculated as follows:

$$(x_1, y_1) + (x_2, y_2) = (x_3, y_3)$$

that $(x_3, y_3) = (\lambda^2 - x_1 - x_2, \lambda(x_1 - x_3) - y_1)$.

The amount of λ is

$$\lambda = \begin{cases} \frac{y_2 - y_1}{x_2 - x_1}, (x_1, y_1) \neq (x_2, y_2) \\ \frac{3x_1^2 + a}{2y_1}, (x_1, y_1) = (x_2, y_2) \end{cases}.$$

Next subsection investigates the advantages of ECC-based algebraic groups that persuaded many researchers to use them.

2.2 Advantages of ECC-based Cryptosystems

In continue to what pointed out in the Sect. 2.1, it is possible to claim that elliptic curves are the basis of a large variety of cryptographic schemes. In compare with RSA-based cryptosystems, the required key size of ECC based is significantly smaller. Tables 1 and 2 represent the suggested key sizes for two mentioned cryptosystems based on the claims of NIST [17] and ECRYPT [18], respectively.

In these tables, the security level refers to the required number of bits of mentioned algebraic group so that the discrete logarithm remains a hard problem.

Table 1 Key sizes of NIST standard [17]

Security level (bits)			
Category of cryptosystems	80	128	256
ECC style	160	256	512
RSA style	1024	3072	15,360

Table 2 Key sizes of ECRYPT standard [18]

Security level (bits)			
Category of cryptosystems	80	128	256
ECC style	160	256	512
RSA style	1248	3248	15,424

3 Review of ID-RSA

In this section, we are going to review the identity-based protocol proposed by Eissa et al. [15]. For the sake of simplicity, we call it ID-RSA. In order to resist RSA cryptanalysis, the authors attempted to ensure that the users' public keys are attainable only by trusted users. Therefore, it is assumed that each user is a part of a coalition. If any user out of one coalition needs public key of a user which is inside the coalition, it should request to the coalition. The detail explanation of this cryptosystem is as followed.

3.1 ID-RSA Phases

It is possible to summarize the ID-RSA in the following main phases.

Setup

The output of this phase is public parameters of the cryptosystem (Params), which are generated by a trusted third party.

$$\text{Params:} \langle G, G_T, q, P, \hat{e}, n, H_1, H_2, H_3 \rangle$$

Here, G is an additive algebraic group as an input of $\hat{e}: G \times G \rightarrow G_T$ which is a bilinear pairing that maps elements of mentioned group (such as P) to an element of the multiplicative algebraic group G_T. The order of the mentioned groups is the same at prime number q. In addition, n represents the number of bits of e and N in the RSA cryptosystem. The rest are one-way hash functions that $H_1:\{0,1\}^* \rightarrow G_1^*$, $H_2:G_T \rightarrow \{0,1\}^n$ and $H_3:\{0,1\}^* \rightarrow \{0,1\}^n$.

Node Initialization

In this phase, some of public/private parameters for current users/coalitions will be generated. Three types of public parameters have been considered; identity-key, general-key and public-key. The first one for users/coalitions "i" can be computed by all available users as $Q_i = H_1(ID_i\|time)$, whereas the rest are just computable by their owners. More precisely, user/coalition "i" chooses $e_i \in_{random} \mathbb{Z}_q^*$ then performs RSA key generation algorithm. Similar to RSA, d_i is the private key of the user/coalition "i" and $\langle e_i, N_i \rangle$ represent the corresponding public key. Finally, the mentioned user/coalition computes and publishes its general key $P_i = (d_i \cdot P)$.

Public key obtaining process

In this phase, a user who needs public key of another user inside a coalition can make a request. As a simple scenario, imagine that user A requires public key of user B in the coalition $ID - RSA_i$, thus by performing the following steps this phase can be done.

Step 1. $A \rightarrow ID - RSA_i : P_A, ID_B$

In this step, A sends a request—consist of its general key and identifier B- to $ID - RSA_i$ coalition. It means, user A needs $\langle e_B, N_B \rangle$.

Step 2. $ID - RSA_i \rightarrow A : \langle U, C, W, Y \rangle$

In this step, if A is in the trusted list, the coalition $ID - RSA_i$ will respond to A tuple $\langle U, C, W, Y \rangle$ that $U = P_i$, $C = e_B \oplus H_2(g_i)$ where g_i is $\hat{e}(Q_A, P_A)^{d_i} \times \hat{e}(d_iQ_i, P_A)$, $W = e_B \cdot P$ and $Y = N_B \oplus H_3(e_B)$.

Step 3. Public key extraction by A

In this step, A can extract $\langle e_B, N_B \rangle$ by a set of computation; $g_A = \hat{e}(d_AQ_A, P_i) \times \hat{e}(Q_i, P_i)^{d_A}$, $e_B = C \oplus H_2(g_A)$, $N_B = Y \oplus H_3(e_B)$ and finally A accepts public key of B if $W = e_B \cdot P$ holds.

ID-RSA correctness

To validate the correctness of ID-RSA, it should be proved that both sides reach the same value through the mentioned computations above. Hence, the value of g_A and g_i should be equal at $\hat{e}(Q_A + Q_i, P)^{d_id_A}$. It can be proven through the following computations that ID-RSA is a correct scheme.

$$g_A = \hat{e}(d_AQ_A, P_i) \times \hat{e}(Q_i, P_i)^{d_A}$$
$$= \hat{e}(Q_A, P)^{d_id_A} \times \hat{e}(Q_i, P)^{d_id_A} = \hat{e}(Q_A + Q_i, P)^{d_id_A}$$
$$g_i = \hat{e}(Q_A, P_A)^{d_i} \times \hat{e}(d_iQ_i, P_A)$$
$$= \hat{e}(Q_A, P)^{d_id_A} \times \hat{e}(Q_i, P)^{d_id_A} = \hat{e}(Q_A + Q_i, P)^{d_id_A}$$

4 Proposed Protocol

This section assigns to introducing our proposed certificateless authenticating public key protocol. Similar to ID-RSA, our scheme is constructed based on three phases named setup, node initialization, and public key obtaining process as followed.

Setup

The public output of our scheme, Params, includes following items:

$$\text{Params:} \langle G, q, P, n, H_1, H_2, H_3 \rangle$$

Here, G is a cyclic elliptic curve group with order q and the generator P. The integer number n is the same as n in IDRSA protocol. In addition, $H_1:\{0,1\}^* \to G^*$, $H_2:G \to \{0,1\}^n$ and $H_3:\{0,1\}^* \to \{0,1\}^n$ are three one-way collision-free hash functions.

Node Initialization

Similar to the ID-RSA scheme, the entities can be user or coalition logically. Here, the public and private parameters of mentioned entities are the same as what introduced in ID-RSA. In addition, each entity such as the entity who possesses ID_i generates the parameter $E_i = e_i P_i$.

Public key obtaining process

In this phase, the scenario is similar to what proposed in ID-RSA except that the details of the steps. Here, these steps are as followed:

Step 1. $A \to \text{Coalition}_i : E_A, P_A, ID_B$

In this step, the second input introduces the node A as the entity who sent the request and the third one refers to the identity of the entity who his public key is requested.

Step 2. $\text{Coalition}_i \to A : \langle E_i, U, C, W, Y \rangle$

In this step, the inputs U, C, W and Y are as followed.
$U = P_i$, $C = e_B \oplus H_2(g_i)$ that g_i is equal to $g_i = d_i(E_A + e_i P_A)$ $W = e_B \cdot P$ and $Y = N_B \oplus H_3(e_B)$.

Step 3. In this step, the node A must be able to extract the public key of B (which are e_B and N_B) and verify its authenticity as follows:

First of all, A computes $g_A = d_A(E_i + e_A P_i)$, then computes $e_B = C \oplus H_2(g_A)$. Clearly, the result of g_A and g_i must be the same. In addition, the entity A computes $N_B = Y \oplus H_3(e_B)$. To verify authenticity of the obtained public key, the entity A investigates the equality of $(W = e_B \cdot P)$ to decide whether accept or reject the obtained public key of B.

Correctness of the Proposed Protocol

Beside of what mentioned above, it is necessary to prove that the computed values of g_A and g_i are the same. The two equalities below will lead to this result:

$$g_A = d_A(E_i + e_A P_i) = d_A e_i P_i + d_A e_A P_i$$
$$= (d_A e_i d_i)P + (d_A e_A d_i)P$$
$$g_i = d_i(E_A + e_i P_A) = d_i e_A P_A + d_i e_i P_A$$
$$= (d_i e_A d_A)P + (d_i e_i d_A)P$$

As a result, the functionality of our proposed protocol is logically correct.

Table 3 Computational costs of operations in type 2 and type 3 bilinear pairings [19]

Group operation	Computational cost	
	Type 2	Type 3
Scholar multiplication in G (SM)	1	1
Point addition in G (A)	Negligible	Negligible
Exponent in G_T (E_T)	3	3
Pairing (P)	21	20

5 Efficiency Comparison

The basis of this section is to compare the cost of computing authentication parts of our proposed scheme and ID-RSA, which are the cost of computing g_A and g_i. High expense of computing bilinear pairings [19] is the main reason that made our proposed protocol more efficient than ID-RSA. In order to improve the efficiency of ID-RSA, we have tried to propose our protocol based on group operations of elliptic curves instead of computing bilinear pairings. Table 3 illustrates the cost of operations based on what depicted in [19].

To compare the cost of mentioned two schemes, we have focused on the expense of g_A or g_i parts of "public key obtaining process," as the core of them. Based on the Table 3, the cost of g_A or g_i parts of ID-RSA is equal to "$E_T + SM + 2P$," while the cost these parts in our proposed scheme is "$2(SM) + A$," which is quite efficient in compare with ID-RSA.

In the growth of the number of requests for the Step 3 in ID-RSA scheme, the output is more effectual. Based on what is illustrated in Table 3, by assuming that "n" is the number of such request, the rate of growth of computational cost for g_A or g_i parts of "public key obtaining process" in our proposed scheme is "$2n$," while this value in IDRSA would be "$46n$" or "$44n$" based on the use of type 2 or type 3 bilinear pairings, respectively.

6 Conclusion

In this paper, we proposed a new certificateless cryptographic protocol to authenticate the public key of other participants. The main contribution of this paper is to improve the computational cost of the ID-RSA protocol by the use of elliptic curve-based algebraic groups instead of multiplicative ones over finite fields as the output of bilinear pairings. Our analysis shows that besides a significant improvement of computational cost, our proposed protocol is much more efficient from the perspective of the rate of computational expense growth in compare with ID-RSA protocol.

References

1. Zhao, S., Akshai, A., Frost, R., Bai, X.: A survey of applications of identity-based cryptography in mobile ad-hoc networks. IEEE Commun. Surv. Tutorials Early Access **14**, 380–400 (2011)
2. Abusalah, L., Khokhar, A., Guizani, M.: A survey of secure mobile ad hoc routing protocols. IEEE Commun. Surv. Tutorials IEEE **10**(4), 78–93 (2008)
3. Sen, C., Salmanian, M., Kellett, M.: A Mobile Ad Hoc Networking Test Bed. DRDC, Defence R&D, Ottawa (2005)
4. Hu, Y., Perrig, A., Johnson,D.: Packet leashes: a defense against wormhole attacks in wireless ad hoc networks. Proceedings of IEEE INFORCOM, 2002
5. Capkun, S., Buttyan, L., Hubaux, J.: Sector: secure tracking of node encounters in multi-hop wireless networks. Proceedings of the ACM workshop on security of ad hoc and sensor networks, 2003
6. Yi, S., Naldurg, P., Kravets, R.: Security-aware ad-hoc routing for wireless networks. Report No. UIUCDCS-R-2002-2290, UIUC, 2002
7. Sanzgiri, K., Dahill, B., Levine, B., Shields, C., Belding-Royer, E.: A secure routing protocol for ad hoc networks. Proceedings of IEEE international conference on network protocols (ICNP), pp. 78–87, 2002
8. Hu, Y., Johnson, D., Perrig, A.: SEAD: secure efficient distance vector routing in mobile wireless ad-hoc networks. Proceedings of the 4th IEEE workshop on mobile computing systems and applications (WMCSA'02), pp. 3–13, 2002
9. Oliveira, L.B., Dahab, R.: Pairing-based cryptography for sensor networks. Presented at IEEE international symposium on network computing and applications, Cambridge, July 2006
10. Gaubatz, G., Kaps, J.-P., Oztruk, E., Sunar, B.: State of the art in ultra-low power public key cryptography for wireless sensor networks. In: Proceedings of Per Sec '05, IEEE, pp. 146–150, 2005
11. Liu, J.K., Baek, J., Zhou, J., Yang, Y., Wong, J.W.: Efficient online/offline identity-based signature for WSN. In: Proceedings of IJIS, pp. 287–296, 2010
12. Shamir, A.: Identity-based cryptosystems and signature schemes. Advances in Cryptology—Crypto. Lecture Notes In Computer Science. Springer, Berlin (1984)
13. Boneh, D., Franklin, M.: Identity based encryption from the weil pairing. Advances in Cryptology—Crypto. Springer, Berlin (2001)
14. Al-Riyami, S.S., Paterson, K.G.: Certificateless public key cryptography. In: Laih, C.S. (ed.) Advances in Cryptology C Asiacrypt. Lecture Notes in Computer Science, pp. 452–473. Springer, Berlin (2003)
15. Eissa, T., Razak, S.A., Ngadi, M.A.: A novel lightweight authentication scheme for mobile ad hoc networks. AJSE **37**, 2179–2192 (2012)
16. Li, L., Wang, Z., Liu, W., Wang, Y.: A certificate less key management scheme in mobile ad hoc networks. 7th international conference on wireless communications, networking and mobile computing, pp 1–4, China, 2011
17. NIST recommendation for key management part 1: general, Nist Special publication 800–57, August 2005
18. ECRYPT yearly report on algorithms and key sizes, 2004
19. Chen, L., Cheng, Z., Smart, N.P.: Identity-based key agreement protocols from pairings. Int. J. Inf. Secur. **6**, 213–241 (2007)

A Resource-Efficient Integrity Monitoring and Response Approach for Cloud Computing Environment

Sanchika Gupta, Padam Kumar and Ajith Abraham

Abstract Cloud computing has an immense need of file monitoring techniques to be applied so that system-specific configuration files are not modified by the virtual machine users or cloud insiders/outsiders for carrying out unwanted operations such as privilege escalation or system misconfiguration. The article critically describes the previous work done in the area of file monitoring and why there is a need for lightweight and efficient mechanism for its deployment in cloud. A centralized cum distributed approach for file monitoring of VMs on a host in cloud and the initial outcomes of its execution are presented in this article. The scheme does not use any database for storing file integrity, which eventually results in increased computational and storage efficiency. The technique is presented for the use by admins for ensuring integrity of specific files at VM hosts but can be upgraded to provide support to user-specified files as well.

Keywords File monitoring · Lightweight · Efficient · Cloud computing · Security · Hosts · Virtual machine · Cloud

S. Gupta (✉) · P. Kumar
Indian Institute of Technology Roorkee, Roorkee, India
e-mail: dr.sanchikagupta@gmail.com

P. Kumar
e-mail: padamfec@iitr.ernet.in

A. Abraham
Machine Intelligence Research Labs, Scientific Network for Innovation and Research Excellence, Auburn, WA, USA
e-mail: ajith.abraham@ieee.org

A. Abraham
IT4Innovations, VSB—Technical University of Ostrava, Ostrava, Czech Republic

© Springer International Publishing Switzerland 2015
A. Abraham et al. (eds.), *Pattern Analysis, Intelligent Security and the Internet of Things*, Advances in Intelligent Systems and Computing 355,
DOI 10.1007/978-3-319-17398-6_31

1 Introduction to Cloud and File Monitoring

Cloud computing is an efficient and scalable technology that is used for providing computing and data storage resources to users as a pay per usage model [1]. As cloud technology is still in its nascent stages, there are many concerns regarding security [2, 3]. We have studied vulnerabilities occurring in the cloud environment [4] and identified that privilege escalation attacks which allow malicious modifications to system-specific configuration files allocated to a VM can lead to severe attacks in cloud. Hence, an emerging concern is how to ensure that the integrity of resources that are provided to VMs are not modified at the user end to carry out unwanted operations. File monitoring is an approach for ensuring integrity of files by storing its content integrity information in normal scenario and using the same to identify, respond, and alert the stakeholders in case of malicious subversions [5]. Ensuring system and configurations, file integrity is a major concern for cloud environment and there are not enough techniques presented or defended yet, which are resource saving and cloud specific in practice. The need for such file monitoring arises from the reason that the file residing on VMs allocated to the users contains a noticeable amount of system-specific sensitive data, for example, firewall rules, network-specific interface setting configuration files, and security policies [6]. These configuration and system-specific files on the VM can be modified in case of privilege escalation attacks, and if performed successfully, these modifications can subvert security and integrity of the cloud as a whole.

The article critically describes the previous techniques that exist for file integrity monitoring and discusses how they are not lightweight and cloud deployment specific in nature. We propose a novel lightweight file monitoring and response approach in the form of a tool which is efficient in terms of computational cost, improves storage efficiency, and provides an easy way for ensuring file integrity of VMs remotely [7]. Thus, a secure file monitoring tool (SFMT) was developed and tested on private cloud environment.

2 Related Works

For file integrity monitoring, all systems and deployment environments might not have the same resources or same level of security requirements. On the basis of this, plenty of techniques have been proposed and practiced. We have categorized them into two well-known types.

2.1 Utility Applications for File Monitoring

These include file monitoring techniques that are implemented as standalone applications. They create a content hash of file, store them in a database, and then perform file monitoring based on stored hash on a regular basis. These techniques are good as they come as standalone applications and require no subtle modifications to the system. But they are heavyweight as continuous file integrity monitoring adds burden to the system and thus decreases its efficiency.

2.2 Kernel Modification or Memory-Based Approaches

These approaches add a hook or an access control or an integrity check inside the kernel module itself. They are different in the sense that they do not run as utility applications rather apply the file integrity monitoring operations during discrete events itself such as file access or modification and are more real time in nature. These techniques save resources in terms of computational power but require kernel modification and are platform dependent in their nature of deployment. Table 1 shows the description of these techniques.

From the available file integrity monitoring tools, it was identified that these techniques are more focused on deployment strategies, hashing techniques, and incident response handling. Very few schemes pay focus on the concerns for environments such as cloud where the major criterion for a file monitoring tool to get accepted for deployment is its efficiency. Moreover, many schemes also do not adhere to the required minimal expectations from file monitoring and response applications so as to be considered as lightweight implementations. Many schemes provide additional set of features such as extensive event logging, policy enforcements, secure communication, and secured databases that eventually increases their computational and storage needs. Such features might be needed for standalone systems or systems where resources are not of concern, but for resource critical systems where it is required that a solution should be managed more easily, then present heavyweight systems cannot be deployed with a higher efficiency expectation. The study reveals that there are not enough proven utility applications for file monitoring that work in reducing the overall resource requirements and managerial efforts needed for systems such as cloud. We have hence critically analyzed all the advantages from the existing utility applications and worked upon so as to include all necessary and essential features with a mindset of reducing the resource needs.

3 Secure File Monitoring Tool—SFMT

SFMT is a lightweight scheme as it does not utilize any database for storing integrity hash information of a file rather stores the integrity of the file in the file itself. It stores the encrypted file integrity and attributes information in well-defined pair of tags, that is, "<SFMT></SFMT>" and "<SFMT_FA></SFMT_FA>" tags. During file integrity monitoring operations, this information is extracted from the defined tags and is used to identify whether the file has been modified or not.

Table 1 File integrity monitoring techniques' abstract description

Name	Type of solution	Specifics	Integrity stored	Response	Cloud specific	Deployment architecture
Tripwire [8]	File integrity application	Policy governed file monitoring tool and takes care of file attributes	In database	Governed by policy file defined by user	No	Client–server
Cim Trak [9]	File integrity application	Stores integrity in good baseline provide good logs for analysis	In database	Governed by policy file	No	Client–server
Samhain [10]	File integrity application	Use PGP-signed files and database for maintaining integrity. Works in stealth mode	In database	Governed by policy file	No	Client–server provides Web-based management console
Stealth	File integrity application	Integrity of client is checked by controller over SSH connection	In database	Changes are written to reports which are mailed regularly to intended recipients	No	Client–server
OSSEC and Osirirs [11]	File and host integrity application	Fully featured file integrity cum host event-based centralized monitoring application	In database	Governed by policy file with extra features for deeper analysis	No	Centralized client–server
ADIE	IDS	Uses the concept of regular expressions to include or exclude the files for monitoring	In database	Governed by policy file	No	Centralized client–server
Radmind	File integrity application	Command line utility tool designed to remotely monitor the file system for integrity of multiple UNIX machines	In database	Feature of reverting back the changes if a change is performed on a monitored file	No	Client–server

(continued)

Table 1 (continued)

Name	Type of solution	Specifics	Integrity stored	Response	Cloud specific	Deployment architecture
Linux IDS	Kernel modification-based approaches	Modifies the Linux access control semantics known as DAC (discretionary access control)	In database	Policy governed in kernel	No	Kernel-based standalone
I3FS [12]	Stackable file system based	On access file integrity checking system and checks the integrity of the file in real time	In database	Policy governed	No	Stackable file system and is designed so that it can be mounted over existing file system such as NFS or EXT3
Linux security module	Kernel hooks provisions	Provides a framework, in kernel hooks that are provided for addition of integrity and file system security operations	In database	Policy governed in kernel hooks	No	Standalone kernel based
XenFIT	XEN hypervisor-based, dynamic memory patching	Instead of managing hash databases, there is concept of dynamically patching the memory of protected virtual machine	No database	Policy governed	Yes	VM cloud based
Flogger [13]	Kernel modification based	Records the file centric access and transfers information from kernel spaces of virtual machines and physical machines in cloud	Database	Policy governed	Yes	Cloud based

The scheme is designed such that admin of cloud environments can monitor the host and virtual machines for file integrity of configuration and system-specific files whose modification/misconfiguration can allow the VM users or malicious insiders or outsiders to carry out unwanted operations such as DDOS attacks by the modification of intrusion detection policies, or of firewall rules and other attacks including privilege escalation. Our proposed technique will allow admins on privilege VMs to execute SFMT as a standalone and managed application on a client VM and specify the files that need to be monitored. The integrity response is very simple, such that it replaces the file with its backup replica and creates logs at the central monitoring system for further analysis of subversion. This response scheme is effective and a valid approach because in case of configuration or system-specific files assigned to a VM by a cloud administrator, the aim is that they should remain consistent and if modified they should be replaced with their authentic replicas to maintain integrity and operability of the system. By using this strategy of alert and response, a huge amount of resources is saved compared to the conventional file monitoring applications. The logs of file monitoring events are collected for each VM at centralized monitoring system. Sections 3.1 and 3.2 give a detailed description and design of the technique.

3.1 Introduction to SFMT

SFMT follows the well-known steps of standard file integrity monitoring approaches with a lightweight implementation mindset [14]. The technique of storing integrity information intelligently in file itself and using the same for file monitoring is a novel approach for saving ample amount of resources required for establishing, maintaining, and updating databases and reducing platform dependencies. This technique is proposed to reduce the amount of resources needed for file integrity monitoring as compared to the known file monitoring applications so as to make them available for integrity monitoring in resource critical environments such as cloud [6]. The technique is focused on environments where integrity and efficiency are major requirements vis-à-vis modification and backtracking. It fits the bill perfectly for an environment where the only need is to have authentic files in place, such as the system-specific configuration files that drive operational characteristics of a system, and the best response that applies in case of their modifications is to alert, log, and replace them with their previously best-known replica. However, it is equally capable of integrity monitoring in its normal usage with user space files. Our developed and deployed technique on XEN environment [15] does integrity monitoring for admin-specified files; however, the same can be upgraded to provide users at each VM the capability to monitor integrity of user-specified files as well. It does a little storage for tracing information which is not needed in environments where replacing the file works and tracing the information of who is responsible for the attacks has little significance. The details of the operations for file monitoring that are implemented by the proposed approach are as follows.

3.1.1 File Integrity Establishment

The proposed technique is centralized cum distributed and hence has benefits of both. It does not become a bottleneck by doing all the activities that require computational resources at a centralized VM. Instead, it provides a synchronous copy of file integrity monitoring tool, that is, SFMT at each VM that executes availing the resources of the host VM. Yet the concept is distributed because only the admin can specify which configuration and system-specific files present at that VM need monitoring, and is the one who is responsible for event log analysis and to provide policies for response in case the integrity is subverted. The design of the approach is based on files; everything from integrity of a file to policy or initialization information resides in the files itself. Through NFS, the host VM's file monitoring script can access the specific file; we call it as initialization file that contains the information of which host files need to be monitored. The file can only be modified by admin, and the structure for it is as follows: Each line of the file has first value as the path of the original file that needs to be monitored and a space separated value that describes the backup replica path for that file that needs to be used for replacement if the original file integrity is lost. When the script is executed over the VM host for file integrity monitoring, the file integrity establishment module reads the information as described in the initialization file, opens the original and backup file replica, and creates integrity hashes using the well-known SHA-256 bit algorithm with added secret value as described by the admin so as to make it irreproducible by an attacker after doing a malicious change to the file. If the attacker does the change knowing the algorithm, he can generate the hash and put it into the file, but use of the secret value removes this easy possibility, making the process more stringent to attacks. After creating the integrity hash for the file contents, the hash is appended in between the well-defined <SFMT></SFMT> tags. A file contains not only the file content integrity information but also the file attribute information. We have identified the minimal and most effective file attributes, namely user mode or permissions, user id, and group id. We identified that apart from the other file attributes such as modification time, number of links, file size, and I-node number whose information is somehow captured in the file, these attributes are quite useful in the identification of permission modifications that cannot be captured in the file integrity information. Hence, with the file integrity information in between the <SFMT></SFMT> tags, we also append the file attribute-specific information in the <SFMT_FA></SFMT_FA> tags. This file attribute-specific information is then utilized during integrity monitoring stage to find out discrepancies if any, in mode, user id, and group id of the files. But to avoid modifications to the value of file attributes to evade detection, the value is concatenated and encrypted with a crypt function (A Perl-specific encryption specification). We have used this one-way function that takes an input storing and the SALT to create cryptographic information for our implementation; however, more secure and robust cryptographic method can be used to encrypt the file attribute information. This makes the information to be safely captured in the file itself.

```
if(tied %dbm){
print "FILE $db_file now open\n";
<>;
$j=0;
foreach my $Key (sort keys %dbm)
{
$lines[$j]=$Key."\t\t\t".$dbm{$Key}."\n";
print $lines[$j]."\n";
$j++;
<>;
}
}
else
{
die " sorry unable to open $db_file\n";
}

untie %dbm;
}
<SFMT>8203a94545f519bdee11d57b8b0b23804d10e3fad46b2b24b77b0eb4485dc746</SFMT><SFMT_FA>33204 500 500</SFMT_FA>
```

Fig. 1 Snapshot of a file on VM host with integrity information added by SFMT

```
/home/sanchika/Desktop/time.pl  /home/sanchika/Desktop/filemonitoring/files/time.pl
/home/sanchika/Desktop/cprogram.c /home/sanchika/Desktop/filemonitoring/files/cprogram.c
/home/sanchika/Desktop/filemonitoring/files/normal/anomalycallexittimer.pl /home/sanchika/Desktop/filemonitoring/files/backup/anomalycallexittimer.pl
/home/sanchika/Desktop/filemonitoring/files/normal/anomalycallprogram.pl /home/sanchika/Desktop/filemonitoring/files/backup/anomalycallprogram.pl
/home/sanchika/Desktop/filemonitoring/files/normal/callexittimer.pl /home/sanchika/Desktop/filemonitoring/files/backup/callexittimer.pl
/home/sanchika/Desktop/filemonitoring/files/normal/callprogram.pl /home/sanchika/Desktop/filemonitoring/files/backup/callprogram.pl
/home/sanchika/Desktop/filemonitoring/files/normal/display.pl /home/sanchika/Desktop/filemonitoring/files/backup/display.pl
/home/sanchika/Desktop/filemonitoring/files/normal/getdata.pl /home/sanchika/Desktop/filemonitoring/files/backup/getdata.pl
/home/sanchika/Desktop/filemonitoring/files/normal/MalGui.pl /home/sanchika/Desktop/filemonitoring/files/backup/MalGui.pl
```

Fig. 2 Initialization file snapshot at a monitored VM

Figure 1 shows the snapshot of the file with added integrity information, and Fig. 2 shows a snapshot of initialization file at a monitored VM.

If the technique finds a file with already added integrity information, it skips that file and does not override the integrity information; however, it brings this event into the notice of the admin through email and event logs so that any misconfigurations or attackers' mischiefs can be handled manually if needed. The technique is intelligent in the sense that addition of the integrity information does not affect any file in its way of operation. It has a list that contains file type and the characters that should be used for appending the integrity information as comments so that they do not create any defect in their operation. For example in case of a .c file, "//" are used as first starting characters before adding the file integrity and attribute information. In this way, the technique is able to store integrity information to any file keeping into account that addition of such information does not affect its operation or functionality.

3.1.2 File Integrity Monitoring, Alert, and Response

File integrity monitoring is periodic in nature. It follows a specific and lightweight mechanism to alert the admins in which the host script sends the mail with the details of the event happened along with the information of the file integrity and the attribute changes done maliciously on a specific file on VMs. It also uploads the event logs periodically to the Web server through FTP so that they can be analyzed by the admin on the fly through management Web server with associated credentials. The approach to file integrity monitoring is lightweight in the sense that neither does it access database for extracting the integrity of the file nor does it manage any key to make database access secure, as all the integrity information exists in the file itself. It extracts the information from between the <SFMT> </SFMT> and <SFMT_FA></SFMT_FA> tags. The integrity monitor extracts the current integrity of the file and its attributes, which is then verified with the current integrity as calculated in the same manner from the file contents. If the current file integrity information matches with extracted integrity from file, then it shows that the file is intact and the tool moves to the next file for integrity checking operations and so on. Again the information of which files need to be monitored is present in the shared read-only initialization file between the admin VM and the host VMs. If in any case the file integrity gets changed, the backup file is scanned by the integrity monitoring tool for identifying whether it is authentic or not. If the backup file is itself modified, an alert with priority 1 is lodged in the event log files and the emails are sent to the admin for manual intervention as SFMT cannot do any operation in this situation. However, if the backup file is authentic, then the original file is replaced by its backup replica from the specified location. SFMT also stores the information of this event in the log file and uploads the information on the specified FTP.

3.2 SFMT Design and Deployment

The deployment strategy used in SFMT is shown in Fig. 3. We have deployed the SFMT tool in all the VMs in centralized cum distributed manner under control of central VM in the cloud environment. We have chosen this deployment strategy as it is lightweight when a privilege VM on each host in a cloud environment will control all the other VMs for their file monitoring operations.

The privilege VM shares the initialization and event log file with each of the VM under its control. The files are properly configured, and the permissions are set in a manner such that the files can only be accessed by SFMT, which runs as privileged process on VMs. The files are shared with NFS. In our deployment scenario, we have kept all the backup files pertaining to a VM host separately under admin VM shared with NFS. The schematic flow diagram of working of SFMT is shown in Fig. 4 which depicts how the SFMT is executed by admin VM user for a specific VM host and how the actions are taken based on operation type which can be either integrity establishment or integrity monitoring. It also depicts what will happen when the integrity of original file or its backup replica is subverted and how the alert and response are taken care of by SFMT.

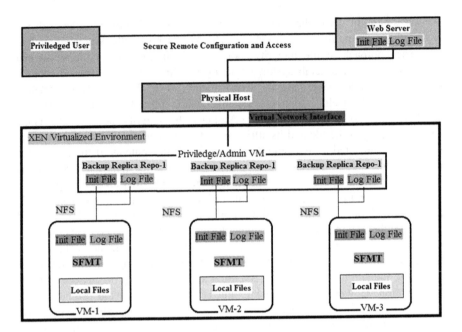

Fig. 3 SFMT deployment strategy

4 Experimental Setup, Results, and Analysis

The scheme was tested in cloud stack management server and XEN as hypervisor on the host. The host and VM run Ubuntu 12.04 and Ubuntu 11.10 releases. The SFMT tool is created as an executable Perl script in Linux. SFMT implementation uses Perl 5.14 active state release. Perl is chosen for implementation as the concept of file integrity establishment and monitoring involves numerous file access and text searching operations and Perl is known to be one of the best languages for regular expression matching. Many of the other modules that are a part of SFMT implementation include response module which incorporates mail sending, event logging alert and reporting components. We have used SMTP authentication implementation for secure transmission of alert emails by the host VM. The network file system between the admin and host VMs allows in sync sharing of the log and initialization file from centralized VM. Secure FTP connection is used for uploading the event log files to the management Web server. We have used as dataset a set of 100 files at each VM that describes the configuration of system and user programs. Admin at the privileged VM defines for each VM the initialization file that the specific VM SFMT can access through NFS with read privileges. The file can be updated anytime by the admin based on the needs. This means that he/she can append or delete the files which need file integrity establishment and monitoring. Admin also shares a log file with other VMs that log the event information. Both email alerts and Web upload-based event logging are introduced to make the system more

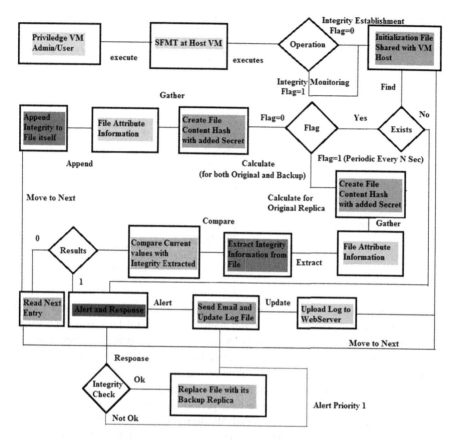

Fig. 4 Workflow of SFMT in cloud environment

robust and responsive. Users at host VMs have no control over the operation of SFMT; it is fully controlled by centralized VM. It is same as all other kind of background processes that run on the host for carrying out regular activities. Admin is responsible for setting the email set that needs to be reported in case file gets subverted. Admin can also choose the time period after which the file monitoring is done over the file set specified in the initialization file. After iterations and analyzing the CPU consumption, we set the value to be 1/10 of a second. This means after each 1/10 of a second, the next file is scanned as given in initialization file and in round robin fashion; hence, if 10 files are specified in the initialization set, it will take 1 s for SFMT to perform the file monitoring operation on the same file again. If the value of time period of monitoring is set too high, it will save resources, but if made too low, then it will result in a high CPU utilization for file monitoring activities. Hence, a proper value needs to be determined based on cost benefit analysis. The average CPU consumption we found in measuring at various time periods is shown in Fig. 5. We have also divided the detection speed discretely in

Fig. 5 Average CPU
consumption versus detection
speed for various time period
values of monitoring

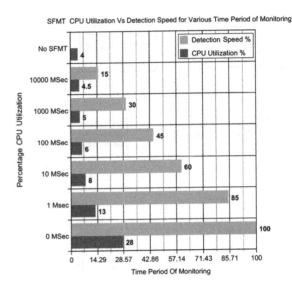

levels that we expect with various time periods. With synchronous run (when SFMT runs round robin on the set of files without delays), we have seen an average CPU utilization on a VM of about 28 % and a real-time detection, hence representing the detection speed at highest level of 100 %. Similarly, we have shown the statistics at other runs when we varied time period of monitoring from 1 ms to every 10,000 ms and a special case when SFMT is not running.

The aim is to set correct values for time period for periodic file monitoring with best CPU utilization and detection speed. We found the value of time period as per 100 ms as best as it gives best ratio of detection speed and CPU utilization among all. The snapshot of CPU utilization without and with SFMT with defined period is as shown in Fig. 6.

We also created the lightest possible database implementation that can counter SFMT. We engaged both of them to do integrity monitoring on a set of 20 files and identified that the time taken by SFMT at various workloads is higher or competitive with its database-based implementation. This is when all optimizations for

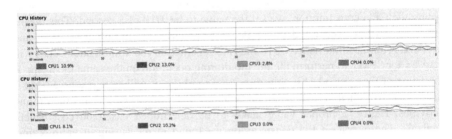

Fig. 6 CPU utilization without and with SFMT running on a VM

Fig. 7 Comparing database- and SFMT-based implementation at various CPU workloads for monitoring a set of files

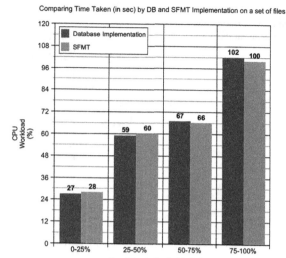

monitoring and response exist in SFMT and DB implementation, which shows that it is a far better and lightweight file monitoring and response tool compared with tools providing ample features for monitoring response and analysis that may not be a need in resource critical systems. Figure 7 describes the statistics of such experimentation.

We have tried a real-time attack on the set of files that are secured by SFMT. We have written a program that does malicious modifications to files that are secured by SFMT every second and in a round robin fashion similar to what SFMT does during file monitoring. We ran the attack script for 10 min and recorded the CPU utilization and efficiency of SFMT in alert and response. SFMT sent emails whenever it found discrepancies in file integrity to admins, checked for integrity-verified backup replica, and if authentic replaced the old file with its backup replica and simultaneously it updated the event logs on the server for further analysis and reporting. The CPU utilization at various time frames of attack is as follows.

The analysis shows that SFMT stabilizes file monitoring and reporting alert and response operations during attack and is very consistent in its attack detection alert and response in rigorous attacks over the file monitored. It also does not overburden the network for alert and logging. Figure 8 describes the behavior of SFMT in alert and response scenario and how it stables down after an attack. It saves a decent amount of resources in storage as we identified that it only adds 100 bytes to a file for appending integrity. We recorded a saving of around 70 % of storage requirements in comparison with a database-based implementation in our experimentation.

Fig. 8 SFMT CPU utilization during alert and response in attack scenarios

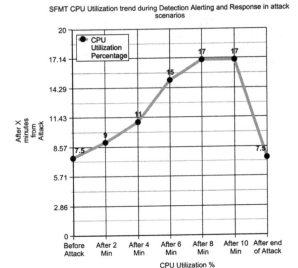

5 Conclusions

The article describes the need for lightweight file monitoring approaches for monitoring system-specific and configuration files on VMs in cloud environment and provides initial results of its deployment on XEN platform. The article also lists techniques and tools available for file monitoring with their description and the gaps that suggest a need for cloud-specific and lightweight solutions. The proposed solution deploys file monitoring without the use of database for storing file integrity information and instead uses the file itself for storing it in an intelligent manner so that it does not affect its manner of operation if any. The technique provides the same level of security and is more resource efficient and computationally superior than database-based implementations. It takes the lightweight measures as taken from previous schemes for response, alert, and reporting and work in a consistent behavior at various CPU workloads. It is highly platform independent and configurable to be centrally controlled. Creation of a commercial version of the tool is in progress.

References

1. Mell, P., Grance, T.: Effectively and Securely Using the Cloud Computing Paradigm (v0.25). http://csrc.nist.gov/organizations/fissea/2009-conference/presentations/fissea09-pmell-day3_cloud-computing.pdf (2009)
2. Brown, E.: NIST issues cloud computing guidelines for managing security and privacy. Natl. Inst. Stand. Technol. Spec. Publ. 800-144, 1–80. http://csrc.nist.gov/publications/nistpubs/800-144/SP800-144.pdf (2012)

3. Khan, K.M., Malluhi, Q.: Establishing trust in cloud computing. IEEE IT Prof. **12**(5), 20–27 (2010)
4. Subashini, S., Kavitha, V.: A survey on security issues in service delivery models of cloud computing. J. Netw. Comput. Appl. **34**(1), 1–11 (2011)
5. Quynh, N.A., Takefuji, Y.: A novel approach for a file-system integrity monitor tool of Xen virtual machine. In: ASIACCS '07 Proceedings of the 2nd ACM Symposium on Information, Computer and Communications Security, ACM, pp. 194–202 (2007)
6. Gupta, S., Sardana, A., Kumar, P.: A light weight centralized file monitoring approach for securing files in cloud environment. In: The 7th International Conference for Internet Technology and Secured Transactions (ICITST-2012), pp. 382–387. IEEE (2007)
7. Gupta, S., Kumar, P., Sardana, A., Abraham, A.: A secure and lightweight approach for critical data security in cloud. In: Fourth International Conference on Computational Aspects of Social Networks (CASoN), pp. 315–320. IEEE (2012)
8. Kim, G.H., Spafford, E.H.: The design and implementation of tripwire: a file system integrity checker. In: Proceedings of the 2nd ACM Conference on Computer and Communications Security, ACM, pp. 18–29 (1994)
9. Cim Trak—file integrity monitoring and compliance management. http://www.locins-consulting.com/file-integrity-monitoring-and-compliance-management.asp; http://www. cimtrak.com (2008)
10. Rainer, W.: Book the Samhain HIDS—Overview of Available Features, pp. 1–11 (2011)
11. Cheng, J.-B., Cid, D.B., Hargrave, V., Starks, M.: OSSEC an Open Source Host-based Intrusion Detection System (2008–2012)
12. Patil, S., Kashyap, A., Sivathanu, G., Zadok, E.: I3FS: an in-kernel integrity checker and intrusion detection file system. In: Proceedings of the 18th USENIX Conference on System Administration, USENIX Association, pp. 67–78 (2004)
13. Ko, R.K.L., Jagadpramana, P., Lee, B.S.: Flogger: a file-centric logger for monitoring file access and transfers within cloud computing environments. In: TRUSTCOM '11 Proceedings of the 2011 IEEE 10th International Conference on Trust, Security and Privacy in Computing and Communications, IEEE Computer Society, pp. 765–771 (2011)
14. Kwon, H., Kim, T., Yu, S., Kim, H.: Self-similarity based lightweight intrusion detection method for cloud computing. In: Nguyen, N., Kim, C.-G., Janiak, A. (eds.) Intelligent Information and Database Systems. Lecture Notes in Computer Science 6592, pp. 353–362. Springer, Berlin (2011)
15. Paul, B., Boris, D., Keir, F., Steven, H., Tim, H., Alex, H., Rolf, N., Ian, P., Andrew, W.: Xen and the art of virtualization. SIGOPS Oper. Syst. Rev. **37**(5), 164–177 (2003)

Bookmarklet-Triggered Literature Metadata Extraction System Using Cloud Plugins

Kun Ma and Ajith Abraham

Abstract In this paper, a bookmarklet-triggered literature metadata extraction system using cloud plugins is designed to find metadata of the publisher Web pages. First, we propose selector-syntax extractors using CSS-like syntax. Furthermore, we deploy them in the cloud. Finally, a bookmarklet-triggered way is proposed to execute cloud script to extract metadata of current Web pages. Compared with current methods, this system works across browser platforms with flexibility and extensibility and without installing additional plugins.

Keywords Metadata extraction · Content negotiation · Bookmarklet · Cloud computing · Plugins

1 Introduction

Literature metadata extraction is a kind of method to extract metadata (author, title, affiliation, email, abstract, keyword, etc.) from semi-structured publishing Web pages. Literature metadata extraction system is a broad class of software applications targeting at extracting information from the literature Web pages. The method addressing the extraction issue interacts with a Web page and extracts data stored in it. The extracted information consists of elements in the page as well as the full text of the page itself. With this method, we can use metadata results to assist the literature entry, literature sharing, and literature recommendation.

K. Ma (✉)
Shandong Provincial Key Laboratory of Network Based Intelligent Computing,
University of Jinan, Jinan, China
e-mail: ise_mak@ujn.edu.cn

A. Abraham
Machine Intelligence Research Labs, Scientific Network for Innovation
and Research Excellence, Auburn, USA
e-mail: ajith.abraham@ieee.org

© Springer International Publishing Switzerland 2015
A. Abraham et al. (eds.), *Pattern Analysis, Intelligent Security
and the Internet of Things*, Advances in Intelligent Systems and Computing 355,
DOI 10.1007/978-3-319-17398-6_32

With regard to the issue of the literature metadata extraction, there have been several main approaches [1, 2], i.e., automatical approach, semi-automatical approach, and manual approach. Among these approaches, automatical and semi-automatical approaches are intelligent methods with a low accuracy. Although the manual approach has a high accuracy, it requires manually writing templates or extraction rules. Some other intelligent methods, such as HMM-based approach [3] and roadrunner [4], require a large training sample to construct the extractor. To further improve the accuracy of these methods is very difficult.

In this paper, we design bookmarklet-triggered literature metadata extraction system using cloud plugins. By implementing selector-syntax extractors and the bookmarklet-triggered way, we aim at a highly flexibility and interoperability. The contributions of this paper are several folds. First, we present how to find elements using selector-syntax extractors that we design. Furthermore, extractors are deployed in the cloud. The update of extractors is not considered by users. Finally, we propose a bookmarklet-based way to trigger selector-syntax extractors, which enables users to extract literature metadata from Web pages of different publishers in an unobtrusive way. Besides, this approach works across browser platforms without installing additional plugins and software except a bookmarklet.

This bookmarklet-triggered literature metadata extraction system is advantageous because it provides the independent trigger way. Users could extract the metadata without installing any software. This is easy to be integrated with current third-party systems. This approach is unified since it is suitable for all the heterogeneous publishers. Besides, as a cache of the extraction result, it is stored with schema-free document stores in the cloud.

The rest of the paper is organized as follows. Section 2 discusses the related work and background, and Sect. 3 presents the architecture, extractor, and triggered way of literature metadata extraction. Section 4 gives the survey result of this system. Brief conclusions are outlined in the last section.

2 Related Work

2.1 Literature Metadata Extraction

With regard to this issue, there have been several main approaches [5].

The first is extracting metadata from none-scanned document, i.e., Office word file, latex source document, and PDF document that conforms to the template. Due to the diversity of the template, some more manual intervention is generally needed with higher failure rate. An example is that CrosssRef provides an open-source tool called PDF-Extract for identifying and extracting semantic metadata from PDFs of scholarly journal articles or conference proceedings [6]. This tool performs structural analysis and categorizes sections to determine column bounds, headers, footers, sections, titles, and so on. Another example is that metadata can be extracted from a PDF file embedding with pre-generated XMP file [7]. However, the disadvantage is that this is not a standard that all content producers will follow, and not all the PDFs contain this information. There is also another method to identify DOI in the PDF first and then obtain the metadata from the DOI content negotiation service.

The second approach to extract literature metadata is using result resolution of academic search engines. For instance, developers can quickly find out information about literature information (researchers, papers, conferences, journals, organizations, and keywords) by using Google Scholar and Microsoft Academic search engine. However, the result resolution is heterogeneous, and not always correct due to lack of APIs.

The third approach to extract literature metadata is using APIs of third-party literature databases. Some third-party organizations provide APIs (e.g., metadata search API by CrossRef, ResearcherID Web Services by Thomson Reuters, and APIs by Scopus and ORCID) to acquire literature metadata by unique literature ID (e.g., DOI, ArXiv, and PMID).

The fourth approach to extract literature metadata is DOI content negotiation [5, 6, and 8]. DOI content negotiation is used to retrieve different representations of a work. It allows the developer to request a particular representation of literature metadata. Instead of accessing this service directly, DOI content negotiation makes a resolution via http://dx.doi.org by specifying the content types of HTTP Accept header. The advantage is that content negotiation enables literature metadata extraction without knowing its origin or specifics of the DOI registration agency.

The fifth approach to extract literature metadata is DOI resolver. It is the interface for a specific registration agency to obtain the bibliographic metadata or Web page from DOI. However, the interface of DOI resolver is heterogeneous, and each of the registration agencies has its own specification. The challenge is how to integrate the resolver together and design a DOI resolver for the new registry agency.

The last approach is data extraction from semi-structured Web pages [1, 2]. This taxonomy is based on the main technique used by each tool to generate a wrapper: wrapper development, wrapper induction tools [9, 10], HTML-aware tools [11], and ontology-based tools [12, 13]. Our methods presented in this paper are efforts toward developing HTML-aware tools using cloud plugins.

After analyzing the three solutions, we select a flexible plugin-based method to extract literature metadata. Compared with current manual approaches, we design a selector-syntax extractor with cloud plugins.

2.2 Triggered Way

Along with Web 2.0 and big data technology, the concepts have led to the development and evolution of Web-based applications. Therefore, we use bookmarklet and NoSQL document stores to benefit the implementation of literature metadata extraction system.

There are mainly two ways to trigger an action in the browser. The first is bookmarklet. A bookmarklet is a bookmark stored in a Web browser that invoking JavaScript commands [8]. For example, a bookmarklet might allow the user to trigger literature metadata extraction, click the bookmarklet, and then present the metadata results in the sidebar. It is designed to add one-click functionality to a browser or Web page. When clicked, a bookmarklet performs some functions.

The second is browser plugin. Generally, browser plugin utilizes ActiveX or Chrome Plugins to trigger a high-privilege action that is not triggered by the Web elements (buttons or links). Compared with bookmarklet, this way need extra software installation to complete the task.

3 Literature Metadata Extraction System

3.1 Architecture

First, we discuss the architecture of our proposed bookmarklet-triggered literature metadata extraction system. As depicted in Fig. 1, the architecture is composed of two points of view in order to reduce the complexity: browser layer and cloud layer. Browser layer provides bookmarklet to trigger literature metadata extraction and a sidebar to display the extraction result. Cloud layer provides selector-syntax extractors using cloud plugins. Each plugin is made by different selectors. After the success of literature metadata extraction, the results are saved in the cache presented by NoSQL document stores.

3.2 Selector-Syntax Extractor

We propose an extractor to find or manipulate matching elements using a CSS-like selector syntax, which allows very powerful and robust queries of the HTML DOM tree. We implement a contextual *select* method, which is available in an element or elements, to filter by selecting from a specific element, or by chaining select calls. In addition, *select* method returns a list of elements, which provides a range of methods to extract and manipulate the results.

Regular Selector We divide regular selectors into two categories: element selector and attribute selector, which is shown in Table 1. Element selector returns elements by tag, ID, and class name (e.g., 1–3), and attribute selector returns elements by attribute value supporting wildcard characters and regular expressions (e.g., 4–8).

Selector Combination Next, we design some new selectors with the combinations of the above, which is shown in Table 2. They are divided into element combinations (e.g., 1–3), iterator combinations (e.g., 4–7), and group combinations (e.g., 8).

Fig. 1 Architecture of bookmarklet-triggered literature metadata extraction system

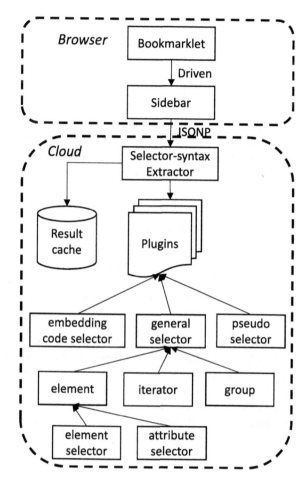

Table 1 Regular selector

No.	Name	Description	Example
1	tagname	Find elements by tag	div
2	#id	Find elements by ID	#title
3	.class	Find elements by class name	.cls
4	[attr=value]	Elements with attribute value	[alt=test]
5	[attr^=value]	Elements with attributes starting with	[href^=/t/]
6	[attr$=value]	Elements with attributes ending with	[href$=/t/]
7	[attr*=value]	Elements with attributes containing with	[href*=/t/]
8	[attr~=value]	Elements with attributes matching RepEx	[src~=/[0–9].jpe?g/]

Table 2 Selector combinations

No.	Name	Description	Example
1	el#id	Elements with ID	div#title
2	el.class	Elements with class	div.cls
3	el[attr]	Elements with attribute	img[alt]
4	ancestor child	Child elements descending from ancestor	div p
5	parent>child	Child elements descending directly from parent	div>p
6	sA+sB	Finds sibling B immediately preceded by sibling A	div.#title+div
7	sA~sX	Finds sibling X preceded by sibling A	h2~a
8	el,el	Group multiple selectors	div#title,div.cls

Table 3 Embedding code selector

No.	Name	Description	Example
1	{has(seletor)}	Find elements matching the selector	div{has(a)}
2	{contains(text)}	Find elements containing given text	div{contains(abc)}
3	{matches(regex)}	Find elements matching regex	div{contains(.jpe?g)}

Embedding Code Selector Next, embedding code selectors (that start with {and end with}) are further extended, which is shown in Table 3. Since we use dependency injection to embed some manual codes and regular selectors, it becomes much cleaner and easier to follow. Some codes of the complicated business process can be embedded in selectors. These selectors can execute some user-defined functions to find elements by boolean expressions. This is a new way of the separation of manual codes and regular selectors.

3.3 Cloud Extraction Plugins

Finally, a selector-syntax extractor is made by the above three selectors. We deploy selector-syntax extractors in the cloud with plugins. This way has the advantage that the browser/client does not maintain the selectors. Once the publisher Web pages upgrade, the only thing to do is to update the selector-syntax extractors in the cloud, which is transparent to the users and developers.

3.4 Bookmarklet-Triggered Way

We present the triggered way to invoke cloud selector-syntax extractor plugins using bookmarklet. Bookmarklet is a bookmark stored in a Web browser that contains JavaScript commands to trigger cloud selector-syntax extractor. We put the following code to execute cloud JavaScript *c.js*.

```
javascript:void((function(){var%20element=document.
    createElement('script');element.setAttribute(\
    'src','http://[cloud]/c.js?url='+window.location.href);
        document.body.appendChild(element);})())
```

After that, cloud Javascript *c.js* will take *URL* as the input and extract literature metadata using selector-syntax extractors. The following code is the segment of *c.js*. Figure 2 shows the screenshot of our bookmarklet-triggered literature metadata extraction system. After clicking the bookmarklet, the metadata is extracted to display in the new layer.

```
var url=$.getUrlVar('url');
$.ajax({dataType: 'jsonp',url:'c.php?url='+url,success:
    function(json){
//display json.title, json.author, json.abstract...
});});
```

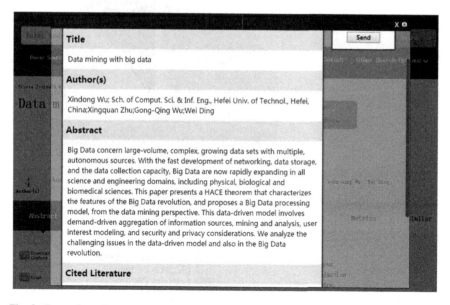

Fig. 2 Screenshot of our bookmarklet-triggered literature metadata extraction system

4 Survey Result

4.1 Bookmarklet Versus Browser Plugin

We now turn to quantitative evaluation of how well our literature metadata extraction system ran in terms of user satisfaction. We made a survey to the researchers of our university. We asked the researchers to rank the two prototype systems that we developed: One is a bookmarklet-triggered and another is plugin-triggered. Figure 3 shows the feedback of user satisfaction. We have received 121 valid feedbacks. A total of 71.9 % of researchers thought that our system is user-friendly, 21.7 % of researchers thought that browser plugin is better than book-marklet, and 7.4 % thought that both systems are OK. Furthermore, more than 80 % of researchers who dislike bookmarklet are not used to both of these systems. We believe that the bookmarklet-triggered way will be better after the understanding of Web 2.0 technology.

Fig. 3 Survey result of bookmarklet and browser plugin

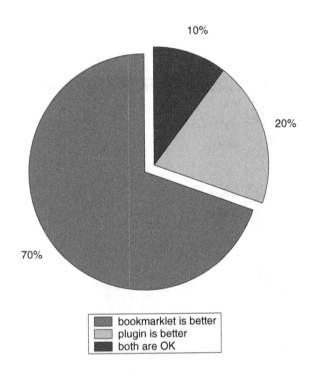

5 Conclusions

In our demonstration, we will present a bookmarklet-triggered literature metadata extraction system using cloud plugins. Selector-syntax extractors and bookmarklet-triggered way have been proposed to make the system flexible and extensible.

Acknowledgments This work was supported by the Doctoral Fund of University of Jinan (XBS1237) and the Shandong Provincial Natural Science Foundation (ZR2014FQ029).

References

1. Laender, A.H., Ribeiro-Neto, B.A., da Silva, A.S., Teixeira, J.S.: A brief survey of web data extraction tools. ACM Sigmod Record **31**(2), 84–93 (2002)
2. Ferrara, E., De Meo, P., Fiumara, G., Baumgartner, R.: Web data extraction, applications and techniques: a survey. arXiv preprint arXiv:1207.0246 (2012)
3. Movshovitz-Attias, D., Cohen, W.W.: Alignment-hmm-based extraction of abbreviations from biomedical text. In: Proceedings of the 2012 Workshop on Biomedical Natural Language Processing, Association for Computational Linguistics (2012) pp. 47–55
4. Crescenzi, V., Mecca, G., Merialdo, P., et al.: Roadrunner: Towards automatic data extraction from large web sites. VLDB. **1**, 109–118 (2001)
5. Ma, K., Yang, B.: A simple scheme for bibliography acquisition using doi content negotiation proxy. Electron. Libr. **32**(6), 806–824 (2014)
6. Ma, K., Yang, B., Chen, G.: Doi proxy framework for automated entering and validation of scientific papers. In: Web-Age Information Management. Springer (2013) pp. 799–801
7. Sullivan, S.J.: An archival/records management perspective on pdf/a. Rec. Manag. J. **16**(1), 51–56 (2006)
8. Ma, K., Zhang, L.: Bookmarklet-triggered unified literature sharing system in the cloud. Int. J. Comput. Appl. **5**(4), 217–226 (2014)
9. Kushmerick, N.: Wrapper induction for information extraction. PhD thesis, University of Washington (1997)
10. Dalvi, N., Kumar, R., Soliman, M.: Automatic wrappers for large scale web extraction. Proc. VLDB Endowment **4**(4), 219–230 (2011)
11. Xu, Z., Yan, D.: Designing and implementing of the webpage information extracting model based on tags. In: International Conference on Intelligence Science and Information Engineering (ISIE), IEEE (2011) pp. 273–275
12. Embley, D.W., Campbell, D.M., Smith, R.D., Liddle, S.W.: Ontology-based extraction and structuring of information from data-rich unstructured documents. In: Proceedings of the Seventh International Conference on Information and Knowledge Management, ACM (1998) pp. 52–59
13. Flesca, S., Furche, T., Oro, L.: Reasoning and Ontologies in Data Extraction. Springer, Berlin (2012)

Printed in the United States
By Bookmasters